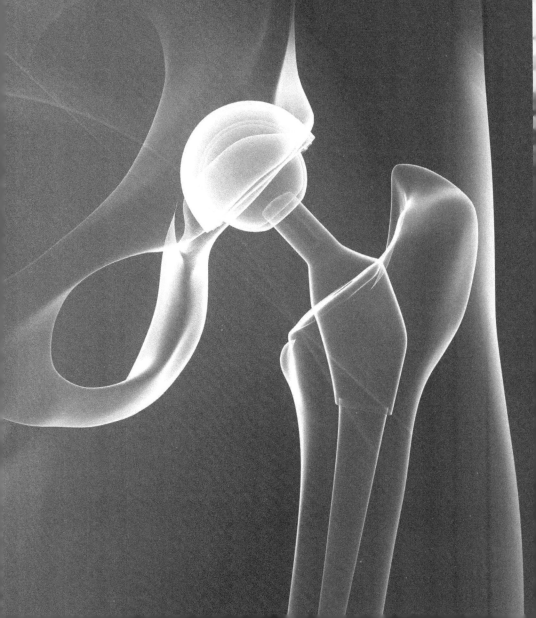
The Hip Joint in Adults

The Hip Joint in Adults
Advances and Developments

K. Mohan Iyer

PAN STANFORD PUBLISHING

Published by

Pan Stanford Publishing Pte. Ltd.
Penthouse Level, Suntec Tower 3
8 Temasek Boulevard
Singapore 038988

Email: editorial@panstanford.com
Web: www.panstanford.com

British Library Cataloguing-in-Publication Data
A catalogue record for this book is available from the British Library.

The Hip Joint in Adults: Advances and Developments
Copyright © 2018 by Pan Stanford Publishing Pte. Ltd.
All rights reserved. This book, or parts thereof, may not be reproduced in any form or by any means, electronic or mechanical, including photocopying, recording or any information storage and retrieval system now known or to be invented, without written permission from the publisher.

For photocopying of material in this volume, please pay a copying fee through the Copyright Clearance Center, Inc., 222 Rosewood Drive, Danvers, MA 01923, USA. In this case permission to photocopy is not required from the publisher.

ISBN 978-981-4774-72-7 (Hardcover)
ISBN 978-1-351-26244-6 (eBook)

In memory of my respected teacher, late Geoffrey V. Osborne

and my colleague, late George S. E. Dowd,
MD, M.Ch. (Orth.), FRCS.

To

My grandson, Vihaan

My wife, Mrs Nalini K. Mohan

My daughter, Deepa Iyer, MBBS, MRCP (UK)

My son, Rohit Iyer, BE (IT)

Contents

Foreword	xvii
Preface	xxi
Acknowledgements	xxv

1. Advances in Slipped Upper Femoral Epiphysis   1
Kishan Gokaraju, Nimalan Maruthainar and M. Zahid Saeed

1.1	Definition		1
1.2	Epidemiology		1
	1.2.1	Demographics	2
	1.2.2	Hip Location	2
	1.2.3	Other Risk Factors	2
1.3	Aetiology		2
1.4	Pathophysiology		4
1.5	Clinical Features		5
1.6	Investigations		6
	1.6.1	First-Line Imaging	6
	1.6.2	Radiographic Signs	7
	1.6.3	Additional Imaging	9
1.7	Classification		10
1.8	Management		10
	1.8.1	Stable SUFE	11
	1.8.2	Unstable SUFE	12
1.9	Surgical Techniques		13
	1.9.1	In Situ Fixation	13
	1.9.2	Open Reduction	14
	1.9.3	Osteotomies	14
	1.9.4	Cheilectomy	15
	1.9.5	Bone-Peg Epiphysiodesis	15
1.10	Postoperative Regime		16
1.11	Complications		16

2. Advances in Fractures and Dislocations of the Hip Joint   21
Thomas Pepper, Philip Ahrens and M. Zahid Saeed

2.1	Hip Fractures	21

	2.1.1	Intracapsular Fractures	22
		2.1.1.1	Classification	22
		2.1.1.2	Clinical features	22
		2.1.1.3	Management	23
		2.1.1.4	Complications	23
	2.1.2	Extracapsular Fractures	24
		2.1.2.1	Classification	24
		2.1.2.2	Clinical features	25
		2.1.2.3	Management	25
		2.1.2.4	Complications	26
2.2	Traumatic Dislocations of the Hip Joint	26
	2.2.1	Posterior Dislocation	26
	2.2.2	Anterior Dislocation	28
	2.2.3	Central Dislocation	29
	2.2.4	Dislocated Hip Prosthesis	29

3. Advances in Surface Replacement of the Hip Joint — 31
Senthil Muthian, Nicholas Garlick and M. Zahid Saeed

3.1	Hip Resurfacing	31
	3.1.1	Introduction	31
	3.1.2	History	32
	3.1.3	Problems with Conventional Total Hip Replacement	32
	3.1.4	Young Patients	33
	3.1.5	Biomechanics	33
	3.1.6	Tribology	34
	3.1.7	Indications	35
	3.1.8	Risk Factors	35
	3.1.9	Contraindications	35
	3.1.10	Cautions	36
	3.1.11	Selection of Ideal Patients	36
	3.1.12	Female Sex	36
	3.1.13	Surgical Considerations	37
		3.1.13.1	Approach	37
		3.1.13.2	Acetabulum	37
		3.1.13.3	Femur	38
	3.1.14	Results	38
		3.1.14.1	Implant-specific results	38
		3.1.14.2	Mechanism of failure	39
		3.1.14.3	Complications	39

	3.1.15	Imaging	40
		3.1.15.1 Radiographic features of impending failure	41
		3.1.15.2 Radiological features requiring closer observation	41
	3.1.16	Revision Surgery	41
	3.1.17	Current Recommendations for Follow-Up	41
	3.1.18	Summary	43

4. Advances in Periprosthetic Fractures of the Hip Joint 47
Ali-Asgar Najefi, Arthur Galea, Nicholas Garlick and M. Zahid Saeed

4.1	Periprosthetic Fractures of the Proximal Femur	47
	4.1.1 Introduction	47
	4.1.2 Risk Factors for PFF	48
	4.1.2.1 Patient factors	48
	4.1.2.2 Local risk factors	50
	4.1.2.3 Surgical factors	50
	4.1.3 Classification	52
	4.1.4 Investigations	55
	4.1.5 Surgical Considerations and Treatment of PFFs	56
	4.1.5.1 Vancouver A_G and A_L fractures	56
	4.1.5.2 Vancouver A1 and A2 fractures	57
	4.1.5.3 Vancouver B1 fractures	57
	4.1.5.4 Vancouver B2 fractures	57
	4.1.5.5 Vancouver B3 fractures	59
	4.1.5.6 Vancouver C fractures	60
4.2	Periprosthetic Acetabular Fractures	60

5. Minimally Invasive Surgery of the Hip Joint 73
Matthew K. T. Seah and Wasim Khan

5.1	Comparing Minimally Invasive Approaches	74
5.2	Quantifying Soft-Tissue Trauma	76
5.3	MIS in Trauma	77
5.4	Proximal Femoral Fractures: Sliding Hip Screw Fixation	77
5.5	Hip Arthroscopy: Current and Projected Trends	78
5.6	Conclusion	80

6. Computer-Assisted Navigation of the Hip Joint   87
Matthew K. T. Seah and Wasim Khan

6.1	Limitations of Conventional Alignment Jigs	88
6.2	Types of Computer Navigation Systems	90
6.3	The Accuracy of Computer-Assisted Navigation in THA	91
6.4	Computer Navigation in Total Hip Resurfacing	92
6.5	Limitations of Computer Navigation	93
6.6	Robotic Surgery: The Future of Navigation?	94
6.7	Conclusion	96

7. Advances in the Treatment of Meralgia Paresthetica in Surgery of the Hip Joint in Adults   101
K. Mohan Iyer

7.1	Aetiology and Risk Factors	103
7.2	Signs and Symptoms	103
7.3	Differential Diagnosis	104
7.4	Treatment for Meralgia Paresthetica	104

8. Advances in Surgery for Bursitis of the Hip Joint in Adults   107
K. Mohan Iyer

8.1	Causes and Symptoms of Trochanteric Bursitis	108
8.2	Investigations	109
8.3	Psoas Bursitis	111
8.4	Treatment	111
8.5	Snapping Hip	111

9. Advances in Surgery of the Hip Joint in Rheumatoid Arthritis in Adults   115
K. Mohan Iyer

10. Advances in Surgery of the Hip Joint in Tuberculosis Arthritis in Adults   119
K. Mohan Iyer

11. Advances in Fractures in the Neck of the Femur in Adults   123
Dayanand Manjunath

11.1	Fracture of the Neck Femur	123

		11.1.1	Introduction	123
			11.1.1.1 Epidemiology	123
			11.1.1.2 Pathophysiology	124
		11.1.2	Anatomy	124
			11.1.2.1 Osteology	124
			11.1.2.2 Blood supply to the femoral head	124
		11.1.3	Classification	125
			11.1.3.1 Garden's classification for the fractured neck of a femur	125
			11.1.3.2 Pauwels classification	125
			11.1.3.3 Anatomic classification	126
		11.1.4	Presentation	126
		11.1.5	Imaging	126
		11.1.6	Treatment	127
			11.1.6.1 Surgical considerations	127
			11.1.6.2 Closed reduction manoeuvres	127
			11.1.6.3 The role of capsulotomy	128
			11.1.6.4 Fixation methods	129
	11.2	Fractured Femur Neck of More Than Three-Week Duration		131
	11.3	Management of a Fractured Femur Neck in People >60 Years Old		132
		11.3.1	Hemiarthroplasty	133
		11.3.2	Complications of Hemiarthroplasty	135
	11.4	Complications		136
	11.5	Other Methods of Osteosynthesis		137
12.	**Advances in Hip Arthroscopy**			**141**
	Prakash Chandran			
	12.1	Introduction		141
	12.2	Surgical Anatomy of the Hip		143
	12.3	Imaging		147
		12.3.1	Radiography	147
		12.3.2	Evaluation of Images	148
		12.3.3	Computer Tomography Scan of the Hip and Pelvis	150
		12.3.4	Ultrasound of the Hip	151
		12.3.5	Magnetic Resonance	151
			12.3.5.1 Labral tears	152

		12.3.5.2	Femoroacetabular impingement	154
		12.3.5.3	Abnormality of the cartilage	156
		12.3.5.4	Intra-articular foreign bodies	157
		12.3.5.5	Abnormality of the ligamentum teres	157
12.4	Physical Examination of the Hip and Pelvis			157
12.5	Surgical Technique			164
	12.5.1	Patient Position		165
	12.5.2	Portals		165
	12.5.3	Theatre Layout and Equipment		165
	12.5.4	Portal Placement		166
	12.5.5	Central Compartment		167
	12.5.6	Peripheral Compartment		167
12.6	Complications			168
12.7	Specific Conditions			169
	12.7.1	Acetabular Labral Tears		169
	12.7.2	Femoroacetabular Impingement		172
	12.7.3	Articular Cartilage		174
	12.7.4	Hip Septic Arthritis		175
	12.7.5	Pigmented Villonodular Synovitis and Synovial Chondromatosis		175
	12.7.6	Arthroscopy in Hip Trauma		176
	12.7.7	Osteonecrosis of the Femoral Head		176
	12.7.8	Painful Hip Arthroplasty		176
	12.7.9	Hip Dysplasia		177
	12.7.10	Slipped Capital Femoral Epiphysis and Perthes Sequelae		177
	12.7.11	Teres Ligament Injuries and Capsule Repair in Cases of Instability		177
	12.7.12	Extra-articular Pathological Conditions around the Hip		178
	12.7.13	External Snapping		178
	12.7.14	Trochanteric Bursitis and Injuries of the Gluteal Muscles		178
	12.7.15	Internal Snapping		179
	12.7.16	Piriformis Syndrome		179
12.8	Rehabilitation			179
12.9	Outcome Assessment			181

13. Mesenchymal Stem Cell Treatment of Cartilage Lesions in the Hip — 185

George Hourston, Stephen McDonnell and Wasim Khan

- 13.1 Introduction — 185
- 13.2 Cartilage Lesions — 186
- 13.3 Traditional Therapeutic Strategies and Limitations — 187
- 13.4 Mesenchymal Stem Cells: A History — 190
- 13.5 Mesenchymal Stem Cells: Clinical Potential — 192
- 13.6 Animal Models of Chondral Lesions — 194
- 13.7 Clinical Trials — 196
- 13.8 Future Prospects — 196
- 13.9 Concluding Remarks — 199

14. Advances in Short-Stem Total Hip Arthroplasty — 211

Karl Philipp Kutzner

- 14.1 Background — 211
- 14.2 Classification of Short Stems — 212
- 14.3 Philosophy of Modern Short-Stem THA — 215
  - 14.3.1 Individualised Positioning: Reconstruction of the Anatomy — 216
  - 14.3.2 Bone- and Soft-Tissue Sparing Implantation — 218
  - 14.3.3 Metaphyseal Anchorage — 219
- 14.4 Osteointegration and Migration Pattern — 220
- 14.5 Indications and Contraindications — 221
- 14.6 Conclusions — 222

15. Advances in Legg–Calvé–Perthes Disease — 227

J. S. Bhamra, P. Singh, S. Madanipour and A. Malhi

- 15.1 Background — 227
- 15.2 Classifications Systems and Pathophysiology of Legg–Calvé–Perthes Disease — 228
- 15.3 Treatment Goals — 228
- 15.4 Current Trends in Total Hip Arthroplasty — 230
- 15.5 Femoroacetabular Impingement and Legg–Calvé–Perthes Disease — 231
- 15.6 Radiological Advances — 232

15.7	Use of Bisphosphonates in Legg–Calvé–Perthes Disease?	233
15.8	Future Therapies: Preclinical Role of Morphogenic Protein-2 in Legg–Calvé–Perthes Disease?	235
15.9	Future Therapies: Interleukin-6 and Synovitis	235
15.10	Future Therapies: The Use of Mesenchymal Stem Cells?	235
15.11	Conclusions	236

16. Advances in Haemophilic Hip Joint Arthropathy — 241

Muhammad Zahid Saeed, Amr Saad, Haroon A. Mann and Nicholas Goddard

16.1	Recent Advances in Haemophilic Hip Joint Arthropathy	241
	16.1.1 Introduction	241
	16.1.2 Epidemiology	242
	16.1.3 Pathophysiology	242
	16.1.4 Clinical Features	243
	16.1.5 Investigations	244
	16.1.6 Differential Diagnosis	245
	16.1.7 Management	246
	16.1.8 Multidisciplinary Team Approach	246
	16.1.9 Preoperative Screening Tests	247
	16.1.10 Surgical Techniques	247
	16.1.11 Perioperative Regime	248
	16.1.12 Complications	249
	16.1.13 Long-Term Prognosis	250
16.2	Conclusions	250

17. Direct Anterior Approach to the Hip Joint — 255

John O'Donnell

17.1	Introduction	255
17.2	Our Technique of Direct Anterior Approach THR Using a Leg Holder	257
17.3	Postoperative Care	264
17.4	Indications for the Direct Anterior Approach	265
17.5	Specific Complications	265
17.6	Conclusion	265

18. Total Hip in a Day: Setup and Early Experiences in Outpatient Hip Surgery — 267

Manfred Krieger and Ilan Elias

- 18.1 New Hip in a Day: Setup and Initial Clinical Experiences in Germany — 267
  - 18.1.1 Introduction — 267
  - 18.1.2 The Course of the Day of Surgery — 269
  - 18.1.3 Discussion — 273

19. Advances in Osteoarthritis of the Hip — 287

Pratham Surya, Sriram Srinivasan and Dipen K. Menon

- 19.1 Introduction — 287
- 19.2 Epidemiology — 289
- 19.3 Aetiology of Primary Hip OA — 290
- 19.4 Basic Science of Cartilage and Changes in Hip Arthritis — 292
- 19.5 Early-Stage OA — 295
- 19.6 Late-Stage OA — 296
- 19.7 Symptoms of OA — 296
  - 19.7.1 Initial Symptoms — 296
  - 19.7.2 Progression of Disease — 296
- 19.8 Diagnosis — 296
- 19.9 Management of OA — 297
- 19.10 Animal Models of OA — 297
  - 19.10.1 Invasive Model — 298
  - 19.10.2 Noninvasive Model — 298
- 19.11 Nonpharmacological Management of Hip OA — 298
- 19.12 Pharmacological or Drug Treatment — 299
- 19.13 Surgical Treatment — 300
- 19.14 Drugs Assessed for Disease-Modifying Potential in OA — 301
- 19.15 New Targets for Treating OA — 301
- 19.16 Conclusion — 302

20. Advances in Primary and Revision Hip Arthroplasty — 305

Shibu P. Krishnan and G. Gopinath

- 20.1 Update on Polyethylene — 306
  - 20.1.1 E-Poly — 307

20.2	Ceramics in Total Hip Arthroplasty		308
	20.2.1	Alumina Matrix Composite (Biolox Delta Ceramic)	308
	20.2.2	Oxidised Zirconium-Bearing Surface (Oxinium)	309
	20.2.3	Hydroxyapatite Coating	309
20.3	Current Status of Metal-on-Metal Bearing Surfaces		309
20.4	Computer-Assisted THA		310
20.5	Noncemented (Biologic) Fixation of Hip Components		311
	20.5.1	Porous Coating	311
	20.5.2	Grit Blasting	311
	20.5.3	Hydroxyapatite Coating	312
	20.5.4	Trabecular Metal Technology	312
20.6	Enhanced Recovery Programmes for Primary THA		313
	20.6.1	Preoperative Measures	313
	20.6.2	Intraoperative Measures	313
	20.6.3	Postoperative Measures	314
20.7	Modular Femoral Stem Designs in THA		314
20.8	Cementless Acetabular Cup Designs		315
20.9	Bearing Surfaces in Modern THA		316
20.10	Dual-Mobility Cup		316

21. Advances in Adult Dysplasia — 319

Kaveh Gharanizadeh

21.1	Adult Hip Dysplasia		319
21.2	Hip Dysplasia		322
21.3	Epidemiology and Natural History		323
21.4	Pathophysiology and Aetiology		324
21.5	Pathomorphology		326
21.6	Femur Morphology		330
21.7	Natural History of Hip Dysplasia		332
21.8	Clinical Examination and Diagnosis		335
21.9	Imaging		336
	21.9.1	Acetabular WB Index or Tonnis Angle	336
	21.9.2	Acetabular Sharp Angle	337
	21.9.3	Acetabular Depth–Width Ratio	337
	21.9.4	Femoral Side Criteria	337

21.9.5	Neck–Shaft Angle	338
21.9.6	Fovea Alta	338
21.9.7	CT Scan	341
21.9.8	MRI	342
21.10	Classification	343
21.11	Management	344
21.12	Rotational Osteotomy	348
21.13	Periacetabular Osteotomy	348
21.14	Indications of Reorientation Osteotomy	349
21.15	Femoral Osteotomy	356
21.16	Criteria of Correct Reorientation after PAO	358
21.17	Outcome of Reorientation Osteotomy	359
21.18	Salvage Osteotomy	361
21.19	Chiari Osteotomy	361
21.20	Results	361
21.21	Neuromuscular Patients	362
21.22	Symptoms	362
21.23	Arthroscopy	364
21.24	PAO Complications	364
21.25	Asymptomatic Dysplastic Patients	365
21.26	Summary	366

22. Advances in Avascular Necrosis of the Hip Joint — 383
Kaveh Gharanizadeh

22.1	Femoral Head Osteonecrosis	383
22.1.1	Introduction	383
22.1.2	Pathology and Aetiology	384
22.1.3	Other Diseases	387
22.1.4	Clinical Presentation and Diagnosis	388
22.1.5	Imaging	388
22.1.5.1	Plain X-ray	388
22.1.5.2	MRI	391
22.1.5.3	CT scan	392
22.1.5.4	Nuclear bone scan	392
22.1.6	Differential Diagnosis	392
22.1.7	Classification	393
22.1.7.1	Collapse	393
22.1.7.2	The lesion extent	396
22.1.7.3	Location of the lesion	396

		22.1.7.4	Depression rate of the head and head flattening	396
		22.1.7.5	Author's preference	396
	22.1.8	Treatment		397
		22.1.8.1	Nonsurgical treatment	397
		22.1.8.2	Core decompression	398
		22.1.8.3	Bone graft	401
		22.1.8.4	Proximal femoral osteotomy	403
		22.1.8.5	Hip replacement surgery	404
	22.1.9	Overview		406

Index   423

Foreword

Ever since the visionary Sir John Charnley came out with his understanding of the hip joint, its biomechanics, its crippling diseases, the approach to its diagnosis and its management more than all, different surgical approaches with innumerable implants to replace the hip have been a bugbear and challenge to every orthopaedic surgeon. The literature and innovations on this subject perhaps far surpass those in any other field of orthopaedic surgery.

In my journey over the past 50 years in orthopaedic surgery, I still remain confused as to the surgical approach, the bearing surface and the variable biomechanism after a stiff, painful hip joint.

I have had the pleasure of knowing Dr K. Mohan Iyer over the past 35 years. His innovative thinking and mastery in solving everyday problems we face in surgery have, indeed, baffled me. He has been working on solutions to different problems we face in hip joint diseases, and his other edited books have shown us the depth of his knowledge on this subject, which he has penned down with illustrations for the benefit of surgeons and learning residents all round the world.

It was indeed a pleasure going through this illustrated book and the quest for new knowledge with co-authors from different parts of the world who have given descriptive details in their chapters, from slipped upper femoral epiphysis (SCFE) to computer-assisted surgery to mesenchymal stem cell therapy for cartilage defects in the hip.

On behalf of the orthopaedic fraternity, I congratulate Dr Iyer for his tenacity of purpose in bringing out such a contribution. He has indeed left behind his footsteps in the sands of time for surgeons to remember him and contribute to the management of different disorders of the hip.

Professor (Dr) M. Shantharam Shetty, MS (Orth), FRCS (Eng), FACS, FICS
Pro-Chancellor, NITTE University
Chairman, Tejasvini Hospital
Ex-President, Indian Orthopaedic Association
Mangalore, India

Preface

Following the launch of my book *The Hip Joint* on October 5, 2016, I thought it was my sincere duty to write about the hip joint in adults, in addition to the advances and developments. I have been 'brought up' in orthopaedics by my respected teacher late Geoffrey V. Osborne, who was also a patron of the Indian Orthopaedic Association (IOA). I still have fond memories of him and cherish them till today. In his honour, I wrote an article that came out in the journal *Neurosurgery* (Granger, A., Sardi, J. P., Iwanaga, J., et al. (2017). Osborne's ligament: a review of its history, anatomy, and surgical importance, *Cureus*, 9(3). doi: 10.7759/Cureus.1080).

I have been extremely fortunate to get Assoc. Prof. John O'Donnell, MBBS, FRACS, FAOrthA, to write the chapter 'Direct Anterior Approach to the Hip Joint', which is the fastest-growing area of hip replacement surgery in Australia and in the world. He is an inaugural board member of the International Society for Hip Arthroscopy (ISHA). He was the president (2013–2014) of the International Society for Hip Arthroscopy. He is also the founding president of the Australian Hip Joint Preservation Surgery Society. After studying the technique in Paris, he started to perform direct anterior approach (DAA) total hip replacement surgery in Australia in 2007. DAA is now the fastest-growing area of hip replacement surgery in Australia.

I also got impressive chapters by Kaveh Gharanizadeh, assistant professor of orthopaedic surgery at the Iran University of Medical Sciences (IUMS) who specialises in adult hip dysplasia and recent advances in avascular necrosis of the femoral head. He was a fellow of hip preserving surgery at Inselspital, Bern University, Switzerland, in 2010 under the supervision of Prof. Siebenrock. He was also Adult Reconstruction Fellow at the University of British Colombia (UBC), Canada, in 2012 and at Charite Hospital, Berlin University, Germany, in 2014. He is currently working as a hip surgeon, doing all kinds of hip surgeries, including hip-preserving surgery and adult reconstruction from childhood to adulthood.

I am extremely thankful to my colleague Mr Nicholas Goddard, MBBS FRCS Di Hand Surgery (Paris) Dip Microsurgery (Paris). Mr Goddard is a consultant orthopaedic surgeon with a particular interest in hand and wrist surgery and surgery for bleeding disorders. He has an international reputation in both fields and is vice chairman of the Musculoskeletal Committee of the World Federation of Haemophilia. He is also editor of a book on the surgical management of haemophilia by E. C. Rodriguez-Merchan, MD (editor) and Christine A. Lee (editor), published in May 2008 by Wiley-Blackwell. Mr Goddard was awarded the Henri Horoszowski Prize 2011 in Dubai by the World Federation of Haemophilia. He has contributed to the chapter 'Advances in Haemophilic Hip Joint Arthropathy'.

Above all, my sincere thanks to Mathys Ltd Bettlach, Switzerland, for its permission for certain figures in the chapter 'Advances in Short-Stem Total Hip Arthroplasty' by Dr Karl Philipp Kutzner from Germany, who is also on the editorial board of the journal *Recent Advances in Arthroplasty (RAA)*. Also my sincere thanks to Medacta for its permission to use the images in Chapter 17, 'Direct Anterior Approach to the Hip Joint.' I also wish to express my thanks to Vijay Kumar Kempanna and Sharad Goyal for their permissions for two line drawings from Chapter 6 of the book *The Hip Joint*.

I would also like to thank Nicholas Garlick, my colleague at the Royal Free London NHS Foundation Trust, whose special interests are in hip and shoulder surgery, and he regularly lectures in orthopaedic surgery. He has contributed to advances in the surface replacement of the hip joint and advances in periprosthetic fractures of the hip joint.

I thank Arthur Galea, who was with me during my tenure at the Royal Free Hospital, for his contribution to 'Advances in Periprosthetic Fractures of the Hip Joint'. I also thank my colleague Philip Ahrens for his contribution to 'Advances in Fractures and Dislocations of the Hip Joint' and Nimalan Maruthainar for his views on the chapter 'Advances in Slipped Upper Femoral Epiphysis'.

In particular, I thank Muhammad Zahid Saeed, FRCS (Tr and Orth) and member of ISAKOS (International Society of Arthroscopy, Knee Surgery and Orthopaedic Sports Medicine), for taking active interest in involving all the specialist registrars working in the department for helping me out in various chapters.

I wish to thank Wasim Khan, Cambridge, UK, who was instrumental in contributing to the chapter 'Computer Navigation in Hip Arthroplasty' in my book *The Hip Joint* and the chapters 'Computer-Assisted Navigation of the Hip Joint', 'Minimally Invasive Surgery of the Hip Joint' and 'Mesenchymal Stem Cell Treatment of Cartilage Lesions in the Hip' for this book, co-written by George Hourston and Stephen McDonnell (Division of Trauma & Orthopaedics, Addenbrooke's Hospital, University of Cambridge, UK).

I am also thankful to Dayanand Manjunath, Bangalore; Prakash Chandran, Warrington and Halton Hospitals, UK; Kaveh Gharanizadeh, Iran; Shibu P. Krishnan, London, UK; Girish Gopinath, Trust Specialty Registrar, Buckinghamshire NHS Trust; Jagmeet Bhamra, Specialist Registrar in Trauma & Orthopaedic Surgery, London Deanery, London, UK; Prashant Singh, Core Surgical Trainee, London Deanery, Trauma & Orthopaedic Surgery Department, University Hospital Lewisham High Street, London; Suroosh Madanipour, Core Surgical Trainee, London Deanery; and Arfan Malhi, Consultant in Trauma & Orthopaedic Surgery, Trauma & Orthopaedic Surgery Department, University Hospital Lewisham High Street, London, for their active contributions to various chapters in this book.

Finally, I thank Pratham Surya, Sriram Srinivasan and Dipen K. Menon, Kettering, UK.

Above all, I am extremely thankful to Dr med. Manfred Krieger and Dr med. Ilan Elias, Wiesbaden, Frankfurt, Germany, for the chapter 'Total Hip in a Day: Setup and Early Experiences in Outpatient Hip Surgery'.

K. Mohan Iyer
M.Ch.Orth. (Liverpool, UK), M.S. Orth. (Bom), F.C.P.S. Orth. (Bom),
D'Orth. (Bom), MBBS (Bom)
Senior Consultant Orthopaedic Surgeon, Bengaluru, India

Acknowledgements

I am extremely grateful to Prof. Dr Shantharam Shetty, senior orthopaedic surgeon in Karnataka, India, for his encouraging foreword for this book and who also gave a foreword for my book *The Hip Joint*.

I am also thankful to John O'Donnell, Nicholas Goddard, Nicholas Garlik, Philip Ahrens, Arthur Galea, Nimalan Maruthainar, Muhammad Zahid Saeed, Amr Saad, Haroon A. Mann, Kishan Gokaraju, Thomas Pepper, Senthil Muthian, Ali-Asgar Najefi, Dayanand Manjunath, Matthew K. T. Seah, Wasim Khan, Prakash Chandran, George Hourston, Stephen McDonnell, Jagmeet Bhamra, Karl Philipp Kutzner, Manfred Krieger, Ilan Elias, Pratam Surya, Sriram Srinivasan, Dipen K. Menon, Prashant Singh, Suroosh Madanipour, A. Malhi, Shibu P. Krishnan, Girish Gopinath and Kaveh Gharanizadeh for their efforts to write and complete their respective chapters on time.

I am thankful to Jenny Rompas and Stanford Chong, Directors and Publishers, and Sarabjeet Garcha, Senior Editorial Manager, Pan Stanford Publishing Pte. Ltd., Singapore, for guiding me throughout the publishing process; Archana Ziradkar, Senior Editor, for her invaluable help while preparing the manuscript; and Naveen Kumar for typesetting the book.

Above all, I highly appreciate the help of my son, Rohit Iyer, in the presentation and publication of this book.

Chapter 1

Advances in Slipped Upper Femoral Epiphysis

Kishan Gokaraju, Nimalan Maruthainar and M. Zahid Saeed
Trauma and Orthopaedic Department, University College London Medical School, Royal Free Hospital, Pond Street, London NW3 2QG, United Kingdom
z.saeed@nhs.net, mzahidsaeed@hotmail.com

1.1 Definition

Slipped upper femoral epiphysis (SUFE), also termed slipped capital femoral epiphysis (SCFE), is a common paediatric hip disorder involving failure of the proximal femoral physis with subsequent displacement.

1.2 Epidemiology

It is the most common adolescent hip disorder, occurring in up to 10 per 100,000 people. Important associated factors are as follows [1].

The Hip Joint in Adults: Advances and Developments
Edited by K. Mohan Iyer
Copyright © 2018 Pan Stanford Publishing Pte. Ltd.
ISBN 978-981-4774-72-7 (Hardcover), 978-1-351-26244-6 (eBook)
www.panstanford.com

1.2.1 Demographics

- Males (male to female ratio 3:2)
- Age 10–16 years (i.e., associated with puberty)
    - Males 12–16 years (average 13.5 years)
    - Females 10–14 years (average 12.2 years)
- Growth spurt (i.e., average ages in boys 13.4 years and girls 12.2 years)
- African Americans or Pacific Islanders

1.2.2 Hip Location

- The left hip is more common than the right one.
- A bilateral SUFE occurs in 17%–63% (if not simultaneous, the second usually occurs within 18 months of initial presentation with the first hip).

1.2.3 Other Risk Factors

- Obesity (single greatest risk factor)
- Trauma
- Femoral retroversion
- Family history (7% risk of a second family member being affected)
- Skeletally immature children
- Endocrine disorders (e.g., hypothyroidism, hyperparathyroidism, parahypopituitarism, hypogonadism, pituitary tumours and growth hormone therapy)
- Metabolic disorders (e.g., rickets and renalosteodystrophy)
- Radiotherapy (i.e., specifically around the hip or pelvis)
- Chemotherapy
- Down syndrome

1.3 Aetiology

The cause of a SUFE is unclear, but it is suspected that an abnormal load exerted on a normal proximal femoral physis or a normal load exerted on an abnormal physis may be the reason for injury.

Normally, the perichondral ring of Lacroix is a band of tissue that surrounds the physis, providing support during growth (Fig. 1.1).

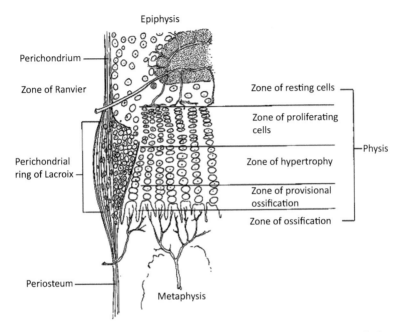

Figure 1.1 Diagram of a physis demonstrating differing zones and the perichondral ring of Lacroix.

During growth spurts, particularly after the onset of puberty, the physis widens and may result in thinning and weakening of the ring. As this occurs in the proximal femoral physis, the risk of a SUFE is increased.

The upper femoral physis is normally horizontally orientated, but a more oblique or even vertical growth plate is mechanically less stable and makes the patient prone to a slip due to increased shear forces. Obesity is associated with a relatively retroverted (mean of 0° anteversion) femoral neck (with decreased head-neck offset), and so increased load transmitted through this abnormally aligned physis significantly increases risk of injury.

The natural growth and development of the proximal femoral physis, as with other growth plates, relies on multiple mechanical, endocrine and metabolic factors. For example, growth hormone is required for longitudinal growth of the physis, but during puberty this process is enhanced, causing weakening of the growth plate. Normally pituitary hormone, which stimulates growth, is balanced by gonadal hormone, resulting in cessation of physeal growth. Imbalance of these hormones places the physis at risk of injury from shear forces.

A SUFE occurring before 10 years of age or in a child below the 50th percentile should raise concern of an underlying endocrine disorder. A SUFE in each hip should be considered, along with investigation of an endocrine cause. One must be aware of the high proportion of patients who develop bilateral slips, usually occurring within 18 months of the contralateral slip. Vitamin D and haemophilia A have also been described as risk factors.

It has also been reported as a rare case report in haemophilia A [2].

1.4 Pathophysiology

The upper femoral epiphysis is the only physis in the body that lies within the joint capsule. The primary blood supply to the femoral head and hence the epiphysis is from the extracapsular circumflex arteries. Blood supply via a branch of the obturator artery through the ligamentum teres is thought to be minimal and begins to diminish in this age group.

A SUFE is a Salter Harris type 1 injury, but despite the terminology, the epiphysis does not actually move. Movement of the metaphysis occurs in relation to the stationary epiphysis, which is held in place by the thick ligamentum teres. While the appearance on pelvic radiographs is such that the femoral head appears to have slipped posteriorly and inferiorly, it is actually the neck of the femur that is displaced anteriorly and superiorly along with external rotation. Less frequently, the opposite occurs, where the metaphysis displaces posteriorly and inferiorly, resulting in an anterosuperior SUFE.

The injury occurs in the weak hypertrophic zone of the growth plate, which ordinarily takes up approximately one-third of its width. Irregularity and widening of the hypertrophic zone and a weakened perichondral ring make the physis susceptible to a slip. During growth spurts, the orientation of the physis is altered to a more oblique configuration, leading to increased shear forces. These forces are worsened in a retroverted hip. Histologically, a disorganised appearance of the physis is seen with ultimately chondrocyte death and reduced irregularly ordered collagen [3].

The displacement in a SUFE may be slow to progress, with minimal symptoms, but alternatively may slip further, increasing pain and affecting function. With increasing metaphyseal displacement, both function of the growth plate and the blood supply to the epiphysis are threatened. If chronic or missed, the natural progression is for a callus to form at the site of the slip within the stripped periosteum and eventually ossify. This hump of bone may impinge on the acetabulum in later life, giving rise to the cam lesion described in femoroacetabular impingement (FAI) [3]. In addition, there is a direct association of chondrolysis and subsequent osteoarthritis with the severity of the slip.

1.5   Clinical Features [1]

The primary symptom of a SUFE is pain in the groin or around the hip. Similarly to other hip joint pathology, pain may be experienced as referred pain in the ipsilateral thigh or knee, along the distribution of the obturator nerve.

In an acute setting, this is likely to present as sudden-onset, severe pain in a previously asymptomatic hip, often preceded by trauma. In a chronic SUFE, the pain may be mild and constant or gradually worsening, over a period of weeks or even months. With an acute-on-chronic slip, the patient may experience sudden increase in pain following a background of milder hip symptoms.

In 40% of the cases the patient may have pain on both sides due to a bilateral SUFE, with symptoms in one hip often starting within a year of onset of pain in the initially affected side. In 20% of all cases, patients present with pain in both hips at the same time due to a simultaneous bilateral SUFE.

As with most displaced hip fractures, the patient may hold his or her leg in a flexed and externally rotated position. Movement may be restricted, and there may be a noticeable limp secondary to pain, necessitating the need for walking aids. Severe pain may understandably result in an inability to weight-bear on the affected side, indicating an unstable injury and a potentially poorer prognosis.

The antalgic gait or inability to weight-bear, walk or stand will be seen at assessment. A displaced SUFE often causes a flexed and externally rotated position of the limb with or without shortening. Chronic slips may be accompanied by wasting of the thigh or gluteal muscles due to disuse atrophy secondary to ongoing pain.

Groin or hip tenderness is common in acute injuries. The range of motion may be reduced with restricted internal rotation, flexion and abduction. With passive flexion, the hip may be forced into external rotation (Drennan's sign). Weakness may be present secondary to pain in an acute SUFE or due to wasting in the chronic condition.

1.6 Investigations [1]

1.6.1 First-Line Imaging

- Plain anteroposterior (AP) pelvis radiograph (Fig. 1.2)

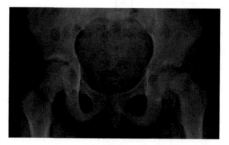

Figure 1.2 Anteroposterior pelvic radiograph demonstrating a left-sided SUFE (the epiphysis remains within the acetabulum, but the metaphysis has displaced superiorly).

Frog leg lateral radiograph of both hips (Fig. 1.3)

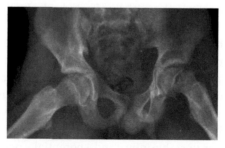

Figure 1.3 Frog leg lateral radiograph of both hips demonstrating a left-sided SUFE.

1.6.2 Radiographic Signs

- **Shenton line** – a curved line drawn along the medial border of the proximal femur extending proximally along the inferior border of the femoral neck and continuing along the inferior border of the superior pubic rami. This line is disrupted with a SUFE (as seen with displaced hip fractures) (Fig. 1.4).

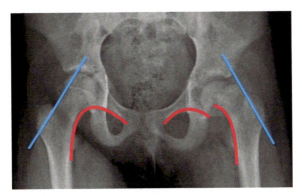

Figure 1.4 Anteroposterior pelvic radiograph demonstrating Shenton lines (red), intact on the normal right side but disrupted on the left side, with the SUFE. Klein's lines (blue) bisect the superior third of the femoral head on the normal right side but miss the femoral head on the left side, with the SUFE, as the neck has displaced superiorly.

- **Klein's line** – a line drawn along the superior border of the femoral neck on an AP pelvic radiograph. Normally Klein's line intersects the superior one-third of the femoral head. With a SUFE, as the head is positioned inferiorly in comparison to the neck, Klein's line will not intersect the femoral head (**Trethowan's sign**) (Fig. 1.4).
- **Epiphysiolysis** – a lysis at the physis that appears as lucency or widening of the growth plate, seen in the early SUFE or preslip (Fig. 1.5).
- **Decreased height of the femoral epiphysis** – the femoral head positioned posteriorly and inferiorly in comparison to the neck, possibly giving a narrower appearance.
- **Metaphyseal blanch of steel** – blurring of borders of the physis, giving a crescent-shaped sclerotic appearance on an

AP radiograph due to overlap of the epiphysis (posteriorly) and the metaphysis (anteriorly) from an established slip (radiographic sign described by H. H. Steel) (Fig. 1.6).

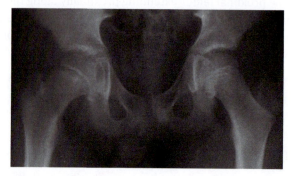

Figure 1.5 Anteroposterior radiograph of both hips demonstrating widening of the right proximal femoral physis in an early SUFE.

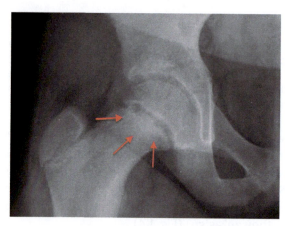

Figure 1.6 Anteroposterior radiograph of the right hip demonstrating increased opacity at the metaphysis due to overlapping of the posteriorly positioned femoral head in relation to the neck (metaphyseal blanching of steel).

- **Southwick's angle (head-shaft angle)** – on a frog leg lateral radiograph, the angle between an axial line drawn along the femur and a perpendicular line to one drawn connecting the superior and inferior tips of the femoral head epiphysis (Fig. 1.7).

Investigations | 9

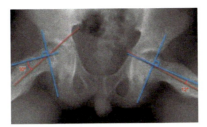

Figure 1.7 Frog leg lateral radiograph of both hips demonstrating the normal head-shaft angle (Southwick's angle) on the right side but an increased angle on the left side due to the displaced SUFE.

Comparison is made to the contralateral angle. Normal is 12 degrees, but it is greater with a SUFE.

- **Superior callus formation in a chronic SUFE** – callous formation at the site of the injured physis possibly demonstrated by chronic slips.

1.6.3 Additional Imaging

- Computed tomography (CT): Can confirm the presence and severity of displacement in an apparently normal radiograph (rarely required) or identify physeal closure in a chronic or missed injury.
- Ultrasound (US): Is useful for identifying early slips and presence of joint effusion.
- Magnetic resonance imaging (MRI): Can help to diagnose a slip or preslip condition if radiographs appear normal. May show physeal widening or metaphyseal oedema (Fig. 1.8).

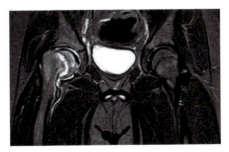

Figure 1.8 Coronal image of a pelvic T2-weighted MRI demonstrating a right-sided displaced SUFE with surrounding signal change due to localised oedema.

- Bone scan: Demonstrates increased uptake in a SUFE or chondrolysis but decreased uptake with subsequent osteonecrosis.

1.7 Classification

Classifications of SUFEs are based on the duration and extent of symptoms experienced by the patients and also the severity of the displacement as seen on imaging modalities.

- **Loder classification**
  Provides prognostic information
  - Stable: Able to weight-bear with or without crutches (risk of avascular necrosis [AVN] < 10%)
  - Unstable: Unable to weight-bear, even with crutches (risk of AVN ≈ 47%)
- **Southwick's classification**
  Measures the difference of the head-shaft angles (described above) between contralateral sides
  - Mild: <30
  - Moderate: 30–60
  - Severe: >60
- **Temporal classification (rarely referred to)**
  - (Preslip)
  - Acute: Symptoms <3 weeks
  - Chronic: Symptoms >3 weeks
  - Acute-on-chronic: Acute exacerbation of chronic symptoms
- **Wilson's classification (grading of slip)**
  - Grade I: 0%–33%
  - Grade II: 34%–50%
  - Grade III: >50%

1.8 Management

Conservative management is associated with a high complication rate, including further displacement, osteonecrosis and chondrolysis. Operative management is therefore indicated in nearly all cases, and nonoperative management is no longer an option. The aims are primarily to stabilise the injury; prevent worsening of the displacement, rather than to reduce the slip; and reduce potential complications. Reduction of an acute slip remains largely controversial, unless it is highly unstable or significantly displaced.

1.8.1 Stable SUFE

Stable injuries are judged by the patient's ability to weight-bear, according to the Loder classification. The duration of symptoms does not have a bearing on stability; however, increased longevity of symptoms with a less severe slip would suggest greater stability. Mild to moderate angles according to Southwick's classification and Grade I–II slips often fall into this group.

For a stable SUFE, in situ pinning is the treatment of choice to prevent further slip. This involves not reducing the slip but fixing it in its position at diagnosis with one (Fig. 1.9) or two cannulated screws [4]. Reduction of the slip is avoided as this increases the risk of disruption to the femoral head blood supply, resulting in an increased incidence of AVN. Pinning in situ increases the likelihood of preservation of whatever blood supply is still intact after the slip.

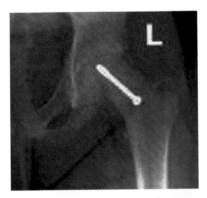

Figure 1.9 Anteroposterior left hip radiograph demonstrating fixation of a left SUFE without reduction using one partially threaded screw.

The use of one versus two screws remains a topic of controversy. Insertion of a screw may result in direct injury to the femoral neck vasculature supplying the femoral head, so insertion of two screws understandably has a greater risk of osteonecrosis than one screw. Similarly, the use of two screws increases the risk of potential joint penetration and subsequent chondrolysis. The theoretical increase in biomechanical and rotational stability of two screws is not significant in a stable SUFE, and so the benefits do not outweigh the risks [1]. However, the use of more than one screw warrants consideration in more severe unstable slips.

1.8.2 Unstable SUFE

A patient with an unstable SUFE cannot weight-bear according to the Loder classification. Grade III slips and severe slips according to the Southwick classification often fall into this group when discussing operative management. This group has a higher rate of complications, and treatment options can be more complex, with certain issues, such as the optimal number of screws inserted and whether to reduce a slip prior to fixation, remaining controversial.

The more severe, unstable slip may require consideration for the use of two screws to provide greater biomechanical support [5] (Fig. 1.10).

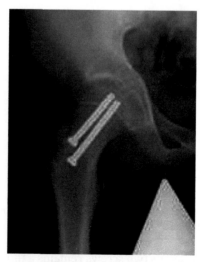

Figure 1.10 Anteroposterior right hip radiograph demonstrating fixation of a SUFE using two partially threaded screws.

The increased risks associated with the use of two screws are still relevant and so require an appropriately experienced and able surgeon. For such injuries identified within 24 hours, closed reduction to improve position prior to fixation may be considered and as evidence suggests, gentle manipulation is not detrimental to the outcome. This can be performed on the table at the time of surgery. If 24 hours have elapsed, soft tissues and vasculature start to become contracted and so manipulation increases the risk of damage to these, with subsequent AVN. In a delayed presentation of over 24 hours, skin traction for three weeks prior to surgery is advised. In any scenario, forceful reduction should not be performed due to the significant risk of vascular disruption [1, 6].

Urgent aspiration of the hip prior to fixation has been discussed as an option to decompress the haemarthrosis and relieve pressure on the femoral head blood supply. Acutely, the decision to perform a closed reduction for a severe slip must be weighed up against the alternative options, including an open reduction and internal fixation or an initial osteotomy.

1.9 Surgical Techniques

1.9.1 In Situ Fixation

With regard to screw fixation, the aim is to prevent a further slip or progression of a stable injury to an unstable one. To achieve this the screw, inserted percutaneously and under X-ray guidance, must cross the physis and attain adequate purchase in the femoral epiphysis. Screws may be partially or fully threaded as long as the threads cross the physis [7]. This is in contrast to fixation of an undisplaced intracapsular hip fracture in a skeletally mature individual where fracture compression is achieved by ensuring the threads of the partially threaded screws are advanced past the fracture.

As the resting position of the femoral head in a SUFE, in relation to the neck, is inferior and posterior, pinning in situ requires appropriate orientation of the screw using intraoperative imaging to get adequate hold in the malpositioned femoral head. To achieve this, the entry point of the screw is anterosuperiorly on the proximal femur and the screw is angled towards the centre of the malpositioned head on both AP and lateral radiographs. Avoidance of the superior and anterior quadrant of the femoral head is important to prevent increased risk of AVN. A minimum of five threads must cross the physis, but, alternatively, when using a partially threaded screw with a 16 mm thread length, ensuring an estimated 50:50 split of threads in the epiphysis and metaphysis would be reasonable to attain adequate purchase in each and prevent further progression. The tip of the screw should be within 5–10 mm of the subchondral bone to get adequate purchase on the epiphysis without penetrating the articular surface as this leads to chondrolysis. To satisfy all criteria, using a partially threaded screw with a 32 mm thread length or a fully threaded screw may be more advisable [1, 7].

1.9.2 Open Reduction

Indications for an open reduction include severe acute or acute-on-chronic slips that cannot be adequately reduced by a closed technique. Options include surgical dislocation of the hip joint with reduction of the slip through an anterior approach or reduction with a proximal osteotomy (see below). Vascular disruption is associated with this technique.

1.9.3 Osteotomies [1, 8]

The aim of treatment by osteotomy is to restore gross alignment, enabling greater function, and reduce risk or delay onset of subsequent arthritis. It also has the potential benefit of relieving tension on the contracted vasculature. It has been used following failure to achieve adequate closed reduction in a severe acute slip, although this is associated with a high risk of complication. Correctional osteotomy of a malunion in missed or chronic slips causing persistent impairment of function have been demonstrated to reduce complication rates. Similarly, acute injuries can be immobilised in traction for three weeks to stabilise circulation to the femoral head prior to osteotomy to maintain a lower incidence of AVN and chondrolysis.

Osteotomy can be performed at differing levels along the femur, from the subcapital down to the subtrochanteric region. The more proximal the level of osteotomy, the greater the possible correction, so these are usually reserved for greater deformities. However, the more proximal the osteotomy, the greater the risk of osteonecrosis due to the closer proximity to the circumflex arteries and their integral ascending branches.

- **Subcapital:** A closing wedge osteotomy may be performed near the physis to correct the deformity. This is associated with a high risk of AVN (2%–100%) and chondrolysis (3%–37%).
- **Midcervical:** Dunn described the cuneiform osteotomy, which entails performing a shortening closing wedge osteotomy distal to the physis in order to reduce the tension of the posterior vessels. A modified Dunn procedure with formation of an epiphyseal vascular flap is also described in the literature.
- **Base of cervix:** Kramer et al. discussed an osteotomy at the base of the neck distal to the vasculature in the posterior retinaculum.
- **Trochanteric:** Abraham detailed a trochanteric osteotomy to correct varus and retroversion.

1.9.4 Cheilectomy

When a bone hump is formed following a missed SUFE, this can be resected to prevent impingement, as in FAI, to improve pain and movement.

1.9.5 Bone-Peg Epiphysiodesis

This technique involves creating a channel along the neck of the femur, across the physis and into the femoral head. As much of the physis is curetted out through this channel a cortico-cancellous bone graft is inserted as a peg to encourage closure of the physis. This is less frequently performed due to greater operative time and significant associated complications (e.g., further slippage, AVN, chondrolysis, infection and heterotopic ossification).

1.10 Postoperative Regime

- **Rehabilitation:** Following in situ fixation of a stable SUFE, patients are allowed to mobilise with early weight bearing as tolerated. However, unstable slips treated with in situ fixation must mobilise non-weight-bearing for a minimum period of six to eight weeks [1]. Osteotomies require a period of protected mobilisation until the osteotomy demonstrates signs of union.

- **Follow-up:** All patients are followed up regularly until the proximal femoral physis of the affected limb closes. Loss of range of motion or worsening pain may indicate AVN or chondrolysis.

- **Contralateral prophylactic fixation [1]:** Indications for prophylactic fixation of the other hip include a symptomatic contralateral hip and known underlying metabolic or endocrine disorders (as these are significant risk factors) or the patient is unreliable and compliance with follow-up is predicted to be poor. A SUFE in the contralateral hip is often diagnosed within one year of the initial SUFE.

1.11 Complications [1]

- **AVN [1]:** If a SUFE is untreated, there is significant incidence of AVN from disruption of the vasculature at the initial trauma, increasing with severity of the slip up to 50% if unstable. Tamponade of the blood supply to the femoral head from a tense haemarthrosis is also a cause of osteonecrosis. The rate of AVN is lowered significantly with in situ fixation due to stabilisation of the injury and vasculature, enabling coagulation. However, there has been evidence to show screw placement in the posterosuperior quadrant of the head may be detrimental to preserving blood supply (Fig. 1.11).

    Iatrogenic disruption of the vulnerable arterial supply may also occur from forced closed reduction of a SUFE or from a more proximally located osteotomy if one is performed. In light of this, it is clear to see why some confusion may be present about whether the initial injury or the subsequent treatment is the primary cause of such complications with this condition.

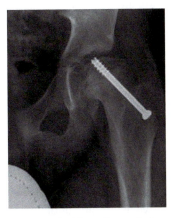

Figure 1.11 Anteroposterior right hip radiograph demonstrating AVN of the femoral head following screw fixation of a SUFE.

- **Chondrolysis [6]:** Lysis of the articular cartilage may occur following a SUFE due to ischaemia and synovial malnutrition and is more common in females. Cartilage damage is primarily due to iatrogenic causes though. Undetected screw penetration of the chondral surface is the most common (Fig. 1.12) cause, but this can also occur following open reduction or osteotomy.

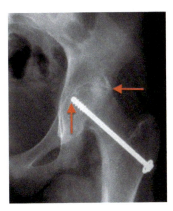

Figure 1.12 Anteroposterior left hip radiograph demonstrating close proximity of the screw to the joint with loss of joint space (chondrolysis).

- **FAI [4]:** Callus formation at the superior aspect of the neck following a SUFE results in a prominent bony lump. This gives rise to a cam-type impingement[1] and associated labral pathology that are often significantly symptomatic. Treatment options include arthroscopic resection of the bony lesion with the aim of relieving symptoms and delay onset of early associated degenerative changes.
- **Osteoarthritis:** Reports suggest significant incidence of osteoarthritic change following a SUFE.
- **Contralateral SUFE:** As discussed, up to 63% of SUFEs can be bilateral and so a high index of suspicion is required with ongoing follow-up until skeletal maturity.
- **Deformity or limb length discrepancy:** This is often a result of the proximal femur failing to remodel. Surgical management involves corrective osteotomy of the proximal femur.
- **Slip progression:** Increase in slip may still occur in 1%–2% of patients following in situ fixation.
- **Others:**
  - Infection 1%–2%
  - Chronic pain 5%–10%
  - Periprosthetic fracture

References

1. Aronsson DD, et al. Slipped capital femoral epiphysis: current concepts. *J Am Acad Orthop Surg*, 2006; **14**(12):666–679.
2. Iyer D, Brueton R. Slipped upper femoral epiphysis with haemophilia A. *Indian J Orthop*, 2007; **41**(3):250–251.
3. Agamanolis DP, et al. Slipped Capital femoral epiphysis: a pathological study. II: An ultrastructural study of 23 cases. *J Paediatr Orthop*, 1985; 5:47–58.
4. Basheer SZ, et al. Arthroscopic treatment of femeroacetabular impingement following slipped capital femoral epiphysis. *Bone Joint J*, 2016; **98-B**(1):21–27.

---

[1]The word "cam" comes from the Dutch word meaning "cog," in which the femoral head is aspherical (not perfectly round). In cam impingement, therefore, the femoral head cannot rotate smoothly inside the acetabulum, forming a bump on the edge of the femoral head that grinds the cartilage inside the acetabulum.

5. Kishan S, et al. Biomechanical stability of single-screw versus two-screw fixation of an unstable slipped capital femoral epiphysis model: effect of screw position in the femoral neck. *J Paediatr Ortop*, 2006; **26**(5):601–605.
6. Phillips SA, et al. The timing of reduction and stabilisation of the acute, unstable, slipped upper femoral epiphysis. *J Bone Joint Surg Br*, 2001; **83-B**(7):1046–1049.
7. Upasani V, et al. Biomechanical analysis of single screw fixation for slipped capital femoral epiphysis: are more threads across the physis necessary for stability? *J Paediatr Orthop*, 2006; **26**(4):474–478.
8. Biring GS, et al. Outcomes of subcapital cuneiform osteotomy for the treatment of severe slipped capital femoral epiphysis after skeletal maturity. *J Bone Joint Surg Br*, 2006; **88-B**:1379–1384.

Chapter 2

Advances in Fractures and Dislocations of the Hip Joint

Thomas Pepper, Philip Ahrens and M. Zahid Saeed
Trauma and Orthopaedic Department, University College London Medical School,
Royal Free Hospital, Pond Street, London NW3 2QG, United Kingdom
z.saeed@nhs.net, mzahidsaeed@hotmail.com

2.1 Hip Fractures

A hip fracture is defined as a fracture occurring in the region between the femoral head and 5 centimetres inferior to the lesser trochanter [1]. It is often a surrogate marker of underlying poor health, and this is reflected in a 10% mortality rate at one month and approximately 33% mortality at one year [1]. Risk factors for hip fracture include advanced age, osteoporosis and recurrent falls – the latter is the commonest mechanism of injury in the older population. Hip fractures may be divided into intracapsular and extracapsular fractures.

The Hip Joint in Adults: Advances and Developments
Edited by K. Mohan Iyer
Copyright © 2018 Pan Stanford Publishing Pte. Ltd.
ISBN 978-981-4774-72-7 (Hardcover), 978-1-351-26244-6 (eBook)
www.panstanford.com

2.1.1 Intracapsular Fractures

2.1.1.1 Classification

These may be further defined as subcapital (immediately beneath the head), transcervical (middle region of the femoral neck) and basal (just proximal to the trochanters). Intracapsular fractures may be classified radiographically according to the displacement of fragments [2]:

- Stage I – an incomplete impacted fracture
- Stage II – a complete but undisplaced fracture
- Stage III – a complete fracture with displacement but some end-to-end bony contact (Fig. 2.1)
- Stage IV – a fracture with no end-to-end bony contact

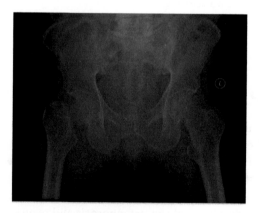

Figure 2.1   Right-sided Garden's stage III fracture.

2.1.1.2 Clinical features

A displaced fracture manifests clinically as a shortened and externally rotated leg. Patients may still be able to weight-bear on impacted fractures, and patients with dementia may not complain of pain. If there is strong clinical suspicion of a fracture but this is not demonstrated radiographically, a magnetic resonance imaging (MRI) or bone scintigraphy scan should be ordered.

2.1.1.3 Management

In stage III and IV fractures the intramedullary and retinacular vascular supply to the femoral head is interrupted, leaving the only supply from the vessels of the ligamentum teres. The latter blood supply is frequently inadequate or absent in the older population, and as a result there is a high risk of nonunion and avascular necrosis. The treatment of choice for displaced intracapsular fractures in advanced age therefore involves prosthetic replacement of the femoral head. Current National Institute for Health and Care Excellence (NICE) guidance suggests total hip replacement for patients who are able to walk independently outdoors with a stick or better, are not cognitively impaired and are medically fit for the procedure [1]. In the remainder, hemiarthroplasty is indicated. Whichever modality is appropriate, cemented prostheses result in improved mobility and less pain [3].

Stage I and II fractures do not cause the same degree of vascular compromise to the femoral neck and may be treated reliably with closed reduction and internal fixation using cannulated screws or a dynamic hip screw. Fixation facilitates early postoperative mobilisation and minimises pulmonary complications and pressure sores, but in the bedbound or demented patients with little pain, nonoperative treatment may be acceptable. If surgery is planned it should be undertaken with urgency in order to minimise ischaemia of the femoral head, although this must be balanced with sufficient preoperative optimisation of the patient.

2.1.1.4 Complications

Complication rates accompanying hip fractures in older patients are high, but most are associated with pre-existing comorbidities, which leave them especially prone to pneumonia and pressure sores. Venous thromboembolism is a great risk in this group of patients, and this may be reduced by appropriate use of compression stockings, perioperative heparin prophylaxis and early mobilisation.

Avascular necrosis of the femoral head occurs in approximately 10% of patients with undisplaced fractures. It is only possible to diagnose this retrospectively, it being visible on bone scintigraphy several weeks postoperatively and radiographically several months postoperatively. Although the fracture may still unite, the joint suffers irreversibly damage. In patients over 45 years total hip replacement is indicated.

Nonunion affects approximately a third of all intracapsular fractures, but the risk is higher in stage IV fractures. The aetiology may be femoral head ischaemia, suboptimal reduction or inadequate fixation, and management of the nonunion depends on addressing the underlying factor, with hemiarthroplasty or total hip replacement remaining viable options.

2.1.2 Extracapsular Fractures

2.1.2.1 Classification

Extracapsular hip fractures comprise intertrochanteric fractures, reverse oblique fractures, trochanteric avulsion fractures (Fig. 2.2) and subtrochanteric fractures. Intertrochanteric fractures (Fig. 2.3) are the most common and may be further classified according to the degree of communition and therefore instability [4][1]. Reverse oblique fractures are generally unstable as weight bearing acts to displace the fragments. Subtrochanteric fractures (Fig. 2.4) may be associated with significant blood loss; if nontraumatic the aetiology may be an osteolytic metastatic deposit.

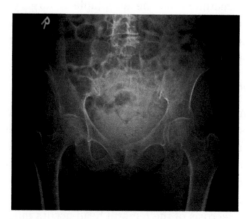

Figure 2.2 Left-sided trochanteric avulsion fracture.

---

[1]Kyle classification of intertrochanteric femoral fractures, OrthoFRACS (http://www.orthofracs.com/adult/trauma/principles/fracture-classification/classification-femur-intertrochanteric.html).

Hip Fractures | 25

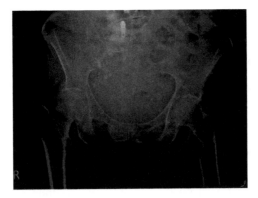

Figure 2.3 Left-sided intertrochanteric fracture.

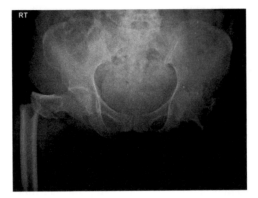

Figure 2.4 Right-sided subtrochanteric fracture.

2.1.2.2 Clinical features

Clinically, extracapsular fractures present in the same way as intracapsular fractures, although the leg may be more externally rotated as the fragments are unsupported by joint ligaments.

2.1.2.3 Management

Extracapsular fractures unite readily, and there are no significant differences in outcome between conservative and operative management [5]; there is a relatively low risk of avascular necrosis of the femoral head as the retinacular vasculature is undisrupted. These fractures are treated by internal fixation nevertheless – using a dynamic compression screw or, if subtrochanteric, an intramedullary nail – in order to facilitate early rehabilitation of the patient.

2.1.2.4 Complications

Although these fractures are less prone to avascular necrosis of the femoral head, patients are subject to the same general postoperative complications as intracapsular fractures – namely pneumonia, venous thromboembolism and pressure sores – all of which may be mitigated by early mobilisation. In addition, there is a risk of metalwork "cut out" of the osteoporotic bone, particularly if the screw has been poorly positioned or there is a suboptimal reduction. Malunion is relatively common, resulting in varus and external rotation deformities, but these are rarely clinically significant. Nonunion is uncommon, but if the fracture has not united after six months, revision is indicated.

2.2 Traumatic Dislocations of the Hip Joint

The hip joint is very stable, and as a result dislocations are relatively rare, requiring extreme force to occur. A dislocated (nonprosthetic) hip is an orthopaedic emergency and should be reduced as quickly as possible under general anaesthesia with a muscle relaxant. It is imperative to test and document neurovascular status prior to attempting reduction. Hip dislocation may be classified according to the direction of movement of the femoral head – posterior, anterior or central.

2.2.1 Posterior Dislocation

This type forms the vast majority of hip dislocations and is frequently associated with a road traffic accident involving a head-on force that causes the knee to strike against the dashboard of the vehicle, forcing the femur backwards. Acetabular injury depends on the position at the time of impact. If the hip is abducted at the time of posterior dislocation, a simultaneous acetabular fracture will occur. However, if the hip is adducted it is unsupported posteriorly and acetabular fracture may be avoided. In isolated posterior hip dislocation, clinically the leg appears shortened and the hip will be held adducted and internally rotated. This picture may be complicated if there is a concomitant femoral shaft fracture. Plain radiographs will show the femoral head to be displaced from and superior to the acetabulum (Fig. 2.5). A computed tomography (CT) scan will better demonstrate any acetabular fracture.

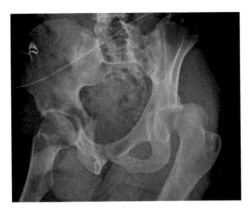

Figure 2.5 Left-sided posterior hip dislocation.

To restore joint congruity the hip and knee are flexed to 90° while traction is applied in the line of the femur and the pelvis is stabilised. A plethora of methods to achieve this have been described (Bigelow [6], Allis [7], Lefkowitz [8] and East Baltimore [9] to name a few). Following reduction the hip should be tested for instability and, if it is present, the patient should be considered for posterior acetabular repair. After reduction the patient should be put into traction for several weeks, with rehabilitation commencing as soon as pain allows. If closed reduction is unsuccessful or postreduction imaging shows intra-articular loose fragments, an operative intervention is necessary.

The sciatic nerve is closely related to the posterior surface of the femoral head, and consequently nerve injury accompanies approximately 10% of posterior hip dislocations [10]. Permanent injury is uncommon, but while nerve recovery occurs, the insensate areas of the limb must be protected from further injury and an ankle dorsiflexion splint should be worn to aid walking. A new sciatic nerve injury following reduction of the hip necessitates exploration of the joint in order to exclude entrapment of the nerve in the joint space.

A concomitant femoral shaft fracture is rare but often leads to late diagnosis of the hip dislocation, as the typical clinical picture is obfuscated. A thorough clinical and radiographic examination will mitigate this.

Avascular necrosis of the femoral head is a late-occurring complication of dislocation, and its incidence rises in proportion to the duration of time dislocated. Ischaemia is thought to result from vessel compression or spasm [11], which is relieved by prompt reduction. If the hip is reduced within 12 hours, the risk of avascular necrosis is approximately 2%–4%; if reduction is delayed past 12 hours, this risk increases by 5.6 times [12]. Osteonecrosis is appreciated earliest on MRI or at 6-12 weeks on plain radiographs. Treatment is operative and often best achieved through total joint replacement.

Secondary osteoarthritis is a potential sequela following hip dislocation and may be a result of avascular necrosis, loose bone fragments in the joint and/or trauma to the articular surface at the time of dislocation.

2.2.2 Anterior Dislocation

Anterior dislocation is much rarer than posterior but in the same way it usually accompanies major vehicular trauma. Clinically, the hip is held abducted – sometimes extremely so – and externally rotated. The leg length is maintained as the femoral head is prevented from moving proximally by the origin of the rectus femoris. The contour of the dislocated head will be visible and palpable in the groin (anteroinferior dislocation) or immediately anterior to the joint (anterosuperior dislocation).

Reduction of an anterior dislocation is achieved in much the same way as a posterior dislocation, except that the hip should be adducted while traction is applied.

Careful examination and documentation of the femoral nerve and vessel function is necessary in anterior dislocations, as these structures are closely related to the anterior hip and are at risk. Avascular necrosis of the femoral head is a possibility but occurs with less frequency than in posterior dislocations.

2.2.3 Central Dislocation

Central dislocation occurs when the femoral head is driven into the pelvis through the floor of the acetabulum. This may be caused by a fall onto the side from a height or in association with another major trauma. Clinical features are variable depending on the aetiology, but the leg is shortened. A CT scan is the most useful imaging modality, but two oblique plain radiographic views should also be sought in addition to anteroposterior and lateral views in order to conduct sufficient initial examination of the acetabulum.

Following resuscitation of the patient and reduction of the dislocation, the hip injury remaining is a complex acetabular fracture. Operative management is indicated for all unstable hips and all displaced acetabular fractures. If there are medical contraindications to surgery or the acetabular fracture is undisplaced and the hip stable, conservative management involving longitudinal traction for six weeks followed by minimal weight bearing for a further six weeks may be appropriate.

2.2.4 Dislocated Hip Prosthesis

This is a relatively common presentation and, similar to a native hip, usually requires reduction under general anaesthetic. However, this need not be undertaken with such urgency as there is no risk of avascular necrosis of the femoral head. Physiotherapy and adherence to hip precautions are the mainstay of prosthetic dislocation prevention, but refractory hips may be improved by replacement with a jumbo prosthesis [13] or by the use of retaining bands.

References

1. National Institute for Health and Care Excellence (NICE). *Hip Fracture: Management*, 2011.
2. Garden R. Low angle fixation in fractures of the femoral neck. *J Bone Joint Surg*, 1961; **43**:647–663.
3. Parker MJ, Gurusamy KS, Azegami S. Arthroplasties (with and without bone cement) for proximal femoral fractures in adults. [Update of *Cochrane Database Syst Rev*, 2006; **3**:CD001706; PMID: 16855974.] *Cochrane Database Syst Rev*, 2010; **6**:CD001706.

4. Kyle RF, Gustilo RB, Premer RF. Analysis of six hundred and twenty-two intertrochantenc hip fractures. *J Bone Joint (Am)*, 1979; **61**:216–221.
5. Handoll H, Parker M. Conservative versus operative treatment for hip fractures in adults: review. *Cochrane Collab*, 2008; 1–31. doi:10.1002/14651858.CD000337.pub2
6. Bigelow H. On dislocation of the hip. *Lancet*, 1878; **111**:860–862.
7. Allis O. *The Hip* (Dorman), 1895.
8. Lefkowitz M. A new method for reduction of hip dislocations. *Orthop Rev*, 1993; **22**:253–256.
9. Schafer SJ, Anglen JO. The East Baltimore Lift: a simple and effective method for reduction of posterior hip dislocations. *J Orthop Trauma*, 1999; **13**:56–57.
10. Cornwall R, Radomisli TE. Nerve injury in traumatic dislocation of the hip. *Clin Orthop Relat Res*, 2000; 84–91. Available at <http://www.ncbi.nlm.nih.gov/pubmed/10943188>
11. Shim SS. Circulatory and vascular changes in the hip following traumatic hip dislocation. *Clin Orthop Relat Res*, 1979; 255–261. Available at <http://www.ncbi.nlm.nih.gov/pubmed/477079>
12. Kellam P, Ostrum RF. Systematic review and meta-analysis of avascular necrosis and posttraumatic arthritis after traumatic hip dislocation. *J Orthop Trauma*, 2016; **30**:10–16.
13. Beaulé PE, Schmalzried TP, Udomkiat P, Amstutz HC. Jumbo femoral head for the treatment of recurrent dislocation following total hip replacement. *J Bone Joint Surg Am*, 2002; **84-A**:256–263.

Chapter 3

Advances in Surface Replacement of the Hip Joint

Senthil Muthian, Nicholas Garlick and M. Zahid Saeed

Trauma and Orthopaedic Department, University College London Medical School, Royal Free Hospital, Pond Street, London NW3 2QG, United Kingdom
z.saeed@nhs.net, mzahidsaeed@hotmail.com

3.1 Hip Resurfacing

3.1.1 Introduction

Hip resurfacing has been an attractive alternative to total hip replacement as it involves less bone resection and reduced risk of dislocation, with an excellent quality of life in a young, active individual with hip arthritis. The primary goal of metal-on-metal hip resurfacing arthroplasty (MOM HRA) is to buy time until an age at which conventional total hip arthroplasty (THA) would be suitable for the patient. If MOM HRA can offer around 10 years of good function without jeopardising the possibility of later conversion to THA, it would be a viable conservative option. These benefits have

---

The Hip Joint in Adults: Advances and Developments
Edited by K. Mohan Iyer
Copyright © 2018 Pan Stanford Publishing Pte. Ltd.
ISBN 978-981-4774-72-7 (Hardcover), 978-1-351-26244-6 (eBook)
www.panstanford.com

been superseded by other concerns recently, limiting the scope of hip resurfacing.

3.1.2 History

Since the development of the first successful THA by Sir John Charnley [1] in the 1960s, the total hip replacement has remained the gold standard of surgical treatment for symptomatic arthritis of the hip. The concept of hip resurfacing was well known at that time but had met with limited success and hence was not popular. Charnley developed Teflon-on-Teflon resurfacing in the 1950s, which was associated with a high failure rate due to poor wear characteristics [2]. Subsequent resurfacing hip arthroplasties by Townley, Muller, Girard and Furuya were recognised but did not gain popularity [3]. There was re-emergence of interest after work done by McMinn and Treacy in the UK, using hybrid metal-on-metal (MOM) bearings with hydroxyapatite-coated shells [4]. The early results were promising, but on long-term follow-up the risk of complications and failure has now resulted in this procedure being used in a limited capacity.

3.1.3 Problems with Conventional Total Hip Replacement

There are several issues with conventional THA that are addressed by a resurfacing arthroplasty, including the following:

- **Dislocation:** This has been reported to be approximately 2% after THA [5]. When using a smaller head than the patient's native femoral head (28–32 mm), dislocation is more common, but when a same-sized head is used as in resurfacing (38–58 mm) the dislocation rate is much lower.
- **Leg lengthening:** This is one of the commonest complications of THA. It is currently recommended that the leg length discrepancy should be less than 10 mm. Leg length discrepancy remains one of the main causes of litigation in the United States following THA. It is very unlikely for a patient to get leg lengthening after a resurfacing arthroplasty.
- **Bone resection:** A significant amount of bone resection is required in THA, and poor bone stock is one of the main

challenges facing the revision hip surgeon. In contrast, resurfacing preserves bone stock and conversion to total hip replacement is relatively simple.

- **Osteolysis:** It presents with a linear pattern in cemented cups and with a cavitary pattern in uncemented acetabular cups. In resurfacing metal cups, use of porous coated metal results in very good fixation and osteolysis is not expected as no polyethylene is present in the system.
- **Stress shielding:** This is a common phenomenon observed causing osteolysis in the proximal femur following hip replacements. This is avoided in resurfacing due to the point of fixation and point of loading being coincident.
- **Revision:** Whilst revision of a conventional THA is a technically demanding and complex procedure with a high complication rate, revising a resurfacing prosthesis is similar to a total hip replacement.

3.1.4 Young Patients

Additional advantages of resurfacing in young and active patients include restoration of normal length, offset and anteversion and better proprioceptive feedback, resulting in a higher functional result and activity level.

3.1.5 Biomechanics

A near-normal anatomy of the proximal femur is maintained after resurfacing [6]. Acetabular components are mainly designed as monoblock implants, which can make subsequent revision difficult. This is due to the smaller ratio of the resurfaced head and relatively thick femoral neck [7]. Impingement of the femoral neck onto the rim of the acetabular component can result in deformation of the bearing surfaces, subluxation, femoral neck fracture and component loosening.

Biomechanical parameters, including neck length, offset, anteversion and head size, are well preserved. Finite element analysis showed that the changes in peak stresses after hip resurfacing arthroplasty (HRA) were low in relation to the failure strength of bone and hence the fracture risk was low [8]. The bone strains produced after HRA are closer to normal than those produced after THA.

Furthermore, proprioceptive feedback is better preserved following resurfacing although this concept has been challenged [9].

Biomechanical testing has revealed that resurfaced hips have a reduced range of motion and a higher rate of neck-to-cup impingement as compared to conventional arthroplasty [6, 10]. Resurfacing systems impinged almost entirely on the femoral neck, while conventional hip arthroplasties had a varied impingement profile [10, 11].

3.1.6 Tribology

Polyethylene cannot be used as the bearing material in hip resurfacing as the inevitable use of a large femoral head size in the resurfacing arthroplasty would lead to excess polyethylene debris, osteolysis, loosening and collapse of femoral heads. Metal bearings would be durable for use in young, active patients when used with a large diameter articulation. In addition, metallic bearings are capable of manufacture as a thin component to avoid excessive resection of valuable bone stock.

Thick film lubrication can occur in both MOM bearings and ceramic-on-ceramic bearings. Thick film lubrication is not possible in a metal-on-polyethylene or ceramic-on-polyethylene bearing because of the high surface roughness of polyethylene [12].

MOM bearings have a low frictional torque through a wide range of head sizes as compared to metal-on-poly bearing surfaces [13].

Prosthetic bearing surfaces for hip resurfacing operations are manufactured from high-carbon (0.20% to 0.25%) cobalt-chromium-molybdenum alloy. The large-diameter MOM components result in very low wear if the diametral clearance (the difference between the diameters of the femoral head and the acetabular cup) is optimised. Deformation is greater in acetabular components manufactured with thin (2 to 4 mm) walls to conserve the pelvic bone. Greater deformation can lead to higher diametral clearance and increased wear. In a 50 mm MOM articulation, at the commonly manufactured clearance of 100 microns the fluid film thickness is increased four times compared to that generated with the same clearance in a

28 mm MOM articulation. This fluid film thickness in the 50 mm bearing is enough to completely separate the two articulating surfaces, resulting in low wear rates [12].

3.1.7 Indications

Resurfacing is an excellent option for young, active patients who place a high mechanical demand on their hips. Current indications include male sex, age less than 60 years and larger patient size.

3.1.8 Risk Factors

Several risk factors have been identified as contributory to early failure and these include:

- Increased acetabular inclination angle (AIA)
- Small femoral component sizes
- Implant design factors
- Developmental dysplasia of the hip (DDH)
- Female sex
- Increased patient age
- Renal disease
- Developmental decreased bone mineral density

3.1.9 Contraindications

- Female patients
- Infection
- Skeletally immaturity
- Severe vascular insufficiency or neuromuscular disease
- Inadequate bone stock, including severe osteoporosis, avascular necrosis (AVN) with loss of >50% of the head and femoral head cysts larger than 1 cm
- Severe renal insufficiency
- Immunosuppression or high-dose corticosteroid therapy
- Severe obesity
- Metal sensitivity (e.g., to jewellery)

3.1.10 Cautions

- Patients requiring a 48 mm femoral head size are at a moderately elevated risk of requiring revision surgery earlier than expected for Birmingham hip resurfacing (BHR, [14]).
- Patients on medications (such as high-dose or chronic aminoglycoside treatment) or with comorbidities (such as diabetes) that increase the risk of future, significant renal impairment should be advised of the possibility of increase in systemic metal ion concentration. Preoperative and postoperative monitoring of renal function should be done for these patients.

3.1.11 Selection of Ideal Patients

The most common indication for MOM HRA is end-stage osteoarthritis (OA) in young, active patients. In these patients, having good bone quality in the femoral head and neck and proper anatomy around the joint produces excellent outcomes if the surgeon has the appropriate skills and training for MOM HRA. Loss of bone stock may compromise the stability and osteointegration of the implant.

Beaulé et al. have developed the Surface Arthroplasty Risk Index (SARI) as a guide for patient selection for HRA [15]. The risk of implant failure is high if the SARI is greater than or equal to 3. The authors applied the SARI in a study of young patients undergoing MOM HRA and showed its impact on clinical outcomes. They found that the SARI was significantly higher in patients with failed implants as compared to patients in whom the implants were performing well. Factors included in the SARI are femoral cysts larger than 1 cm, activity level, previous surgery and weight.

3.1.12 Female Sex

A Canadian Arthroplasty Society report found that the female sex is an independent risk factor for loosening and early failure [16]. In a systematic review, Haughom [17] reported a higher rate of failure in women but argued that this is not necessarily an independent risk factor, due to various confounding factors, including smaller component sizing, gender differences in ligamentous laxity, bone quality and a higher prevalence of DDH leading to higher combined anteversion.

3.1.13 Surgical Considerations

3.1.13.1 Approach

The commonest approach used in the UK is the posterior approach. This provides excellent exposure but risks the deep branch of the medial femoral circumflex artery, which may result in AVN, collapse and failure. The trochanteric slide approach (surgical dislocation) aims to preserve the vascularity of the femoral head but can result in trochanteric nonunion and painful hardware. The direct anterior approach is associated with increased risk of intraoperative femoral neck fracture and injury to the lateral femoral cutaneous nerve. There is no definite evidence to support any particular approach, and all approaches have been reported to have excellent outcomes [18]. A recent Canadian investigation of 550 consecutive patients reported a five-year survivorship of 94.5%, with no statistical differences in patient outcomes or reoperation rates between anterior, trochanteric slide, lateral or posterior approaches [19].

3.1.13.2 Acetabulum

High metal ion levels and wear rates are related to increased acetabular inclination and higher degrees of combined anteversion. Liu and Gross [20] described a safe zone for acetabular component positioning (relative acetabular inclination limit [RAIL]) based on implant size and AIA. When the AIA fell below the RAIL, there were no adverse wear failures or dislocations. Additionally, the risk of having metal ion levels above 10 μg/L was <1% when the AIA fell below the RAIL, regardless of the patient's gender.

Amstutz [21] in 2012 reported that the only variable associated with revision is the contact patch-to-rim distance (CPRD). The CPRD is described as the distance between the centre of the contact patch (the contact area of the femoral articular surface with the acetabular component during any and all functions) and the acetabular rim. Cups with a higher AIA result in a shortened CPRD, which is thought to increase edge loading, accelerate wear and lead to elevated ion levels. Amstutz [21] concluded that the preferred cup orientation is an abduction angle of approximately 42° ± 10° and an anteversion angle of approximately 15° ± 10°.

The CPRD is an indirect predictor for increased metal ion levels and abnormal wear and a value less than or equal to 10 mm has a 37-fold increased risk of having elevated serum cobalt levels and an 11-fold increased risk of having elevated serum chromium levels [22].

3.1.13.3 Femur

Care must be taken to avoid notching of the femoral neck and varus malposition of the femoral component. Intraoperative notching of the neck increases the risk of postoperative fracture [23]. A 2 mm notch reduces the neck strength by 25%, and a 5 mm notch reduces the neck strength by 50%. Varus component malposition increases the strain at the superior neck, with a resultant increased risk of fracture and loosening. A smaller head size has been found to have an increased rate of wear, adverse local tissue reaction (ALTR) and component failure in both clinical and biomechanical settings. As the femoral head diameter increases, the articulation is more likely to promote fluid film lubrication, subsequently improving wear characteristics [23].

3.1.14 Results

Daniel [24] reported in 2014 on a 15-year follow-up and found that men have better implant survival of 98.0% at 15 years than women, with an overall survival of 91.5%. In women <60 years, the overall survival is of 90.5%. Patients under 50 years with OA fare best (99.4%), with no failures in men in this group.

Van der Straeten [25], in 2013, reported a 92.4% survival rate at a 13.2-year follow-up.

3.1.14.1 Implant-specific results

BHR [14] and Conserve Plus (Wright Medical) are Food and Drug Administration (FDA) approved and have been commonly used in the UK (BHR) and in the United States. They have a survival rate of 88%–99% at 10 years.

Two devices have been recalled from the market. In 2010, the articular surface replacement (ASR; DePuy, Warsaw, IN, US) was recalled because of high rates of early failure secondary to poor acetabular design, which resulted in an unacceptably low clearance, cup deflection and increased wear. The Durom acetabular component (Zimmer Inc., Warsaw, IN, US) had higher early failure rates than other designs, with no single design flaw indicated, and was also recalled.

3.1.14.2 Mechanism of failure

MOM HRA failures fall into one of two categories: mechanical failures (such as femoral neck fracture) and bearing-related failures (such as soft-tissue reactions and osteolysis). The most commonly reported reasons for failure of MOM HRA requiring revision include femoral neck fracture, collapse of the femoral head and component loosening [26]. Approximately one in five MOM hip replacements will need revision 10–13 years after they are implanted.

3.1.14.3 Complications

- **Generic complications:** Such as infection, nerve damage, deep vein thrombosis and pulmonary embolism (DVT/PE) and haematoma.
- **Specific failure:** Femoral neck fracture and osteolysis.
- **Local metal ion toxicity:** Soft-tissue inflammatory reactions to metal debris are a recognised complication of MOM hip arthroplasty. These reactions, which are grouped under the umbrella term 'adverse reactions to metal debris' (ARMD), have been called inflammatory pseudotumour, aseptic lymphocytic vasculitis associated lesion (ALVAL) and metallosis. The spectrum of ARMD is extensive and ranges from small asymptomatic cysts to large soft-tissue masses (pseudotumours). ALVAL is a histological diagnosis that describes the unique cellular changes that occur in response to metal particles, namely cobalt and chromium ions. The mechanism is believed to be a T lymphocyte–mediated type IV hypersensitivity reaction, with tissue damage occurring as a result of cytotoxic T cells and activated monocytes/macrophages. The presence of this reaction is thought to be

proportional to the amount of wear debris released but has also been observed in patients with smaller amounts of wear debris [27].

- **Systemic metal ion toxicity:** Concerns regarding long-term risk of cancer, risk in renal failure patients and risk in pregnant women [27].

3.1.15 Imaging

- **Plain radiography:** It should be done in all symptomatic patients with MOM HRA. Plain films tend to underestimate the prevalence of pseudotumours and can therefore falsely reassure clinicians.
- **Ultrasound scan:** Cost effective and safe, this imaging modality provides better visualisation of joint effusions and tendinous pathologies compared with magnetic resonance imaging (MRI). Because it is less sensitive for detection of pseudotumours it has been suggested that ultrasound only be used when metal artefact reduction sequence (MARS) MRI is poorly tolerated, contraindicated or unavailable. Furthermore, as ultrasound is operator dependent, an experienced musculoskeletal radiologist is required to accurately interpret the images.
- **Computed tomography:** While computed tomography (CT) gives a better bony resolution than plain radiography, because osteolysis is a late presentation in ARMD, CT is not suitable as an imaging modality for screening patients.
- **MRI:** With the advent of MARS, MRI has quickly become one of the most popular imaging modalities for visualising pseudotumours and muscle atrophy. When used as a screening tool, it can also detect soft-tissue inflammatory reactions in asymptomatic patients. Due to its high specificity and sensitivity for detection of these reactions and its versatility in assisting with preoperative planning and longitudinal comparison, MARS MRI has been recommended over ultrasound as the first-line modality for assessment of periprosthetic soft tissues in patients with MOM HRA.

3.1.15.1 Radiographic features of impending failure

- **Femoral peg-shaft angle:** A progressive varus indicating osteonecrosis, loosening, fracture and femoral head collapse
- **Femoral component:** A progressive reduction in the distance from the tip of the femoral peg to the lateral femoral cortex, indicating implant subsidence, femoral head collapse and component loosening
- **Peg radiolucent lines:** Component loosening when progressive and >2 mm

3.1.15.2 Radiological features requiring closer observation

- Femoral neck nonprogressive narrowing of <10%
- Femoral neck initial rapid or extensive narrowing of >10%
- Femoral neck scalloping at the neck-rim junction
- Femoral neck osteolysis

3.1.16 Revision Surgery

Recurrence of ARMD and dislocation are the two most common complications with revision surgery for failed resurfacing. When revising MOM HRA implants, it is recommended to utilise a non-MOM articulation such as ceramic-on-ceramic or ceramic-on-polyethylene articulation to reduce local metal ion release and hence the chances of recurrence of ARMD [28]. Surgeons must ensure the entire ARMD lesion is excised and the metal debris completely removed. Failure to do so may contribute to persistence or recurrence of ARMD after revision. Resection of ARMD and pseudotumours have been likened to an oncological procedure where the surgeon should aim for clear resection margins to reduce the risk of recurrence. This is often difficult, however, as pseudotumours can pass through tissue planes, extend into the pelvis and even involve vital neurovascular structures. In these cases, assistance should be sought from experienced vascular surgeons and/or pelvic reconstruction specialists [26].

3.1.17 Current Recommendations for Follow-Up

The Medicines and Healthcare Regulatory Agency (MHRA) has issued guidelines for following up MOM HRA patients as below.

	MOM hip resurfacing (no stem)	
	Symptomatic patients	Asymptomatic patients
Patient follow-up	Annually of the life of the implant	According to local protocols
Imaging: MARS MRI or ultrasound	Recommended in all cases	No - unless concern exists for cohort or patient becomes symptomatic
1st blood metal ion level test	Yes	No - unless concern exists for cohort or patient becomes symptomatic
Results of 1st blood metal ion level test	Blood metal ion level > 7ppb indicates potential for soft-tissue reaction	
2nd blood metal ion level test	Yes - 3 months after 1st blood test if result was >7ppb	
Results of 2nd blood metal ion level test	Blood metal ion level >7ppb indicates potential for soft-tissue reaction especially if greater than previously	
Consider need for revision	If imaging is abnormal and/or blood metal ion levels rising	

Guidance notes:
- This chart provides general guidance to the management of patients with MOM hips. However, each patient must be judged individually.
- Fluid collection alone around the joint in an asymptomatic patient can safely be observed with interval scanning.
- Imaging should carry more weight in decision-making than blood ion levels. Muscle or bone damage on the MRI scan is of utmost concern.
- Seven parts per billion (ppb) equals 119 nmol/L cobalt or 134.5 nmol/L chromium.

3.1.18 Summary

Hip resurfacing is now very limited in practice with most UK surgeons, and very careful patient selection is required, along with meticulous surgical technique and prolonged follow-up, in order to have a successful long-term outcome with MOM HRA. Further studies are required on the systemic effects of metal ion toxicity in patients undergoing ARMD revision in order to make definitive recommendations to improve the outcome in these situations.

References

1. Charnley J. Total hip replacement by low-friction arthroplasty. *Curr Orthop Pract* [Internet], 2014; **25**:105–113. Available from: http://ovidsp.ovid.com/ovidweb.cgi?T=JS&PAGE=reference&D=emed12&NEWS=N&AN=2014329203
2. Griffith MJ, Seidenstein MK, Williams D, Charnley J. Socket wear in Charnley low friction arthroplasty of the hip. *Clin Orthop Relat Res* [Internet], 1968; **1**(137):37–47. Available from: http://www.ncbi.nlm.nih.gov/pubmed/743841
3. Pritchett JW. Curved-stem hip resurfacing: minimum 20-year follow-up. *Clin Orthop Relat Res*, 2008; **466**(5):1177–1185.
4. Daniel J, Ziaee H, Kamali A, Pradhan C, Band T, Mcminn DJW, et al. Ten-year results of a double-heat-treated metal-on-metal hip resurfacing. *J Bone Joint Surg Br*, 2010; **92**:20–27.
5. Dargel J, Oppermann J, Brüggemann GP, Eysel P. Dislocation following total hip replacement. *Dtsch Arztebl Int*, 2014; **111**(51–52):884–890.
6. Girard J. Biomechanical reconstruction of the hip: a randomised study comparing total hip resurfacing and total hip arthroplasty. *J Bone Joint Surg Br* [Internet], 2006; 721–726. Available from: http://www.bjj.boneandjoint.org.uk/cgi/doi/10.1302/0301-620X.88B6.17447
7. Bader R, Klüß D, Gerdesmeyer L, Steinhauser E. Biomechanische aspekte zur implantatverankerung und kinematik von oberflächenersatzhüftendoprothesen. *Orthopade*, 2008; 634–643.
8. Little JP, Taddei F, Viceconti M, Murray DW, Gill HS. Changes in femur stress after hip resurfacing arthroplasty: response to physiological loads. *Clin Biomech*, 2007; **22**(4):440–448.

9. Larkin B, Nyazee H, Motley J, Nunley RM, Clohisy JC, Barrack RL. Hip resurfacing does not improve proprioception compared with THA. *Clin Orthop Relat Res*, 2014; 555–561.

10. Ganapathi M, Vendittoli PA, Lavigne M, Günther KP. Femoral component positioning in hip resurfacing with and without navigation. *Clin Orthop Relat Res*, 2009; 1341–1347.

11. Ellison P. Theoretical relationships between component design, patient bone geometry and range-of-motion post hip resurfacing. *Proc Inst Mech Eng H*, 2012; **226**(3):246–255.

12. Jin ZM, Dowson D, Fisher J. Analysis of fluid film lubrication in artificial hip joint replacements with surfaces of high elastic modulus. *Proc Inst Mech Eng H* [Internet], 1997; **211**(3):247–256. Available from: http://www.ncbi.nlm.nih.gov/pubmed/9256001

13. Shimmin A, Beaulé PE, Campbell P. Metal-on-metal hip resurfacing arthroplasty. *J Bone Joint Surg Am* [Internet], 2008; **90**(3):637–654. Available from: http://content.wkhealth.com/linkback/openurl?sid=WKPTLP:landingpage&an=00004623-200803000-00024

14. Robinson E, Richardson JB, Khan M. *Minimum 10 Year Outcome of Birmingham Hip Resurfacing (BHR), a Review of 518 Cases from an International Register*. Oswestry Outcome Centre, Oswestry, UK.

15. Beaulé PE, Dorey FJ, Le Duff MJ, LeDuff M, Gruen T, Amstutz HC. Risk factors affecting outcome of metal-on-metal surface arthroplasty of the hip. *Clin Orthop Relat Res* [Internet], 2004; (418):87–93. Available from: http://www.ncbi.nlm.nih.gov/pubmed/15043098

16. Powell JN. The Canadian Arthroplasty Society's experience with hip resurfacing arthroplasty: an analysis of 2773 hips. *Bone Joint J*, 2013; **95-B**(8):1045–1051.

17. Haughom BD, Erickson BJ, Hellman MD, Jacobs JJ. Do complication rates differ by gender after metal-on-metal hip resurfacing arthroplasty? A systematic review. *Clin Orthop Relat Res* [Internet], 2015; **473**(8):2521–2529. Available from: http://dx.doi.org/10.1007/s11999-015-4227-8

18. Sershon R, Balkissoon R, Della Valle CJ. Current indications for hip resurfacing arthroplasty in 2016. *Curr Rev Musculoskelet Med*, 2016; **9**(1):84–92.

19. Zylberberg AD, Nishiwaki T, Kim PR, Beaulé PE. Clinical results of the conserve plus metal on metal hip resurfacing: an independent series. *J Arthroplasty*, 2015; **30**(1):68–73.

20. Liu F, Gross TP. A safe zone for acetabular component position in metal-on-metal hip resurfacing arthroplasty: winner of the 2012 HAP Paul award. *J Arthroplasty*, 2013; **28**(7):1224–1230.
21. Amstutz HC, Le Duff MJ, Johnson AJ. Socket position determines hip resurfacing 10-year survivorship. *Clin Orthop Relat Res*, 2012; 3127–3133.
22. Yoon JP, Le Duff MJ, Johnson AJ, Takamura KM, Ebramzadeh E, Amstutz HC. Contact patch to rim distance predicts metal ion levels in hip resurfacing hip. *Clin Orthop Relat Res*, 2013; **471**(5):1615–1621.
23. Davis ET, Olsen M, Zdero R, Papini M, Waddell JP, Schemitsch EH. A biomechanical and finite element analysis of femoral neck notching during hip resurfacing. *J Biomech Eng* [Internet], 2009; **131**(4):041002. Available from: http://www.ncbi.nlm.nih.gov/pubmed/19275431
24. Daniel J, Pradhan C, Ziaee H, Pynsent PB, McMinn DJW. Results of birmingham hip resurfacing at 12 to 15 years: a single-surgeon series. *Bone Joint J*, 2014; **96-B**:1298–1306.
25. Van Der Straeten C, Van Quickenborne D, De Roest B, Calistri A, Victor J, De Smet K. Metal ion levels from well-functioning birmingham hip resurfacings decline significantly at ten years. *Bone Joint J*, 2013; **95-B**(10):1332–1338.
26. Sehatzadeh S, Kaulback K, Levin L. Metal-on-metal hip resurfacing arthroplasty: an analysis of safety and revision rates. *Ont Health Technol Assess Ser*, 2012; **12**(19):1–63.
27. Drummond J, Tran P, Fary C. Metal-on-metal hip arthroplasty: a review of adverse reactions and patient management. *J Funct Biomater* [Internet], 2015; **6**:486–499. Available from: http://www.mdpi.com/2079-4983/6/3/486/
28. Drummond J, Tran P, Fary C. Metal-on-metal hip arthroplasty: a review of adverse reactions and patient management. *J Funct Biomater* [Internet], 2015; **6**(3):486–499. Available from: http://www.mdpi.com/2079-4983/6/3/486/pdf%5Cnhttp://ovidsp.ovid.com/ovidweb.cgi?T=JS&CSC=Y&NEWS=N&PAGE=fulltext&D=emed13&AN=2015403995%5Cnhttp://oxfordsfx.hosted.exlibrisgroup.com/oxford?sid=OVID:embase&id=pmid:&id=doi:10.3390%2Fjfb6030486&issn=2079-4983&

Chapter 4

Advances in Periprosthetic Fractures of the Hip Joint

Ali-Asgar Najefi, Arthur Galea, Nicholas Garlick and
M. Zahid Saeed

*Trauma and Orthopaedic Department, University College London Medical School,
Royal Free Hospital, Pond Street, London NW3 2QG, United Kingdom*
z.saeed@nhs.net, mzahidsaeed@hotmail.com

4.1 Periprosthetic Fractures of the Proximal Femur

4.1.1 Introduction

A periprosthetic fracture of the proximal femur (PFF) is a rare but devastating complication after total hip arthroplasty (THA) [1]. The main treatment goal is to obtain a stable prosthesis and fixation of the fracture, thus allowing early rehabilitation and thereby avoiding complications associated with immobilisation [2]. They represent a challenge with regard to patient management and recovery, and treatment requires profound expertise in both trauma surgery and revision arthroplasty techniques [3, 4].

The Hip Joint in Adults: Advances and Developments
Edited by K. Mohan Iyer
Copyright © 2018 Pan Stanford Publishing Pte. Ltd.
ISBN 978-981-4774-72-7 (Hardcover), 978-1-351-26244-6 (eBook)
www.panstanford.com

With the aging population, rates of revision THA are increasing [5, 6]. There are a greater number of complications after primary or revision THA, including PFF [5–11]. PFF is especially important because it is expensive and associated with a poor outcome, high mortality rate and incomplete functional recovery [11–13].

The incidence for postoperative PFFs after THA for all causes is 1.1% [14]. It ranges from 0.6% after primary THA to 2.4% after revision THA [15]. According to the National Joint Registry (NJR) in the UK, 9.6% of single-stage revisions are attributable to PFF [16].

4.1.2 Risk Factors for PFF

Risk factors can be divided into intra- and postoperative. Risk factors for intraoperative PFF are osteoporosis, rheumatoid arthritis, femoral-stem preparation and surgical technique for inserting the rasp or femoral component, press-fit cementless stem and revision THA [17]. In the case of postoperative PFF, advanced age, female gender, posttraumatic osteoarthritis, osteoporosis, rheumatoid arthritis, proximal femur deformities, previous surgery of the affected hip, implant type (especially cementless stem and press-fit implantation), technical errors such as cortical perforation, cortical stress risers, low-energy trauma, osteolysis, loosening and revision are significant predisposing factors.

We have broken down the risk factors into subgroups. These include patient factors, local factors and surgical factors.

4.1.2.1 Patient factors

Patient factors such as age, gender and body mass index (BMI) contribute to the risk of PFF [4, 18]. PFF is more common amongst female patients, probably due to its association with osteoporosis and remaining structural bone [19, 20]. Older patients are at greater risk of postoperative fractures, as age is associated with other problems, such as osteoporosis and falls [19]. In elderly patients in particular, sufficient stabilisation of such fractures may be difficult in the presence of poor bone stock, previous revision surgeries and high prevalence of medical comorbidities. Most frequently, periprosthetic fracture is the result of a low-energy trauma from the sitting or standing position, which has been shown to account for 75% of fractures in primary THA and for 56% of fractures after revision THA [4, 19, 21, 22]. Patients receiving a hip arthroplasty due to a fracture of the femoral neck constitute a group with an increased risk of sustaining a PFF [23]. The hip fracture patient is generally older and frailer than the osteoarthritis hip patient; in addition, osteoporosis is common among the hip fracture patients [24]. In younger recipients, high-energy trauma can cause PFF [25].

Postoperative activities have a lot of influence on the interfacial micromotion and stress distribution. Excessive stress induces osteoblast apoptosis and osteoclast activation, which may reduce osseointegration. Regional stress increases can also lead to thigh pain or femoral fracture [17]. Higher activity levels in younger patients over a sustained period of time put these patients at higher risk of PFF [26].

Systemic conditions such as rheumatoid arthritis and osteoporosis are risk factors [17, 19, 27]. Treatment with corticosteroids and alcoholism or substance abuse are also known to be associated with increased risk of fracture [28]. Some neuromuscular disorders, such as poliomyelitis or parkinsonism, and certain neurological and medical comorbidities are associated with an increased risk of trauma (due to falls) and may contribute to fractures [25, 28].

There may be higher susceptibility to fracture in patients with secondary osteoarthritis related to developmental dysplasia of the hip (DDH) [7]. DDH leads to alterations in the hip anatomy that additionally result in functional changes, which are more prone to fracture [7]. The most important factors affecting the risk of fracture include increased neck–shaft angle (coxa valga) and antetorsion angle, stenosis of the medullary canal and shortening of the affected extremity. Because of the major deformity of the proximal femoral metaphysis, appropriate fixation of the stem is more difficult and can cause abnormal loads across the bone and consequently the fracture [29]. PFFs have also been found to occur more commonly in patients with a hip fracture prior to arthroplasty [7, 19]. Nonunion, delayed union or malunion, bone remodelling and, ultimately, limb malalignment are often seen in posttraumatic osteoarthritis. This condition leads to the formation of areas of higher loads during implant fixation and may result in a fracture [7]. Another important risk factor for PFF is the occurrence of areas of diminished mechanical properties, cortical perforations and bone defects that may occur after removal of implants used to stabilise the primary fracture or in the fracture line itself [7].

4.1.2.2 Local risk factors

Patients with osteolysis at the greater trochanter are at greatest risk of developing intraoperative and postoperative PFF [4, 30]. Stress shielding (redistribution of load and remodelling of the femur) can occur in response to an altered mechanical environment following a hip replacement and is a predisposing factor for these fractures. Cemented stems are associated with less stress shielding than uncemented stems [31, 32]. The Swedish registry showed that 70% of fractures involved loose prostheses, with 23% known to be loose and 47% first identified as loose at the time of surgery [6, 33].

Anatomic abnormalities, tumour and previous surgery of the affected hip increase the risk of intraoperative PFF, as previous hardware from a prior fracture fixation or osteotomy can cause higher torsional moments when dislocating or trialling [21, 29]. Infection and recurrent hip dislocation are also predisposing factors for postoperative PFF [8, 17].

4.1.2.3 Surgical factors

The risk of intraoperative PFF is affected by implant type and varies with the type of instrumentation for inserting the component [4, 12, 29]. Fractures during both primary and revision surgeries are much more common around uncemented than around cemented implants [18, 29, 34, 35]. In particular, intraoperative rates are higher for uncemented compared with cemented THA as these fractures usually occur during femoral canal broaching or implant insertion while the person is trying to obtain a tight press fit [19, 36, 37]. The goal of cementless stem fixation is to achieve immediate component stability so that bony ingrowth or ingrowth occurs. When a person is trying to achieve this, the strength of the underlying bone can give way, resulting in an intraoperative or early postoperative fracture, as cracks in the proximal femur due to press-fit stem implantation are often only recognised after the patient increases weight bearing [14, 25, 29].

As long as the stem has not established biological fixation, uncemented stems may be more prone to fractures should the patient sustain trauma to the hip [28]. Since the majority of the PFFs are associated with low-energy trauma, such as a simple fall, bone cement may act to reinforce the osteoporotic proximal femur, improve load distribution and hence protect against a shaft fracture. This may explain the low incidence of PFF with cemented prostheses [38].

Intraoperative and early postoperative fractures are mainly related to technical errors at the time of the operation [20, 39]. An important risk factor for PFF is the occurrence of areas of diminished mechanical properties, cortical perforations and bone defects that may occur after removal of the implant used for the primary THA [7]. Accidental cortical perforation can occur during channel preparation and drilling [21, 25, 38]. Fractures during bone preparation can be due to torque generated by power reamers and forceful femoral preparation [4]. Moreover, proximal fracture during insertion of the prosthesis is usually the result of mismatched prosthesis and bone (underreaming) dimensions [17].

There is an elevated risk of intraoperative fractures during dislocation or reduction of the hip during THA [14, 40, 41]. Preparation of the proximal femur with a calcar mill has likewise been correlated with intraoperative fractures [29]. THA with femoral osteotomy for proximal femoral deformity is also associated with a high risk of intraoperative fracture [42]. Pre-existing stress risers, such as cortical windows or perforations, are risk factors for PFF in the early postoperative period [34]. Other risk factors are extruded cement and varus stem position [21].

Revision of the femoral component has a high incidence of intraoperative and early postoperative PFF compared with primary THA [29]. The use of long, straight reamers, underreaming of the femoral canal, use of a longer and large-diameter femoral stem that engages the bow of the native femur, a low ratio between cortical and canal diameters and use of large-diameter stems all seem to be associated with a higher risk of fracture [20, 29]. Fracture during uncemented implant insertion is more common in revision than in primary surgery because of poorer bone quality, distortions of femoral geometry and the need for longer implants [14]. Revision THA of femoral stems tend to transfer forces of weight bearing to the tip of the stem, which may be an independent risk factor for the development of postoperative PFF [4].

In recent years, there has been an increase in intraoperative PFF along with the evolution of minimally invasive surgery (MIS) [13]. Difficulties in exposure and adequate visualisation of the proximal femur (particularly with anterior approaches) were identified as contributing factors to this complication [13, 29, 43].

4.1.3 Classification

The historical classification systems [44–47] for periprosthetic fractures of the femur have been superseded by the Vancouver classification system (VCS), which has become universally accepted [5, 48] (see Tables 4.1 and 4.2). The VCS describes the position of the femoral fracture relative to the prosthetic tip. Femoral component stability within the proximal fragment is the cornerstone of this classification.

Table 4.1 Vancouver classification and treatment of postoperative PFFs

Type	Fracture pattern	Recommended treatment
A	Fractures involving the greater (AG) or the lesser trochanteric (AL) region	Nonoperative or operative
B1	Fractures around a well-fixed femoral stem or just below it	ORIF using long cable-locking plates +/− strut allografts
B2	Fractures around a loose femoral stem or just below it Good femoral bone stock	Revision of the femoral component using an uncemented long-stem component bypassing the fracture site with the distal fixation technique
B3	Fractures around a loose femoral stem or just below it, along with poor quality or severely comminuted proximal bone stock	Revision using a long femur stem with distal fixation and strut cortical allografts or a proximal femoral replacement prosthesis
C	Fractures occurring well distal to the tip of the femoral stem such that their treatment is not influenced by the hip arthroplasty prosthesis	ORIF using a plate and screws, avoiding stress raisers at the tip of the stem

Table 4.2 Vancouver classification and treatment of intraoperative PFFs

Type	Fracture pattern	Recommended treatment
A1	Metaphyseal cortical breach	Bone graft from acetabular reamings
A2	Nondisplaced linear crack	Stabilisation with cerclage wire, screws, or trochanteric claw plate
A3	Displaced/unstable fracture of the greater trochanter/proximal femur	Cerclage wire, cable plate, or trochanteric claw plate before cementation or uncemented implant with a fully coated stem
B1	Diaphyseal cortical perforation	Revision of the femoral component bypassing by 2 cortical diameters +/− cortical strut graft/plate
B2	Nondisplaced linear crack	Wiring for stable implant or bypass with a stem +/− cortical allograft +/− plate and screws for unstable implant
B3	Displaced fracture of the midfemur	ORIF of the fracture and bypass with a longer stem
C1	Cortical perforation	Grafted or bypass with a longer stem and cortical strut graft
C2	Nondisplaced crack extending just above the knee	Cerclage wire +/− cortical strut graft
C3	Diaphyseal fracture of the distal femur that cannot be bypassed with a femoral stem	ORIF with plate +/− cortical strut graft

Type A fractures occur in the trochanteric region and are divided into two different subtypes: A_G fractures occur in the greater trochanter, and A_L fractures occur in the lesser trochanter. Most Vancouver A_G and A_L fractures are associated with particle-induced osteolysis from wear of conventional polyethylene [49]. In 2014, Capello et al. classified a new fracture pattern of the trochanteric region called the 'clamshell' fracture [50]. This refers to fractures of the medial cortex of the femur that include the residual neck, the calcar and the lesser trochanter. The authors classified these fractures as A1 if the stem is well fixed and A2 if the stem is loose.

Table 4.3 UCS classification and treatment options

Type		Acetabulum/pelvis	Femur, proximal
A Apophyseal/ extra-articular	A1 Avulsion of	Anterior Inferior/superior iliac spine	Greater trochanter
	A2 Avulsion of	Ischial tuberosity	Lesser trochanter
B Bed of Implant	B1 Prosthesis stable, good bone	Acetabular rim or good bone	Stem stable, good bone Surface replacement: femoral neck
	B2 Prosthesis loose, good bone	Loose cup, good bone	Loose stem, good bone Surface replacement: loose implant, no proximal bone loss
	B3 Prosthesis loose, poor bone stock	Loose cup, poor bone, bone defect, pelvic discontinuity	Loose stem, poor bone, defect Surface replacement: loose implant, bone loss
C Clear of the implant		Pelvic/acetabular fractures distant to the implant	Distal to the implant and cement mantle
D Dividing the bone between 2 implants or interprosthetic/intercalary		Pelvic fracture between bilateral THRs	Between hip and knee joint replacements, close to the hip
E Each of the 2 bones supporting one arthroplasty		Pelvis and femur	
F Facing and articulating with a hemiarthroplasty		Fracture of the acetabulum articulating with the femoral hemiarthroplasty	–

Type B fractures involve the metaphyseal or diaphyseal femur around a prosthetic stem. Type B1 fractures occur around a stable stem, B2 fractures involve a loose stem with good bone quality and B3 fractures involve bone loss. Vancouver B1 fractures are reported to be at least one-third of all the periprosthetic femoral fractures [51]. According to some series, Vancouver B2– and B3–type fractures account for 65% of all PFFs following primary surgery and 41% of PFFs following revision surgery [23]. Type C fractures are significantly distal to the stem. The reliability and validity of the VCS for periprosthetic femoral fractures after hip replacement have been confirmed [52, 53].

In 2014, Duncan et al. introduced the unified classification system (UCS) to expand upon and update the VCS and apply treatment principles to all periprosthetic fractures [54] (see Table 4.3). The UCS incorporates the previous VCS but is expanded to include types D, E and F [54, 55]. Type D is a fracture affecting one bone that supports two replacements. The most common example involves the femur after hip and knee replacement. Type E involves two bones supporting one replacement. The most common example involves the acetabulum and femur after hip replacement. Type F is an uncommon fracture involving a joint surface that is not resurfaced or replaced but is directly articulating with an implant. The most common example of this type involves the acetabulum following hemiarthroplasty of the hip. This type of fracture can affect the glenoid, the lateral humeral condyle, the acetabulum or the patella [54].

4.1.4 Investigations

Plain radiographs of the hip and whole femur are often all that is necessary for obtaining adequate information on these fractures. Special views such as Judet's views of the acetabulum may be required to evaluate the integrity of columns in cases of acetabular fractures. CT scans, including 3D reconstruction, could give additional information on the acetabular bone stock, integrity of the columns and loosening of the femoral stem. In case of an intrapelvic component, computed tomography (CT) angiography is recommended for preoperative planning to evaluate the risks of potential vascular injuries [56]. Blood inflammatory markers as well as aspiration of the hip joint might be indicated to exclude infection.

4.1.5 Surgical Considerations and Treatment of PFFs

Please see Fig. 4.1.

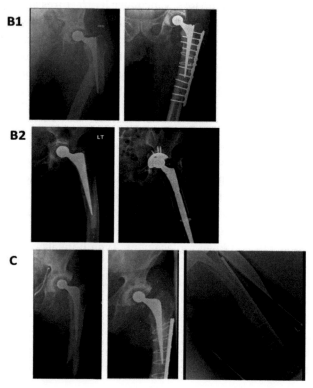

Figure 4.1 Examples of PFF and subsequent fixation. The top image represents a B1 fracture with a stable stem, fixed with open reduction and internal fixation (ORIF). The second row shows a B2 fracture with an unstable stem replaced with a TMFT. The third row shows a type C fracture repaired with an ORIF (the stem was stable).

4.1.5.1 Vancouver A_G and A_L fractures

Vancouver A_G and A_L are usually treated nonoperatively [57]. Surgical management typically aims at addressing the underlying problem. This usually involves an exchange of polyethylene to eliminate the particle generator and treatment of osteolytic lesions with bone grafting procedures. Occasionally, the trochanter is fixed to the femur with a tension band wiring technique [49]. These fractures are successfully treated with a conservative approach if the trochanter is displaced less than 2 cm [58]. It is often advisable to allow acute, minimally displaced A_G fractures six to eight weeks of healing prior to revision surgery in order to minimise the risk of further trochanteric displacement intraoperatively. Regardless of the treatment modality, full weight bearing and active hip abduction should be avoided until fracture healing, which is usually about 6 to 12 weeks [58].

4.1.5.2 Vancouver A1 and A2 fractures

Vancouver A1 and A2 fractures usually occur intraoperatively [59] and are more commonly associated with the use of tapered, proximally coated, cementless stems [60]. Complications include subsidence, fracture propagation and stem loosening. As such, Van Houwelingen et al. recommended surgical management of these fractures, while Capello et al. suggested differentiating such fractures as stable and loose [50, 60, 61]. Fractures with a stable stem can be managed nonoperatively, like Vancouver A_L fractures [58]. In cases where the stem is considered loose, operative intervention is required.

4.1.5.3 Vancouver B1 fractures

Historically, this type of fracture was treated nonoperatively [46, 61] or with skeletal traction [62]. Given the poor results of nonoperative treatment, including nonunion, malunion and the medical complications associated with prolonged bed rest, the standard of care is currently open reduction and internal fixation (ORIF). Most fractures associated with a stable femoral stem can be managed in this way [30, 63–65]. Stability of the stem should be assessed intraoperatively [63]. ORIF for a suspected Vancouver B1 fracture in which the stem is loose is associated with an extremely high failure rate [33, 66].

4.1.5.4 Vancouver B2 fractures

In a B2 fracture, the stem is loose but the bone stock is adequate. The lack of metaphyseal support requires revision arthroplasty stems that bypass the defect [23, 66]. The goals of such a procedure are to restore long-term implant stability and allow fracture healing [40].

An extensively porous-coated stem [67–69] or titanium modular fluted tapered (TMFT) stem [70–74] are used. Cemented stems can also be utilised [57]. With both extensively porous-coated and TMFT stems, the diaphysis is engaged, giving better support. However, extensively porous-coated stems have a higher reported rate of intraoperative and postoperative complications [75]. TMFT stems have better reported outcomes in terms of intraoperative fracture rate, modularity, immediate stability, stress shielding and thigh pain [70–74]. In addition to rotational stability, proximal fractures can be bypassed, with acceptably low complication rates [76–78].

The chosen implant needs to be stable and able to minimise the risk of further fracture, which means that stress risers should be bypassed by at least two femoral cortical diameters. When utilising a TMFT stem, the implant is exposed through the fracture line or along a modified Wagner osteotomy [14]. After this, the previous stem is removed, a prophylactic cerclage is placed 1 cm distal to the osteotomy or fracture and the TMFT stem is placed. The proximal fracture fragments are closed in a soft-tissue-preserving fashion, with two to three cables or wires. Maintenance of the osseous vascularity is far more important than anatomic reconstruction of the proximal femur. In managing inherently unstable transverse fracture patterns it may be advantageous to use onlay cortical strut grafts to augment the intramedullary fixation achieved by the implant [75].

Modular stems offer an acceptable alternative to revision surgery in the presence of large bone defects, reaching 93% survival at seven years [79]. Abdel et al. described 44 patients with Vancouver B_2 and B_3 periprosthetic femur fractures with a 98% union rate, while an additional 11% of the patients had a reoperation for instability at a mean follow-up of 4.5 years. Radiographic results were encouraging, with a 96% rate of stem osseointegration [70]. Munro et al. demonstrated survivorship of 96% at a mean of 54 months. The study showed maintenance or improvement of bone stock in 89% of the cases, with high rates of femoral union [74]. The rate of new revision surgery in cases of PFF is between 1.4% and 5% and that of intraoperative fractures stands between 1% and 18.6% [74].

Subsidence is a complication of modular stems. Hernandez-Vaquero showed some degree of subsidence in 50% of the stems, although the mean did not exceed 4 mm [75]. Subsidence may appear in the first six months and always remain a radiographic finding in the absence of symptoms [80, 81]. By contrast, in some series [82, 83], subsidence was so significant that it resulted in revision surgery in almost 10% of the cases. The rates are higher in those with a higher weight and femoral stem press-fit distance of less than 2 cm.

4.1.5.5 Vancouver B3 fractures

Implant stability, fracture healing and bone stock management are all factors in such fractures [40]. Not only is the stem loose, but the quality of the remaining bone stock is compromised, needing more advanced techniques to bypass or replace the deficient bone stock [75]. Bone loss encountered at the time of surgery is likely to be much greater than that predicted from radiographs [15].

Revision solutions are similar to those described for Vancouver B2 fractures (i.e., extensively porous-coated stems, TMFT stems and cemented stems). It is essential to have secure distal fixation with complex reconstruction of the deficient proximal femur, segmental substitution of the proximal femur with a tumour prosthesis, an allograft–prosthesis composite (APC) and a distally fixed prosthesis with scaffold reconstruction of the proximal femur around the modular device [84]. The use of ancillary bone grafting with either autograft or allogenic bone is recommended to facilitate bony ingrowth in all of these fractures. Other treatment modalities are important to consider, such as impaction bone grafting [85, 86], resection arthroplasty [87], APCs [88, 89] and proximal femoral replacements.

Impaction bone grafting has been used in large metaphyseal bone defects and stem diameters of 17 mm and above with good results [85, 90]. However, subsidence of the stem with consequent loosening and implant instability remains a concern. As such, synthetic bone substitutes with allograft have been recommended to give better structural support to the stem [87, 91]. APCs can be performed utilising a cemented, an uncemented or a partially cemented technique [89]. This procedure is technically demanding and has a high complication rate, including allograft nonunion, graft resorption and disease transmission [92]. In addition, the availability of proximal femoral allografts is rare and expensive, with a 10-year reported survival of 65%–86% [93]. Proximal femoral replacements are considered in those situations where there is severe proximal bone loss (often compromising the trochanter) and in low-demand patients [94].

4.1.5.6 Vancouver C fractures

Vancouver C fractures are well below the stem. The stem is stable, and ORIF without exposing the hip joint can be performed. Corten et al. reported that the fracture can be treated with a plate and screws only if the fracture allows for four bicortical screws distal to the stem and it is at least 2 cm distal to the stem [95]. Abdel et al. described a method to span the length of the femur, with at least four bicortical screws distal to the fracture and a combination of bicortical screws, unicortical screws and cables proximal to the fracture [57]. These fractures are distal enough to the stem to be considered as isolated fractures.

4.2 Periprosthetic Acetabular Fractures

Periprosthetic fractures of the acetabulum are less common than periprosthetic fractures of the femur but are often severe [96]. Diagnosis is established by evaluating clinical signs and symptoms such as pain during hip mobilisation and weight bearing, range of motion (ROM) reduction and lower-limb-length discrepancy [56].

Press-fit impaction of cementless elliptical shells are more likely to cause fracture of the acetabulum than is a true hemispherical cup [97, 98]. Risk factors include underreaming >2 mm, elliptical modular cups, high body mass index (BMI), osteoporosis, cementless acetabular components, dysplasia and radiation therapy [99, 100].

One of the first classifications of periprosthetic fractures of the acetabulum was reported by Peterson and Lewallen. They distinguished two types of fractures according to the cup stability: in type I, the cup was radiographically stable; in type II, the acetabular component was clearly displaced or radiographically loose [101]. Another classification system was proposed by Callaghan et al. in 1999, which divided fractures according to the time of presentation as intraoperative or postoperative and according to implant stability as fixed or loose [102]. In addition, this classification system described both peripheral rim and transverse fractures.

The most widely used classification system is that proposed by Paprosky et al. (see Table 4.4) [103]. Davidson et al. offered a simplified version of this classification that recognised three types of fractures: type 1 is undisplaced fractures not compromising the stability of reconstruction, type 2 is undisplaced fractures that may compromise the stability of reconstruction and type 3 are displaced fractures [20]. The UCS classification can also be applied to acetabular fractures (see Table 4.3) [54].

Table 4.4 The Paprosky classification of periprosthetic acetabular fractures

Problems encountered during component insertion	A = Recognised stable component indicating an undisplaced fracture	B = Recognised displaced fracture indicating that the cup is unstable C = not recognised intraoperatively
Problems encountered during removal	A = with less than 50% bone stock	B = with more than 50% bone stock
Traumatic	A = with a stable component	B = with an unstable component
Spontaneous	A = less than 50% bone stock loss	B = more than 50% bone stock loss
Pelvic discontinuity	A = less than 50% bone stock loss	B = more than 50% bone stock loss C = associated with pelvic radiation

The goals of surgical management are structural integrity, restoration of columns (anterior and posterior) and bone stock and to obtain a stable and well-positioned new cup [104]. The columns must be sufficiently stable to support an acetabular implant and to prevent micro- and macromotion at the bone-implant interface [56].

Treatment depends upon fracture complexity and component stability. Type 1 fractures with a stable prosthesis can be managed nonoperatively with restricted weight bearing. Later, revision of the acetabular component could be technically easier when the fracture has healed [56]. In type 2 and 3 fractures, the loose acetabular component would require revision and the technical complexity will depend on the type of acetabular deficiency and the integrity of the acetabular columns. A range of acetabular reconstruction options are available, such as bone-grafting the defects, using a multihole uncemented component stabilised using screws, using acetabular augments, using 'jumbo cups' with stabilisation of fracture in a distracted position, ilioischiatic cages, plate fixation of structural allografts, triflange cups, cemented acetabular components, reinforcement rings, oblong cups and cup-cage reconstruction [102, 103, 105]. Results of revision THA for acetabular discontinuity have generally been poor [106–110]. This is due to a combination of both patient and surgical factors [56].

References

1. Sadoghi P, Liebensteiner M, Agreiter M, Leithner A, Böhler N, Labek G. Revision surgery after total joint arthroplasty: a complication-based analysis using worldwide arthroplasty registers. *J Arthroplasty*, 2013; **28**(8):1329–1332.

2. Inngul C, Enocson A. Postoperative periprosthetic fractures in patients with an Exeter stem due to a femoral neck fracture: cumulative incidence and surgical outcome. *Int Orthop*, 2014; **39**(9):1683–1688.

3. Young SW, Pandit S, Munro JT, Pitto RP. Periprosthetic femoral fractures after total hip arthroplasty. *ANZ J Surg*, 2007; 424–428.

4. Della Rocca GJ, Leung KS, Pape H-C. Periprosthetic fractures: epidemiology and future projections. *J Orthop Trauma* [Internet], 2011; **25** Suppl 2(6):S66–S70. Available from: http://www.ncbi.nlm.nih.gov/pubmed/21566478

5. Duncan CP, Masri BA. Fractures of the femur after hip replacement. *Instr Course Lect*, 1995; **44**:293–304.

6. Shah RP, Sheth NP, Gray C, Alosh H, Garino JP. Periprosthetic fractures around loose femoral components. *J Am Acad Orthop Surg* [Internet], 2014; **22**(8):482–490. Available from: http://www.ncbi.nlm.nih.gov/pubmed/25063746

7. Nowak M, Kusz D, Wojciechowski P, Wilk R. Risk factors for intraoperative periprosthetic femoral fractures during the total hip arthroplasty. *Pol Orthop Traumatol* [Internet], 2012; **77**:59–64. Available from: http://www.ncbi.nlm.nih.gov/pubmed/23306288

8. Tsiridis E, Krikler S, Giannoudis PV. Periprosthetic femoral fractures: current aspects of management. *Injury*, 2007; **38**(6):649–650.

9. Cook RE, Jenkins PJ, Walmsley PJ, Patton JT, Robinson CM. Risk factors for periprosthetic fractures of the hip: a survivorship analysis. *Clin Orthop Relat Res*, 2008; **466**(7):1652–1656.

10. Lindahl H. Epidemiology of periprosthetic femur fracture around a total hip arthroplasty. *Injury*, 2007; **38**(6):651–654.

11. Katz JN, Wright EA, Polaris JJ, Harris MB, Losina E. Prevalence and risk factors for periprosthetic fracture in older recipients of total hip replacement: a cohort study. *BMC Musculoskelet Disord*, 2014; **15**(1):168.

12. Marsland D, Mears SC. A review of periprosthetic femoral fractures associated with total hip arthroplasty. *Geriatr Orthop Surg Rehabil*, 2012; **3**(3):107–120.

13. Dumont GD, Zide JR, Huo MH. Periprosthetic femur fractures: current concepts and management. *Semin Arthroplasty*, 2010; **21**(1):9–13.

14. Berry DJ. Treatment of Vancouver B3 periprosthetic femur fractures with a fluted tapered stem. *Clin Orthop Relat Res*, 2003; (417):224–231.

15. Learmonth ID. Aspects of current management the management of periprosthetic fractures around the femoral stem. *J Bone Joint Surg Br*, 2004; **8686**(1):13–19.

16. Green M, Howard P, Porter M, Price A, Wilkinson M, Wishart N. 13th Annual report. *Natl Joint Regist*, 2016.

17. SHAR. Annual report 2012. Register SHA, ed. http://www.shpr.se/sv/Publications/DocumentsReports.aspx.

18. Sidler-Maier CC, Waddell JP. Incidence and predisposing factors of periprosthetic proximal femoral fractures: a literature review. *Int Orthop*, 2015; 1673–1682.

19. Jakubowitz E, Seeger JB, Lee C, Heisel C, Kretzer JP, Thomsen MN. Do short-stemmed-prostheses induce periprosthetic fractures earlier than standard hip stems? A biomechanical ex-vivo study of two different stem designs. *Arch Orthop Trauma Surg*, 2009; **129**(6):849–855.

20. Franklin J, Malchau H. Risk factors for periprosthetic femoral fracture. *Injury*, 2007; **38**(6):655–660.

21. Davidson D, Pike J, Garbuz D, Duncan CP, Masri BA. Intraoperative periprosthetic fractures during total hip arthroplasty. Evaluation and management. *J Bone Joint Surg Am* [Internet], 2008; **90**(9):2000–2012. Available from: http://www.ncbi.nlm.nih.gov/pubmed/18762663

22. Thien TM, Chatziagorou G, Garellick G, Furnes O, Havelin LI, Mäkelä K, et al. Periprosthetic femoral fracture within two years after total hip replacement: analysis of 437,629 operations in the Nordic Arthroplasty Register Association database. *J Bone Joint Surg Am* [Internet], 2014; **96**(19):e167. Available from: http://jbjs.org/content/96/19/e167.abstract

23. Schwarzkopf R, Oni JK, Marwin SE. Total hip arthroplasty periprosthetic femoral fractures: a review of classification and current treatment. *Bull Hosp Jt Dis* [Internet], 2013; **71**(1):68–78. Available from: http://www.ncbi.nlm.nih.gov/pubmed/24032586

24. Moazen M, Jones AC, Jin Z, Wilcox RK, Tsiridis E. Periprosthetic fracture fixation of the femur following total hip arthroplasty: a review of biomechanical testing. *Clin Biomech*, 2011; **26**(1):13–22.

25. Savin L, Barharosie C, Botez P. Periprosthetic femoral fractures-- evaluation of risk factors. *Rev Med Chir Soc Med Nat Iasi* [Internet], 2012; **116**(3):846–852. Available from: http://ovidsp.ovid.com/ovidweb.cgi?T=JS&CSC=Y&NEWS=N&PAGE=fulltext&D=medl&AN=23272540%5Cnhttp://sfx.nottingham.ac.uk:80/sfx_local?genre=article&atitle=Periprosthetic+femoral+fractures--evaluation+of+risk+factors.&title=Revista+Medico-Chirurgicala+a+Societ

26. Singh JA, Jensen MR, Harmsen SW, Lewallen DG. Are gender, comorbidity, and obesity risk factors for postoperative periprosthetic fractures after primary total hip arthroplasty? *J Arthroplasty* [Internet], 2013; **28**(1):126-31.e1-2. Available from: http://www.sciencedirect.com/science/article/pii/S0883540312001568

27. Zhu Y, Chen W, Sun T, Zhang X, Liu S, Zhang Y. Risk factors for the periprosthetic fracture after total hip arthroplasty: a systematic review and meta-analysis, *Scand J Surg*, 2015; **104**(3):139–145.

28. Mayle RE, Della Valle CJ. Intra-operative fractures during THA: see it before it sees us. *J Bone Joint Surg Br*, 2012; **94-B**:26–31.

29. Lindahl H, Malchau H, Odén A, Garellick G. Risk factors for failure after treatment of a periprosthetic fracture of the femur. *J Bone Joint Surg Br*, 2006; **88**(1):26–30.

30. Samuelsson B, Hedström MI, Ponzer S, Söderqvist A, Samnegård E, Thorngren K-G, et al. Gender differences and cognitive aspects on functional outcome after hip fracture--a 2 years' follow-up of 2,134 patients. *Age Ageing* [Internet], 2009; **38**(6):686–692. Available from: http://www.ncbi.nlm.nih.gov/pubmed/19767316

31. Moazen M, Mak JH, Etchels LW, Jin Z, Wilcox RK, Jones AC, et al. Periprosthetic femoral fracture: a biomechanical comparison between

vancouver type B1 and B2 fixation methods. *J Arthroplasty*, 2014; **29**(3):495–500.

32. Marqués Lopez F, Muñoz Vives JM. Intraoperative periprosthetic hip fractures. *Eur Orthop Traumatol*, 2013; **4**(2):89–92.

33. Dattani R. Femoral osteolysis following total hip replacement. *Postgrad Med J* [Internet], 2007; **83**(979):312–316. Available from: http://www.pubmedcentral.nih.gov/articlerender.fcgi?artid=2600070&tool=pmcentrez&rendertype=abstract

34. Lindahl H, Malchau H, Herberts P, Garellick G. Periprosthetic femoral fractures: classification and demographics of 1049 periprosthetic femoral fractures from the Swedish National Hip Arthroplasty Register. *J Arthroplasty*, 2005; **20**(7):857–865.

35. Sarvilinna R, Huhtala HSA, Sovelius RT, Halonen PJ, Nevalainen JK, Pajamäki KJK. Factors predisposing to periprosthetic fracture after hip arthroplasty. *Acta Orthop*, 2004; **75**(1):16–20.

36. Hu K, Zhang X, Zhu J, Wang C, Ji W, Bai X. Periprosthetic fractures may be more likely in cementless femoral stems with sharp edges. *Ir J Med Sci*, 2010; **179**(3):417–421.

37. Holley K, Zelken J, Padgett D, Chimento G, Yun A, Buly R. Periprosthetic fractures of the femur after hip arthroplasty: an analysis of 99 patients. *HSS J*, 2007; **3**(2):190–197.

38. Rupprecht M, Sellenschloh K, Grossterlinden L, Püschel K, Morlock M, Amling M, et al. Biomechanical evaluation for mechanisms of periprosthetic femoral fractures. *J Trauma* [Internet], 2011; **70**(4):E62–E66. Available from: http://www.ncbi.nlm.nih.gov/pubmed/21613972

39. Foster AP, Thompson NW, Wong J, Charlwood AP. Periprosthetic femoral fractures: a comparison between cemented and uncemented hemiarthroplasties. *Injury*, 2005; **36**(3):424–429.

40. Lee S-R, Bostrom MPG. Periprosthetic fractures of the femur after total hip arthroplasty. *Instr Course Lect*, 2004; **53**:111–118.

41. Garbuz DS, Toms A, Masri BA, Duncan CP. Improved outcome in femoral revision arthroplasty with tapered fluted modular titanium stems. *Clin Orthop Relat Res*, 2006; **453**:199–202.

42. Khan MA, O'Driscoll M. Fractures of the femur during total hip replacement and their management. *J Bone Joint Surg Br*, 1977; **59**:36–41.

43. Haddad FS, Masri BA, Garbuz DS, Duncan CP. The prevention of periprosthetic fractures in total hip and knee arthroplasty. *Orthop Clin*

North Am [Internet], 1999; **30**(2):191–207. Available from: http://www.ncbi.nlm.nih.gov/pubmed/10196421

44. Masri BA, Meek RMD, Duncan CP. Periprosthetic fractures evaluation and treatment. *Clin Orthop Relat Res*, 2004; (420):80–95.

45. Rüdiger HA, Betz M, Zingg PO, McManus J, Dora CF. Outcome after proximal femoral fractures during primary total hip replacement by the direct anterior approach. *Arch Orthop Trauma Surg*, 2013; **133**(4):569–573.

46. Johansson JE, McBroom R, Barrington TW, Hunter GA. Fracture of the ipsilateral femur in patients wih total hip replacement. *J Bone Joint Surg Am*, 1981; **63**(9):1435–1442.

47. Bethea JS, DeAndrade JR, Fleming LL, Lindenbaum SD, Welch RB. Proximal femoral fractures following total hip arthroplasty. *Clin Orthop Relat Res* [Internet], 1982; (170):95–106. Available from: http://www.ncbi.nlm.nih.gov/pubmed/7127971

48. Jensen JS, Barfod G, Hansen D, Larsen E, Linde F, Menck H, et al. Femoral shaft fracture after hip arthroplasty. *Acta Orthop Scand* [Internet], 1988; **59**(1):9–13. Available from: http://www.ncbi.nlm.nih.gov/pubmed/3354328

49. Roffman M, Mendes DG. Fracture of the femur after total hip arthroplasty. *Orthopedics* [Internet], 1989; **12**(8):1067–1070. Available from: http://ovidsp.ovid.com/ovidweb.cgi?T=JS&PAGE=reference&D=med3&NEWS=N&AN=2771826

50. Capello WN, D'Antonio JA, Naughton M. Periprosthetic fractures around a cementless hydroxyapatite-coated implant: a new fracture pattern is described. *Clin Orthop Relat Res*, 2014; **472**(2):604–610.

51. Brady OH, Garbuz DS, Masri BA, Duncan CP. Classification of the hip. *Orthop Clin North Am*, 1999; **30**(2):215–220.

52. Haidukewych GJ, Langford JR, Liporace FA. Revision for periprosthetic fractures of the hip and knee. *Instr Course Lect* [Internet], 2013; **62**:333–340. Available from: http://eutils.ncbi.nlm.nih.gov/entrez/eutils/elink.fcgi?dbfrom=pubmed&id=23395038&retmode=ref&cmd=prlinks%5Cnpapers2://publication/uuid/5934E0A9-B250-413A-B864-EB85F15DFB2A

53. Capello WN, D'Antonio JA, Naughton M. Periprosthetic fractures around a cementless hydroxyapatite-coated implant: a new fracture pattern is described. *Clin Orthop Relat Res*, 2014; 604–610.

54. Yasen AT, Haddad FS. The management of type B1 periprosthetic femoral fractures: when to fix and when to revise. *Int Orthop*, 2014; 1873–1879.

55. Naqvi GA, Baig SA, Awan N. Interobserver and intraobserver reliability and validity of the vancouver classification system of periprosthetic femoral fractures after hip arthroplasty. *J Arthroplasty*, 2012; **27**(6):1047–1050.

56. Rayan F, Haddad F. Periprosthetic femoral fractures in total hip arthroplasty: a review. *Hip Int*, 2010; **20**(4):418–426.

57. Duncan CP, Haddad FS. The Unified Classification System (UCS): improving our understanding of periprosthetic fractures. *Bone Joint J*, 2014; **96-B**(6):713–716.

58. Vioreanu MH, Parry MC, Haddad FS, Duncan CP. Field testing the Unified Classification System for peri-prosthetic fractures of the pelvis and femur around a total hip replacement: an international collaboration. *Bone Joint J*, 2014; **96-B**(11):1472–1477.

59. Abdel MP, Cottino U, Mabry TM. Management of periprosthetic femoral fractures following total hip arthroplasty: a review. *Int Orthop*, 2015; **39**(10):2005–2010.

60. Pritchett JW. Fracture of the greater trochanter after hip replacement. *Clin Orthop Relat Res* [Internet], 2001; (390):221–226. Available from: http://www.ncbi.nlm.nih.gov/pubmed/11550869

61. Berend KR, Lombardi AV. Intraoperative femur fracture is associated with stem and instrument design in primary total hip arthroplasty. *Clin Orthop Relat Res*, 2010; 2377–2381.

62. Van Houwelingen AP, Duncan CP. The pseudo A(LT) periprosthetic fracture: it's really a B2. *Orthopedics* [Internet], 2011; **34**(9):e479–e481. Available from: http://www.ncbi.nlm.nih.gov/pubmed/21902137

63. McElfresh EC, Coventry MB. Femoral and pelvic fractures after total hip arthroplasty. *J Bone Joint Surg Am*, 1974; **56**:483–492.

64. Scott RD, Turner RH, Leitzes SM, Aufranc OE. Femoral fractures in conjunction with total hip replacement. *J Bone Joint Surg Am*, 1975; **57**(4):494–501.

65. Pike J, Davidson D, Garbuz D, Duncan CP, O'Brien PJ, Masri BA. Principles of treatment for periprosthetic femoral shaft fractures around well-fixed total hip arthroplasty. *J Am Acad Orthop Surg*, 2009; **17**(11):677–688.

66. Müller FJ, Galler M, Füchtmeier B. Clinical and radiological results of patients treated with orthogonal double plating for periprosthetic femoral fractures. *Int Orthop.* 2014; **38**:2469–2472.

67. Ehlinger M, Bonnomet F, Adam P. Periprosthetic femoral fractures: the minimally invasive fixation option. *Orthop Traumatol Surg Res*, 2010; **96**(3):304–309.

68. Haidar SG, Goodwin MI. Dynamic compression plate fixation for postoperative fractures around the tip of a hip prosthesis. *Injury.* 2005; **36**(3):417–423.

69. Kato T, Otani T, Sugiyama H, Hayama T, Katsumata S, Marumo K. Cementless total hip arthroplasty in hip dysplasia with an extensively porous-coated cylindrical stem modified for Asians: a 12-year follow-up study. *J Arthroplasty* [Internet], 2015; **30**(6):1014–1018. Available from: http://www.sciencedirect.com/science/article/pii/S0883540315000480

70. Shen B, Huang Q, Yang J, Zhou ZK, Kang PD, Pei FX. Extensively coated non-modular stem used in two-stage revision for infected total hip arthroplasty: mid-term to long-term follow-up. *Orthop Surg*, 2014; **6**(2):103–109.

71. Thomsen PB, Jensen NJF, Kampmann J, Hansen TB. Revision hip arthroplasty with an extensively porous-coated stem: excellent long-term results also in severe femoral bone stock loss. *Hip Int*, 2013; **23**(4):352–358.

72. Abdel MP, Lewallen DG, Berry DJ. Periprosthetic femur fractures treated with modular fluted, tapered stems. *Clin Orthop Relat Res*, 2014; 599–603.

73. Van Houwelingen AP, Duncan CP, Masri BA, Greidanus NV, Garbuz DS. High survival of modular tapered stems for proximal femoral bone defects at 5 to 10 years follow-up hip. *Clin Orthop Relat Res*, 2013; 454–462.

74. Amanatullah DF, Howard JL, Siman H, Trousdale RT, Mabry TM, Berry DJ. Revision total hip arthroplasty in patients with extensive proximal femoral bone loss using a fluted tapered modular femoral component. *Bone Joint J* [Internet], 2015; **97-B**(3):312–317. Available from: http://www.bjj.boneandjoint.org.uk/content/97-B/3/312

75. Rodriguez JA, Fada R, Murphy SB, Rasquinha VJ, Ranawat CS. Two-year to five-year follow-up of femoral defects in femoral revision treated with the link MP modular stem. *J Arthroplasty*, 2009; **24**(5):751–758.

76. Munro JT, Garbuz DS, Masri BA, Duncan CP. Tapered fluted titanium stems in the management of vancouver B2 and B3 periprosthetic femoral fractures. *Clin Orthop Relat Res*, 2014; 590–598.

77. Hernandez-Vaquero D, Fernandez-Lombardia J, de los Rios JL, Perez-Coto I, Iglesias-Fernandez S. Treatment of periprosthetic femoral fractures with modular stems. *Int Orthop*, 2015; **39**(10):1933–1938.

78. Kwong LM, Miller AJ, Lubinus P. A modular distal fixation option for proximal bone loss in revision total hip arthroplasty: a 2- to 6-year follow-up study. *J Arthroplasty*, 2003; **18**(3 Suppl 1):94–97.

79. Schuh A, Werber S, Holzwarth U, Zeiler G. Cementless modular hip revision arthroplasty using the MRP Titan Revision Stem: outcome of 79 hips after an average of 4 years' follow-up. *Arch Orthop Trauma Surg*, 2004; **124**(5):306–309.

80. Tamvakopoulos GS, Servant CTJ, Clark G, Ivory JP. Medium-term follow-up series using a modular distal fixation prosthesis to address proximal femoral bone deficiency in revision total hip arthroplasty. A 5- to 9-year follow-up study. *Hip Int*, 2007; **17**(3):143–149.

81. Drexler M, Dwyer T, Chakravertty R, Backstein D, Gross AE, Safir O. The outcome of modified extended trochanteric osteotomy in revision THA for Vancouver B2/B3 periprosthetic fractures of the femur. *J Arthroplasty*, 2014; **29**(8):1598–1604.

82. Lakstein D, Backstein D, Safir O, Kosashvili Y, Gross AE. Revision total hip arthroplasty with a porous-coated modular stem: 5 to 10 years follow-up. *Clin Orthop Relat Res*, 2010; **468**(5):1310–1315.

83. Hartman CW, Garvin KL. Femoral fixation in revision total hip arthroplasty. *Instr Course Lect* [Internet], 2012; **61**:313–325. Available from: http://eutils.ncbi.nlm.nih.gov/entrez/eutils/elink.fcgi?dbfrom=pubmed&id=22301243&retmode=ref&cmd=prlinks%5Cnhttp://www.ncbi.nlm.nih.gov/pubmed/22301243

84. Restrepo C, Mashadi M, Parvizi J, Austin MS, Hozack WJ. Modular femoral stems for revision total hip arthroplasty. *Clin Orthop Relat Res*, 2011; 476–482.

85. Stimac JD, Boles J, Parkes N, Gonzalez A, Valle D, Boettner F, et al. Revision total hip arthroplasty with modular femoral stems. *J Arthroplasty*, 2014; **29**:2167–2170.

86. Tangsataporn S, Safir OA, Vincent AD, Abdelbary H, Gross AE, Kuzyk PRT. Risk factors for subsidence of a modular tapered femoral stem used for revision total hip arthroplasty. *J Arthroplasty*, 2014; **30**:1030–1034.

87. Patel PD, Klika AK, Murray TG, Elsharkawy KA, Krebs VE, Barsoum WK. Influence of technique with distally fixed modular stems in revision total hip arthroplasty. *J Arthroplasty*, 2010; **25**(6):926–931.

88. Ohly NE, Whitehouse MR, Duncan CP. Periprosthetic femoral fractures in total hip arthroplasty. *Hip Int* [Internet], 2014; **24**(2):556–567. Available from: http://www.ncbi.nlm.nih.gov/pubmed/24970324

89. Wimmer MD, Randau TM, Deml MC, Ascherl R, Nöth U, Forst R, et al. Impaction grafting in the femur in cementless modular revision total hip arthroplasty: a descriptive outcome analysis of 243 cases with the MRP-TITAN revision implant. *BMC Musculoskelet Disord* [Internet], 2013; **14**(1):19. Available from: http://www.ncbi.nlm.nih.gov/pubmed/23311769%5Cnhttp://www.pubmedcentral.nih.gov/articlerender.fcgi?artid=PMC3556053%5Cnhttp://www.pubmedcentral.nih.gov/articlerender.fcgi?artid=3556053&tool=pmcentrez&rendertype=abstract

90. Ten Have BL, Brouwer RW, Van Biezen FC, Verhaar JA, Viller RN. Erratum: Femoral revision surgery with impaction bone grafting: 31 hips followed prospectively for ten to 15 years (*Bone Joint J*, 2012; **94**:615–618). *Bone Joint J*, 2013; **95**:286.

91. Oshima S, Yasunaga Y, Yamasaki T, Yoshida T, Hori J, Ochi M. Midterm results of femoral impaction bone grafting with an allograft combined with hydroxyapatite in revision total hip arthroplasty. *J Arthroplasty*, 2012; **27**(3):470–476.

92. Mayle Jr RE, Paprosky WG. Massive bone loss: allograft-prosthetic composites and beyond. *J Bone Joint Surg Br*, 2012; **94-B**(Suppl-A):61–64.

93. Min L, Peng J, Duan H, Zhang W, Zhou Y, Tu C. Uncemented allograft-prosthetic composite reconstruction of the proximal femur. *Indian J Orthop* [Internet], 2014; **48**(3):289–295. Available from: http://ovidsp.ovid.com/ovidweb.cgi?T=JS&CSC=Y&NEWS=N&PAGE=fulltext&D=prem&AN=24932036%5Cnhttp://sfx.nottingham.ac.uk:80/sfx_local?genre=article&atitle=Uncemented+allograft-prosthetic+composite+reconstruction+of+the+proximal+femur.&title=Indian+Journal+of+

94. Sheth NP, Nelson CL, Paprosky WG. Femoral bone loss in revision total hip arthroplasty: evaluation and management. *J Am Acad Orthop Surg* [Internet], 2013; **21**(10):601–612. Available from: http://eutils.ncbi.nlm.nih.gov/entrez/eutils/elink.fcgi?dbfrom=pubmed&id=24084434&retmode=ref&cmd=prlinks%5Cnpapers3://publication/doi/10.5435/JAAOS-21-10-601

95. Howie DW, McGee MA, Callary SA, Carbone A, Stamenkov RB, Bruce WJ, et al. A preclinical study of stem subsidence and graft incorporation after femoral impaction grafting using porous hydroxyapatite as a bone graft extender. *J Arthroplasty*, 2011; **26**(7):1050–1056.

96. Babis GC, Sakellariou VI, O'Connor MI, Hanssen AD, Sim FH. Proximal femoral allograft-prosthesis composites in revision hip replacement: a 12-year follow-up study. *J Bone Joint Surg Br*, 2010; **92-B**(3):349–355.

97. Parvizi J, Rapuri VR, Purtill JJ, Sharkey PF, Rothman RH, Hozack WJ. Treatment protocol for proximal femoral periprosthetic fractures. *J Bone Joint Surg Am*, 2004; **86-A** Suppl:8–16.

98. Parvizi J, Vegari DN. Periprosthetic proximal femur fractures: current concepts. *J Orthop Trauma* [Internet], 2011; **25** Suppl 2(6):S77–S81. Available from: http://www.ncbi.nlm.nih.gov/pubmed/21566480

99. Blackely, H, Gross A. Proximal femoral allografts for reconstruction of bone stock in revision arthroplasty of the hip. A nine to fifteen-year follow-up. *J Bone Joint Surg Br*, 2002; **84**:133.

100. Corten K, Vanrykel F, Bellemans J, Frederix PR, Simon J-P, Broos PLO. An algorithm for the surgical treatment of periprosthetic fractures of the femur around a well-fixed femoral component. *J Bone Joint Surg Br* [Internet], 2009; **91**(11):1424–1430. Available from: http://www.ncbi.nlm.nih.gov/pubmed/19880884

101. Laflamme GY, Belzile EL, Fernandes JC, Vendittoli PA, Hébert-Davies J. Periprosthetic fractures of the acetabulum during cup insertion: posterior column stability is crucial. *J Arthroplasty*, 2015; **30**(2):265–269.

102. Benazzo F, Formagnana M, Bargagliotti M, Perticarini L. Periprosthetic acetabular fractures. *Int Orthop*, 2015; **39**(10):1959–1963.

103. Sharkey PF, Hozack WJ, Callaghan JJ, Kim YS, Berry DJ, Hanssen AD, et al. Acetabular fracture associated with cementless acetabular component insertion: a report of 13 cases. *J Arthroplasty*, 1999; **14**(4):426–431.

104. Haidukewych GJ, Jacofsky DJ, Hanssen AD, Lewallen DG. Intraoperative fractures of the acetabulum during primary total hip arthroplasty. *J Bone Joint Surg Am* [Internet], 2006; **88**(9):1952–1956. Available from: http://www.ncbi.nlm.nih.gov/pubmed/16951110

105. Desai G, Ries MD. Early postoperative acetabular discontinuity after total hip arthroplasty. *J Arthroplasty*, 2011; **26**(8):1570.e17-9.

106. Zwartelé RE, Witjes S, Doets HC, Stijnen T, Pöll RG. Cementless total hip arthroplasty in rheumatoid arthritis: a systematic review of the literature. *Arch Orthop Trauma Surg*, 2012; **132**(4):535–546.

107. Peterson CA, Lewallen DG. Periprosthetic fracture of the acetabulum after total hip arthroplasty. *J Bone Joint Surg Am* [Internet], 1996; **78**(8):1206–1213. Available from: http://ovidsp.ovid.com/ovidweb.cgi?T=JS&CSC=Y&NEWS=N&PAGE=fulltext&D=med4&AN=8753713%5Cnhttp://sfx.nottingham.ac.uk:80/sfx_local?genre=article&atitle=Periprosthetic+fracture+of+the+acetabulum+after+total+hip+arthroplasty.&title=Journal+of+Bone+%26+Joint+Surg

108. Callaghan JJ, Yong Sik Kim, Pedersen DR, Brown TD. Periprosthetic fractures of the acetabulum. *Orthop Clin North Am*, 1999; 221–234.

109. Paprosky WG, Sporer SM, O'Rourke MR. The treatment of pelvic discontinuity with acetabular cages. *Clin Orthop Relat Res*, 2006; **453**:183–187.

110. Chatoo M, Parfitt J, Pearse MF. Periprosthetic acetabular fracture associated with extensive osteolysis. *J Arthroplasty*, 1998; **13**(7):843–845.

111. Stiehl JB, Saluja R, Diener T. Reconstruction of major column defects and pelvic discontinuity in revision total hip arthroplasty. *J Arthroplasty* [Internet], 2000; **15**(7):849–857. Available from: http://www.sciencedirect.com/science/article/pii/S088354030029222X

Chapter 5

Minimally Invasive Surgery of the Hip Joint

Matthew K. T. Seah and Wasim Khan
*Department of Trauma and Orthopaedics, Addenbrooke's Hospital,
Cambridge University Hospitals NHS Foundation Trust, Hills Road,
Cambridge CB2 0QQ, United Kingdom*
wasimkhan@doctors.org.uk

Over the past few decades, minimally invasive surgery (MIS) has become increasingly popular in all fields of surgery, and orthopaedics is no exception. Minimally invasive hip surgery is a poorly defined, heterogeneous group of procedures that aim to limit soft-tissue dissection, and while some centres define 'minimal incision' surgery as surgery involving a wound less than 10 cm [1], most would argue that the underlying dissection is more critical and advocate the term 'tissue preserving surgery', especially in the context of minimally invasive approaches to arthroplasty [2]. The belief is that a smaller incision should lead to limited tissue trauma and is therefore associated with reduced patient morbidity, lower blood loss, decreased cost of care, better scar cosmesis and improved functional recovery [3–6]. However, these purported

---

The Hip Joint in Adults: Advances and Developments
Edited by K. Mohan Iyer
Copyright © 2018 Pan Stanford Publishing Pte. Ltd.
ISBN 978-981-4774-72-7 (Hardcover), 978-1-351-26244-6 (eBook)
www.panstanford.com

benefits have to be weighed against the potential disadvantages, which may include reduced visualisation contributing to a possible increased risk of iatrogenic injury and component malpositioning in hip replacements.

Most surgeons would state that as experience guides the surgeon to more accurate incision placement, more precise dissection and more skilful mobilisation of structures, the need for large incisions and extensive dissection decreases. This evolution is certainly seen in the total hip replacement, where the initial approach to the hip described by Charnley required trochanteric osteotomy for good exposure. It became apparent over time that nonunion of the osteotomy as well as painful trochanteric hardware could be problematic, and surgical approaches were developed, eventually demonstrating that the procedure could be performed quite adequately without an osteotomy. Hip replacements are currently being performed via a range of minimalist modifications of the standard hip approaches as well as by nontraditional approaches. These are variably referred to as minimally invasive, but this term has no real specificity or agreed definition. Broadly speaking, minimally invasive surgery can refer to (i) skin incision length, (ii) soft-tissue invasion and (iii) bone and joint preservation. Here, we discuss the literature with regard to MIS for total hip replacement, trauma and hip arthroscopy.

5.1 Comparing Minimally Invasive Approaches

Minimally invasive total hip arthroplasty (MITHA) has often been the subject of debate in recent years. No clear definition exists for what constitutes MITHA, but there is a relative consensus that hip arthroplasties performed with any incision less than 10 cm can be included. Proponents of MIS believe that this approach leads to a faster functional recovery, quicker hospital discharge and increased patient satisfaction. Opponents argue that minimally invasive (MI) approaches are associated with increased iatrogenic nerve injury, prosthesis malposition and increased revision rates, all of these due to the limited field of vision during the surgery [7–9]. During total hip arthroplasty (THA) surgery, the position of the prosthesis is directly related to the efficacy of the procedure, and the bony landmarks can be clearly visualised in traditional approaches while any MIS approach is more challenging in terms of exposure and may theoretically contribute to a component malpositioning. MI surgeries for THAs have been described in five categories: posterior, anterolateral (modified Watson–Jones approach) [10], lateral and single anterior MI approach [11], two-incision muscle-sparing approach [12] and posterior-lateral MI approach [13].

The literature remains inconclusive as to which MIS approach has the best outcomes in THA. Leuchte et al. found that the MIS posterior approach offered a significant advantage in functional ability, symmetry indices of stance, loading rates and single-limb stance in the first six weeks after THA when compared to a MIS Watson–Jones approach [14] but there were no significant differences between the two groups by 13 weeks. In another small study, the MIS Watson–Jones approach was favoured over the lateral transgluteal approach with regard to function, gait and Harris hip score at 6 and 12 weeks after surgery [15]. Meneghini et al. showed no difference between the three MIS THA approaches with regard to discharge, functional recovery or outcome scores over the first year [16]. An aggressive rapid rehabilitation protocol was used in all cases, and most patients were reportedly discharged a day after surgery. Goebel et al. found only a temporary reduction in initial postoperative pain levels after a MIS anterior approach compared with a lateral approach [17]. There was also an improvement in time to attain a range of movement and shorter hospital stay [17].

The various types of MIS have spawned debate over which technique is the most advantageous in accelerating recovery. Pospischill et al. found that there was no difference in the gait analysis of a patient who underwent THA via a MIS lateral approach versus a traditional transgluteal approach [18]. This finding was corroborated by the gait kinematic studies performed by Bennett et al. [19], and they also showed no difference in the immediate postoperative walking ability after MIS compared with standard incision hip replacement [20]. Ogonda et al. [21] and Lawlor et al. [22] published their data showing no enhancement in early walking ability or functional outcome by the sixth week when comparing a MIS posterior approach with a standard posterior approach. Interestingly, Pagnano et al. reported a slower recovery in patients undergoing MIS in a randomised clinical trial [23].

A recent Level III study by van Driessche et al. [24] looked at the MI anterior and Röttinger (between tensor fasciae latae and gluteus medius) approaches as they are deemed to be muscle-sparing. Given that balance is ensured by these muscles, it could be expected that postural control would be better conserved by these two approaches than by the posterior approach. However, no significant differences were found after the operation among patients who underwent surgery via the different approaches.

5.2 Quantifying Soft-Tissue Trauma

One of the arguments for MIS is that a smaller incision and approach equates to less soft-tissue trauma and therefore a faster recovery. The general surgeons were amongst the first to study this, and they noted reductions in acute phase cytokines when comparing laparoscopic and open cholecystectomies [25]. Since then, many studies have tried to replicate this finding in MI arthroplasties, looking at various biochemical markers such as C-reactive protein (CRP), creatinine kinase, myoglobin, aldolase, lactate dehydrogenase, glutamic oxaloacetic transaminase and creatinine. Wohlrab et al. failed to demonstrate a difference in the CRP levels when comparing a MIS Watson–Jones approach with a standard lateral transgluteal approach [15]. Similarly, Ogonda et al. failed to demonstrate a difference in CRP levels when comparing MIS and standard posterior approaches to the hip [21].

MIS has driven the development and refinement of improved and less invasive surgical instrumentation. The introduction of MIS in arthroplasty has also been accompanied by improvements in clinical pathways and the adoption of fast-track protocols. In addition to the MIS literature, there are accounts of fast-track pathways using conventional surgical techniques having achieved similar reductions in length of hospital stay and speed of recovery. Therefore, at present, there are too many confounding factors to show convincingly that MIS in isolation results in accelerated recovery for patients compared with conventional surgery. While not being conclusive in helping us to decide whether MIS is advantageous over conventional surgical techniques, the literature does confirm the value of fast-track pathways in accelerating patient recovery. Optimised preoperative preparation, anaesthesia, analgesia and rehabilitation are certainly aspects of the MIS pathways described in the literature that are here to stay.

5.3 MIS in Trauma

The experimental basis for MIS in trauma surgery is the work of Rhinelander, who showed that two-thirds of the blood supply to the bone is via the nutrient artery and one-third comes from the inflow of the periosteum [26, 27]. Plating methods championed by Farouk and Krettek [28, 29] limit dissection of the periosteum in order to limit devascularisation of the bone. Perren introduced limited-contact plates or point contact fix plates to achieve these same goals [30]. Minimal additional surgical trauma and flexible fixation allow prompt healing when blood supply to the bone is maintained. The biomechanical aspects principally address the degree of instability that may be tolerated by fracture healing under different biological conditions. The strain theory of fracture healing offers an explanation for the maximum instability that can be tolerated and the minimal degree required for induction of callus formation. The biological aspects of damage to the periosteal blood supply, cortical necrosis and temporary cortical porosity help to explain the importance of avoiding extensive contact of the implant with the bone [30].

5.4 Proximal Femoral Fractures: Sliding Hip Screw Fixation

At present, sliding hip screws (SHSs) are still widely regarded as the most conventional devices for the fixation of intertrochanteric femoral fractures and are associated with failure rates between 5% and 20% [31]. Surgical fixation of femoral intertrochanteric fractures can be undertaken with standard devices (SHSs or trochanteric nails) using a MI approach [32], specifically designed MI implants [33] or MIS techniques [34]. Gotfried et al. [33] published the first clinical and biomechanical results of the percutaneous compression plate (PCCP), a MI sliding screw plate implant, used for pertrochanteric fractures with favourable clinical complications rates and biomechanical results [35]. In addition to the PCCP, the newer MI screw plate implant, the Minimally Invasive Screw System (MISS; Lépine, Lyon, France), was developed (double small-diameter screw fixation into the femoral head for PCCP; larger for MISS) and required a different surgical approach. The biomechanical properties of these two MI screw plate implants were demonstrated to be as favourable as conventional hip screws [36]. Loading and mode of failure of these two implants were found to be similar. Future clinical and biomechanical comparisons of these two implants with standard devices will provide further information. MISS and PCCP may improve clinical outcomes and reduce the risk of comorbidities associated with unstable trochanteric fractures without increased risk of mechanical failure.

Kuzyk et al. [37] looked at data from 14 randomised controlled trials to perform a meta-analysis and calculate pooled relative risks of failure of fixation, blood transfusion and mortality. Other outcome measures that were extracted from 17 comparative studies are reported as a systematic review. They found that although significant heterogeneity exists between pooled studies, MI hip fracture plating, nailing or external fixation was associated with a significant decrease in transfusion rate as compared to standard SHS. There was no significant difference for the other comparisons, including mortality between MI plating, nailing or external fixation and standard insertion of a SHS.

5.5 Hip Arthroscopy: Current and Projected Trends

Hip arthroscopy is one of the most rapidly expanding fields in orthopaedics. The most common pathologies treated with hip arthroscopy are labral tears, femoroacetabular impingement (FAI) and chondral lesions. Robust evidence for these treatments is still lacking, but a considerable amount of literature (albeit consisting mainly of noncontrolled cohort studies) demonstrates that 68%–96% of patients show an improvement in symptoms after surgery [38, 39]. Studies of the cost effectiveness of these procedures are yet to be performed.

Less common therapeutic indications for hip arthroscopy include ischiofemoral impingement [40]; sciatic nerve entrapment [41]; synovial chondromatosis and pathology of the psoas tendon, iliotibial band and ligamentum teres [42]. Applications continue to expand, and hip arthroscopy is also used to facilitate hip reduction in infants with development dysplasia of the hip [43] and for the treatment of extracapsular pathology, including iliopsoas tendon and iliotibial band pathology [44].

The use of hip arthroscopy as a solely diagnostic procedure is controversial, as hip magnetic resonance imaging (MRI) has a high specificity and sensitivity (93% and 90%, respectively) for diagnosis of labral tears, although it is considerably less specific and sensitive for the detection of early osteoarthritis (66% and 75%, respectively). MRI is also used to detect most soft-tissue pathologies, as well as bony conditions such as FAI. There is some support for its use as a tool for staging osteoarthritis [44], but its overall role will continue to decline as imaging techniques become more powerful at providing accurate diagnoses of intra-articular hip pathology. In the meantime it has been suggested that diagnostic hip arthroscopy should be considered for pain persisting for six months or more.

According to data from the Hospital Episode Statistics (HES) database, a total of 11,329 hip arthroscopies were performed in National Health Service (NHS) hospitals in England between 2002 and 2013. There was a statistically significant increase of 727% ($p < 0.0001$) during this period, and Palmer et al. projected an increase of 1388% by 2023 [45]. The HES database contains data on patients attending NHS hospitals in England for treatment. Each entry represents a single episode of care, and codes are assigned to diagnoses and procedures during that episode. The study likely underestimates the number of hip arthroscopy procedures performed in England as it does not include those undertaken in private healthcare, but the overall trend is projected to continue despite limited evidence of long-term efficacy or superiority of these procedures over nonoperative measures, and these trends mirror those of subacromial decompression and rotator cuff repair of the shoulder, again in the absence of robust evidence to demonstrate clinical effectiveness [46]. Increasingly questions are also being raised over the efficacy of arthroscopic knee surgery [47, 48]. In the last few years, Swedish and Danish hip arthroscopy registers have been set up to evaluate and report the demographic and outcome data of patients undergoing hip arthroscopy. Similarly, the British Hip Society has set up the Non Arthroplasty Hip Surgery Register, which will provide a record for audit and research. Currently, a standardised approach to defining the pathology (impingement, arthritis, etc.) addressed by hip arthroscopy remains necessary to ensure patients with adequate pathology are indicated for this procedure, thereby attempting to ensure significant clinical improvement after this procedure and enabling comparison of results across nations.

According to current data, the overall risk of an adverse event following hip arthroscopy is 4%–7% [49]. Serious complications, such as infection, thromboembolism, neurovascular damage and hip fracture or dislocation, are rare (<1%) [50]. However, a small number of complications specific to hip arthroscopy have been reported, including abdominal compartment syndrome, perineal and genital trauma, pudendal nerve neuralgia and superficial nerve damage. The overall reoperation rate is 6% and occurs on average 16 months after the primary procedure [51].

5.6 Conclusion

Robust evidence of efficacy is required from clinical studies, and research priorities should reflect the changing nature of healthcare provision. An increasing demand for MI hip surgery may also have implications for physiotherapy rehabilitation and orthopaedic training, and this is particularly salient given that MIS has a steep learning curve and will influence how we deliver training and services to patients.

References

1. Goldstein WM, Branson JJ, Berland KA, Gordon AC. Minimal-incision total hip arthroplasty. *J Bone Joint Surg Am*, 2003; **85-A**(Suppl 4):33–38.
2. Belsky MR. What's new in orthopaedic surgery. *J Am Coll Surg*, 2003; **197**:985–989.
3. Sculco TP. Minimally invasive total hip arthroplasty: in the affirmative. *J Arthroplasty*, 2004; **19**:78–80.

4. DiGioia AM, III, Blendea S, Jaramaz B, Levison TJ. Less invasive total hip arthroplasty using navigational tools. *Instr Course Lect*, 2004; **53**:157–164.
5. Berger RA, Jacobs JJ, Meneghini RM, Della VC, Paprosky W, Rosenberg AG. Rapid rehabilitation and recovery with minimally invasive total hip arthroplasty. *Clin Orthop Relat Res*, 2004; 239–247.
6. Bhandari M, Matta JM, Dodgin D, Clark C, Kregor P, Bradley G, Little L. Outcomes following the single-incision anterior approach to total hip arthroplasty: a multicenter observational study. *Orthop Clin North Am*, 2009; **40**:329–342.
7. Nakamura S, Matsuda K, Arai N, Wakimoto N, Matsushita T. Mini-incision posterior approach for total hip arthroplasty. *Int Orthop*, 2004; **28**:214–217.
8. Woolson ST, Mow CS, Syquia JF, Lannin JV, Schurman DJ. Comparison of primary total hip replacements performed with a standard incision or a mini-incision. *J Bone Joint Surg Am*, 2004; **86-A**:1353–1358.
9. Graw BP, Woolson ST, Huddleston HG, Goodman SB, Huddleston JI. Minimal incision surgery as a risk factor for early failure of total hip arthroplasty. *Clin Orthop Relat Res*, 2010; **468**:2372–2376.
10. Bertin KC, Rottinger H. Anterolateral mini-incision hip replacement surgery: a modified Watson-Jones approach. *Clin Orthop Relat Res*, 2004; 248–255.
11. Siguier T, Siguier M, Brumpt B. Mini-incision anterior approach does not increase dislocation rate: a study of 1037 total hip replacements. *Clin Orthop Relat Res*, 2004; 164–173.
12. Berger RA. The technique of minimally invasive total hip arthroplasty using the two-incision approach. *Instr Course Lect*, 2004; **53**:149–155.
13. Goldstein WM, Branson JJ. Posterior-lateral approach to minimal incision total hip arthroplasty. *Orthop Clin North Am*, 2004; **35**:131–136.
14. Leuchte S, Riedl K, Wohlrab D. [Immediate post-operative advantages of minimally invasive hip replacement: results of symmetry and load from the measurement of ground reaction force]. *Z Orthop Unfall*, 2009; **147**:69–78.
15. Wohlrab D, Droege JW, Mendel T, Brehme K, Riedl K, Leuchte S, Hein W. [Minimally invasive vs. transgluteal total hip replacement. A 3-month follow-up of a prospective randomized clinical study]. *Orthopade*, 2008; **37**:1121–1126.

16. Meneghini RM, Smits SA. Early discharge and recovery with three minimally invasive total hip arthroplasty approaches: a preliminary study. *Clin Orthop Relat Res*, 2009; **467**:1431–1437.
17. Goebel S, Steinert AF, Schillinger J, Eulert J, Broscheit J, Rudert M, Noth U. Reduced postoperative pain in total hip arthroplasty after minimal-invasive anterior approach. *Int Orthop*, 2012; **36**:491–498.
18. Pospischill M, Kranzl A, Attwenger B, Knahr K. Minimally invasive compared with traditional transgluteal approach for total hip arthroplasty: a comparative gait analysis. *J Bone Joint Surg Am*, 2010; **92**:328–337.
19. Bennett D, Ogonda L, Elliott D, Humphreys L, Beverland DE. Comparison of gait kinematics in patients receiving minimally invasive and traditional hip replacement surgery: a prospective blinded study. *Gait Posture*, 2006; **23**:374–382.
20. Bennett D, Ogonda L, Elliott D, Humphreys L, Lawlor M, Beverland D. Comparison of immediate postoperative walking ability in patients receiving minimally invasive and standard-incision hip arthroplasty: a prospective blinded study. *J Arthroplasty*, 2007; **22**:490–495.
21. Ogonda L, Wilson R, Archbold P, Lawlor M, Humphreys P, O'Brien S, Beverland D. A minimal-incision technique in total hip arthroplasty does not improve early postoperative outcomes. A prospective, randomized, controlled trial. *J Bone Joint Surg Am*, 2005; **87**:701–710.
22. Lawlor M, Humphreys P, Morrow E, Ogonda L, Bennett D, Elliott D, Beverland D. Comparison of early postoperative functional levels following total hip replacement using minimally invasive versus standard incisions. A prospective randomized blinded trial. *Clin Rehabil*, 2005; **19**:465–474.
23. Pagnano MW, Trousdale RT, Meneghini RM, Hanssen AD. Slower recovery after two-incision than mini-posterior-incision total hip arthroplasty. A randomized clinical trial. *J Bone Joint Surg Am*, 2008; **90**:1000–1006.
24. Van DS, Billuart F, Martinez L, Brunel H, Guiffault P, Beldame J, Matsoukis J. Short-term comparison of postural effects of three minimally invasive hip approaches in primary total hip arthroplasty: direct anterior, posterolateral and Rottinger. *Orthop Traumatol Surg Res*, 2016; **102**:729–734.
25. Grande M, Tucci GF, Adorisio O, Barini A, Rulli F, Neri A, Franchi F, Farinon AM. Systemic acute-phase response after laparoscopic and open cholecystectomy. *Surg Endosc*, 2002; **16**:313–316.

26. Rhinelander FW. The normal circulation of bone and its response to surgical intervention. *J Biomed Mater Res*, 1974; **8**:87–90.

27. Rhinelander FW. Tibial blood supply in relation to fracture healing. *Clin Orthop Relat Res*, 1974; 34–81.

28. Farouk O, Krettek C, Miclau T, Schandelmaier P, Guy P, Tscherne H. Minimally invasive plate osteosynthesis: does percutaneous plating disrupt femoral blood supply less than the traditional technique? *J Orthop Trauma*, 1999; **13**:401–406.

29. Krettek C, Schandelmaier P, Tscherne H. [New developments in stabilization of dia- and metaphyseal fractures of long tubular bones]. *Orthopade*, 1997; **26**:408–421.

30. Perren SM. Evolution of the internal fixation of long bone fractures. The scientific basis of biological internal fixation: choosing a new balance between stability and biology. *J Bone Joint Surg Br*, 2002; **84**:1093–1110.

31. Ahrengart L, Tornkvist H, Fornander P, Thorngren KG, Pasanen L, Wahlstrom P, Honkonen S, Lindgren U. A randomized study of the compression hip screw and Gamma nail in 426 fractures. *Clin Orthop Relat Res*, 2002; 209–222.

32. Alobaid A, Harvey EJ, Elder GM, Lander P, Guy P, Reindl R. Minimally invasive dynamic hip screw: prospective randomized trial of two techniques of insertion of a standard dynamic fixation device. *J Orthop Trauma*, 2004; **18**:207–212.

33. Gotfried Y. Percutaneous compression plating of intertrochanteric hip fractures. *J Orthop Trauma*, 2000; **14**:490–495.

34. Moroni A, Faldini C, Pegreffi F, Hoang-Kim A, Vannini F, Giannini S. Dynamic hip screw compared with external fixation for treatment of osteoporotic pertrochanteric fractures. A prospective, randomized study. *J Bone Joint Surg Am*, 2005; **87**:753–759.

35. Gotfried Y, Cohen B, Rotem A. Biomechanical evaluation of the percutaneous compression plating system for hip fractures. *J Orthop Trauma*, 2002; **16**:644–650.

36. Ropars M, Mitton D, Skalli W. Minimally invasive screw plates for surgery of unstable intertrochanteric femoral fractures: a biomechanical comparative study. *Clin Biomech (Bristol, Avon)*, 2008; **23**:1012–1017.

37. Kuzyk PR, Guy P, Kreder HJ, Zdero R, McKee MD, Schemitsch EH. Minimally invasive hip fracture surgery: are outcomes better? *J Orthop Trauma*, 2009; **23**:447–453.

38. Clohisy JC, St John LC, Schutz AL. Surgical treatment of femoroacetabular impingement: a systematic review of the literature. *Clin Orthop Relat Res*, 2010; **468**:555–564.

39. Kemp JL, Collins NJ, Makdissi M, Schache AG, Machotka Z, Crossley K. Hip arthroscopy for intra-articular pathology: a systematic review of outcomes with and without femoral osteoplasty. *Br J Sports Med*, 2012; **46**:632–643.

40. Larson CM, Stone RM. Current concepts and trends for operative treatment of FAI: hip arthroscopy. *Curr Rev Musculoskelet Med*, 2013; **6**:242–249.

41. Martin HD, Shears SA, Johnson JC, Smathers AM, Palmer IJ. The endoscopic treatment of sciatic nerve entrapment/deep gluteal syndrome. *Arthroscopy*, 2011; **27**:172–181.

42. Shetty VD, Villar RN. Hip arthroscopy: current concepts and review of literature. *Br J Sports Med*, 2007; **41**:64–68.

43. Eberhardt O, Fernandez FF, Wirth T. Arthroscopic reduction of the dislocated hip in infants. *J Bone Joint Surg Br*, 2012; **94**:842–847.

44. Ilizaliturri VM, Jr., Camacho-Galindo J. Endoscopic treatment of snapping hips, iliotibial band, and iliopsoas tendon. *Sports Med Arthrosc*, 2010; **18**:120–127.

45. Palmer AJR, Malak TT, Broomfield J, Holton J, Majkowski L, Thomas GER, Taylor A, Andrade AJ, Collins G, Watson K, Carr AJ, Glyn-Jones S. Past and projected temporal trends in arthroscopic hip surgery in England between 2002 and 2013. *BMJ Open Sport Exerc Med*, 2016; **2**:e000082.

46. Judge A, Murphy RJ, Maxwell R, Arden NK, Carr AJ. Temporal trends and geographical variation in the use of subacromial decompression and rotator cuff repair of the shoulder in England. *Bone Joint J*, 2014; **96-B**:70–74.

47. Carr A. Arthroscopic surgery for degenerative knee: overused, ineffective, and potentially harmful. *Br J Sports Med*, 2015; **49**:1223–1224.

48. Carr A. Arthroscopic surgery for degenerative knee. *BMJ*, 2015; **350**:h2983.

49. Kowalczuk M, Bhandari M, Farrokhyar F, Wong I, Chahal M, Neely S, Gandhi R, Ayeni OR. Complications following hip arthroscopy: a systematic review and meta-analysis. *Knee Surg Sports Traumatol Arthrosc*, 2013; **21**:1669–1675.

50. Chan K, Farrokhyar F, Burrow S, Kowalczuk M, Bhandari M, Ayeni OR. Complications following hip arthroscopy: a retrospective review of the McMaster experience (2009–2012). *Can J Surg*, 2013; **56**:422–426.
51. Harris JD, McCormick FM, Abrams GD, Gupta AK, Ellis TJ, Bach BR, Jr., Bush-Joseph CA, Nho SJ. Complications and reoperations during and after hip arthroscopy: a systematic review of 92 studies and more than 6,000 patients. *Arthroscopy*, 2013; **29**:589–595.

Chapter 6

Computer-Assisted Navigation of the Hip Joint

Matthew K. T. Seah and Wasim Khan

Department of Trauma and Orthopaedics, Addenbrooke's Hospital, Cambridge University Hospitals NHS Foundation Trust, Hills Road, Cambridge CB2 0QQ, United Kingdom
wasimkhan@doctors.org.uk

Total hip arthroplasty (THA) is an increasingly common procedure, and this comes as no surprise as estimates suggest that around 8.5 million people in the UK are affected by joint pain secondary to arthritis. This number continues to rise as our population ages [1]. According to data from the National Joint Registry for England and Wales, over 179,000 total hip and knee arthroplasties were carried out in 2010, an increase of nearly 300% compared to six years earlier. There has also been a 49% rise in revision THR, and with 35% of hip and knee replacements now carried out in patients below the age of 65, and 12% below the age of 55, this revision burden will grow exponentially. In 2010, payments to patients by the NHS Litigation Authority totalled £863 million and 15% were orthopaedic related. De Palma et al. [2] analysed the financial impact

---

The Hip Joint in Adults: Advances and Developments
Edited by K. Mohan Iyer
Copyright © 2018 Pan Stanford Publishing Pte. Ltd.
ISBN 978-981-4774-72-7 (Hardcover), 978-1-351-26244-6 (eBook)
www.panstanford.com

of complications following hip arthroplasty and found the additional cost of a dislocation within six weeks of surgery was 342% of the primary cost.

Hip arthroplasty is suited to computer navigation as specific targets for component positioning have been defined and the orientation of the acetabular component is probably the most important factor in successful THA [3–6]. Unsurprisingly, computer navigation for acetabular component positioning has received the most attention in the literature. Acetabular component malorientation is associated with increased rates of dislocation, impingement, polyethylene wear and loosening [3]. Positioning of the acetabular component may be inaccurate with the use of traditional jigs [5], and computed tomography–based and imageless navigation may improve the surgeon's ability to insert the acetabular component in the optimal orientation [1]. Restoration of leg length is also important for the outcome of patients undergoing THA, and different techniques using Kirschner wires and Steinmann pins have been described with different success rates for intraoperative measurement and leg length restoration. In current clinical practice, computer-assisted navigation appears to be the most accurate method to obtain leg length equality [7, 8].

Hip resurfacing arthroplasty has gained popularity as an alternative to THA in younger patients due to preservation of the femoral bone stock, improved hip stability and proprioception, and a lower risk for leg length discrepancy. However, the performance and complications of metal-on-metal arthroplasties have questioned its role compared with THA. Moreover, failure of the femoral prosthesis is more common than that of the acetabulum, making the accuracy of fixation of the femoral prosthesis more important. Computer-assisted navigation allows for accurate reconstruction of the femoral bone and optimal fixation of the femoral component.

6.1 Limitations of Conventional Alignment Jigs

Most surgeons use intraoperative alignment jigs to place the acetabular component referencing of the position of the patient on the operating table. In the lateral position vertical orientation of the cup is usually judged from the floor and anteversion from the patient's superior shoulder. However, it is now known that the exact position of the patient's pelvis on the table is difficult to judge with drapes and particularly in obese patients. Additional technical challenges include dysplasia and altered patient anatomy, absence of usual landmarks as commonly encountered in revision cases, poor surgical exposure and obscured view with haemorrhage.

In current practice, there are various ways by which surgeons aim to achieve acetabular component placement within a predetermined 'safe zone'. The most common method utilises a simple mechanical alignment rod, and the surgeon usually judges the component anteversion compared with the patient's superior shoulder, and the position of inclination when compared with the floor, with the additional visual assistance of an alignment rod that can be attached to the cup impactor. However, the surgeon should not assume that the orientation of the pelvis is in line with the table or the patient's body. In the lateral position the lumbar lordotic curve flattens and the pelvis may be flexed forward as much as 35° [9]. In addition, the superior aspect of the acetabulum may be tilted towards the foot of the table by 10°–15°. This means that cups placed with alignment jigs may actually be excessively retroverted and too vertical. Multiple studies that show that surgeons using this manual referencing system often place the acetabular components outside the Lewinneck-defined 'safe zone' and there is significant variability in final cup placement [10, 11]. An analysis of 1952 total hip arthroplasties (THAs) by Callanan et al. indicated that only 62% of cases were placed within their desired inclination of between 30° and 45°, suggesting that surgeons cannot rely on the assumption that the patient's pelvic position is orientated in line with the floor or the long axis of the body [12].

Some surgeons suggest that the transverse acetabular ligament (TAL) is a good reference point for acetabular component positioning and an adjunct to the conventional alignment rod method. Archbold et al. [13] claimed to have identified the TAL in 99.7% of 1000 consecutive THAs; and when using the TAL as the reference for acetabular component placement, the clinical outcome was satisfactory, with only a 0.6% dislocation rate (however, this paper lacked supporting data on postoperative assessment of cup position); and Epstein et al. suggest the TAL is identifiable in less than half of THA operations [14]. Magnetic resonance imaging (MRI) studies have shown that the natural anteversion of the TAL in healthy subjects ranges from 5.3° to 36.1°, which would therefore render this landmark useless in judging acetabular component placement [15]. Furthermore, the natural acetabular position in the native hip is inconsistent and known to change with osteoarthritis, dysplasia and osteonecrosis.

6.2 Types of Computer Navigation Systems

Following advances in 3D sensor technology, it became possible to develop computer navigation systems that use optical or magnetic sensors to determine and track the position of bones, surgical instruments and implants. The preoperative or intraoperative data obtained allow the system to build a 3D model of the pelvis to guide component placement. The imaging systems that are used during scanning can be divided into image-based and imageless systems. Image-based systems require the collection of morphological information by preoperative computerised tomography (CT) or MRI scans or by means of intraoperative fluoroscopy. Imageless systems use a virtual anatomical model that is embedded in the software and is supplemented by intraoperative registration data of anatomical landmarks. The most common means of doing this is using infrared-light-emitting diodes or reflective markers on surgical instruments and at anatomical bony landmarks, including the anterior superior iliac spine and pubic tubercle. Infrared cameras that are coupled with an emitter and the navigation computer track the movement of the markers; however, an uninterrupted line of sight must be maintained. The line of sight does not appear to be a problem with magnetic sensors, although they are less commonly used because electromechanical devices in the theatre may interfere with their accuracy. Imageless systems appear to be the most popular systems in clinical use judging by our experience and the published literature on navigation. The accuracy, however, depends on the ability to identify bony landmarks, and this may be limited by soft-tissue thickness. Image-guided systems are more accurate but involve preoperative planning, associated costs of preoperative CT/MRI imaging and radiation [16, 17].

6.3 The Accuracy of Computer-Assisted Navigation in THA

Proponents of navigated hip arthroplasty have suggested that it may increase the precision of acetabular component placement. There is an increasing body of evidence, including randomised controlled trials, meta-analyses and systematic reviews, that has been published on navigated hip arthroplasty, and the most consistent finding from these studies is that computer navigation improves the accuracy of acetabular cup positioning and minimises outliers in both THA and hip resurfacing rather than significantly improving the acetabular component orientation per se [18]. The accuracy of femoral component placement in hip resurfacing can also be improved with navigation and errors during the surgeon learning curve minimised.

Sariali et al. [19] hypothesised that the use of three-dimensional visualisation tools to identify the planned acetabular component position relative to the acetabular edge intraoperatively would increase the accuracy of cup orientation when compared to regular insertion of the acetabular component. In a prospective randomised controlled study of two groups of 28 patients each, they found that acetabular component anteversion was more accurate in the 3D planning group (mean difference from the planned angle, $-2.7° \pm 5.4°$) compared with the freehand placement group ($6.6° \pm 9.5°$). Although component abduction was also restored with greater accuracy in the 3D planning group, this advantage was not found to be significant when compared to the regular group. There were fewer outliers for both parameters in the 3D planning group, and the operative time did not differ significantly. The conclusion that the use of navigation can improve acetabular component position in THA by reducing the percentage of outliers is shared by Parratte et al. [20], who found lower variations in the mean inclination and anteversion angles in the computer-assisted group when compared to regular insertion. However, the differences in mean inclination and anteversion angles were not significant.

However, the reduced invariance in most studies does not seem to be accompanied by improved clinical outcomes in navigated hips. In a Level III study, Sugano et al. retrospectively studied a cohort of navigated hips and suggested that CT-based navigation reduces the rates of dislocation and impingement-related mechanical complications, leading to revision in cementless THA using ceramic-on-ceramic bearing couples [21]. A more recent Level II study suggests that computer-assisted surgery (CAS) used for acetabular component placement does not confer any substantial advantage in function, wear rate or survivorship at 10 years after THA [22]. Their study noted that CAS is associated with added costs and surgical time, and further studies therefore need to identify what clinically relevant advantages it offers, to justify its continued use in THA. There is a general reluctance of the surgical community to accept navigation as a routine part of the arthroplasty process, but if we assume that optimal component positioning leads to better outcomes, it is logical to predict a resurgence in interest in computer-navigated hip arthroplasty. The results of robust long-term clinical outcome studies comparing navigated with non-navigated hips are awaited.

6.4 Computer Navigation in Total Hip Resurfacing

Hip resurfacing has a place as a bone conserving alternative to THA in the young and active adult with degenerative hip disease. The larger head diameters offer greater postoperative stability and may decrease wear rates. However, hip resurfacing is inherently more difficult to perform than traditional THA because of the limited femoral resection, which makes acetabular visualisation more difficult. Preparing the acetabulum prior to addressing the femoral head creates limitations in terms of exposure and mobilisation of tissues, thus posing a technical challenge to the surgeon and increasing the risk of component malalignment.

As with THA, the orientation of the acetabular component is probably the most important factor in successful hip resurfacing. It has been shown that increasing the inclination of the acetabular component above 55° in hip resurfacing leads to an 'edge loading' effect with a much greater release of metal ions [23, 24]. Elevated serum metal ions have been associated with local pseudotumour formation and have unknown systemic effects, which may include carcinogenic potential [25].

Errors in component positioning during the surgeon learning curve are common in hip resurfacing. Navigation of the acetabular cup in hip resurfacing follows an identical procedure to THA. Imageless navigation in hip resurfacing has been shown to help avoid component malposition during the surgeon learning curve. There is a risk of femoral neck fracture associated with hip resurfacing, with a prevalence reported from 0% to 17% [26, 27]. The notching of the superolateral aspect of the femoral neck and varus placement of the femoral component have also been identified as factors that may increase the likelihood of femoral neck fracture [28, 29]. Preparation of the femoral side is therefore an attractive target for computer guidance. The key step for navigated preparation of the femoral side is the guide wire insertion into the femoral head to determine implant orientation. Online display of the actual wire position in three dimensions allows for immediate correction and best match with the preplanned pin alignment. The wire is then overdrilled and replaced by the instrumentation guide for femoral head preparation. A meta-analysis based on seven studies, 520 patients and 555 hip resurfacing arthroplasties concluded that computer navigation systems make the femoral component positioning in hip resurfacing arthroplasty easier and more precise [30].

6.5 Limitations of Computer Navigation

Reliance on imageless computer navigation systems on the anterior pelvic plane (APP) has been touted as a potentially inaccurate method for determining pelvic position due to registration errors and its lack of accounting for pelvic tilt. The palpation of anatomic landmarks, which is necessary to determine the pelvic coordinate system, can be influenced by uneven soft-tissue distribution, and an ultrasound study assessed the thickness of skin and fat over the three reference points and found that this is potentially a source of error with anteversion analysis [31]. This is particularly problematic in obese patients. Wolf et al. [32] have demonstrated this theoretical effect, showing that inaccurate registration of the APP results in incorrectly placed acetabular components.

Concern over registration has led some groups to use more invasive methods. Dorr et al. [33] used skin puncture over the bony landmarks to improve registration; however, there is concern about donor site morbidity with theoretical, but not reported, increased risk of infection, bleeding, fracture and mechanical pull out. Aside from registration issues, the reliability of APP has been questioned by some. Barbier et al. [34] conducted a prospective, single-centre study of 44 patients who underwent navigated THA, imaged them using a 3D imaging system three months after surgery and found significant differences between operative navigation records and postoperative imaging. Anteversion appeared to show the weakest correlation, with a mean intraoperative anteversion of 20.9° but 29.5° in postoperative imaging. This was thought to be due to both the difficulty in registering APP preoperatively and the interobserver variation intraoperatively. This problem is compounded as extremes of pelvic tilt (which is dynamic) can affect the relationship between the acetabular anatomy and the APP.

Other obstacles to performing CAS relate to theatre logistics. Additional equipment is needed in the operating theatre, which may have to be modified. A technician may also be needed specifically for the navigation equipment. The additional navigation steps throughout the procedure add time to the case, which can be important from the anaesthetic point of view and also in terms of getting through the operating list. Surgical approaches have to be considered carefully as imageless techniques require registration with the patient supine. The patient may then need to be repositioned into the lateral decubitus position. Personnel in the operating theatre need to think carefully about moving so as not to interrupt the line of sight of the sensors for imageless techniques. Both the additional equipment and the possible increase in operating time contribute to the extra immediate and ongoing cost of computer-navigated surgery, which may explain its slow uptake in the National Health Service (NHS).

6.6 Robotic Surgery: The Future of Navigation?

Different computer-assisted surgical systems have been developed for preoperative planning, for surgical navigation and to assist in performing surgical procedures. Robotically assisted surgical

(RAS) devices are one type of computer-assisted surgical system. Sometimes referred to as robotic surgery, RAS devices enable the surgeon to use computer technology to control and move surgical instruments through one or more tiny incisions in the patient's body for a variety of surgical procedures. The benefits of a RAS device may include its ability to facilitate minimally invasive surgery and assist with complex tasks in confined areas of the body.

Indeed there is already evidence that robotic-assisted acetabular cup placement outperforms conventional methods of cup placement [35]. In a matched-pair control study of 100 patients undergoing robotic-assisted THA versus conventional THA, postoperative radiographic analysis showed that robotic-assisted surgery resulted in 100% of cups within Lewinneck's safe zone. This was significantly better than acetabular component placement in conventional THA, which had an average of 80% of the cups in the safe zone.

Domb et al. [36] performed a multisurgeon retrospective analysis of 1980 THA patients, looking at modes of guidance, including a mechanical alignment rod, TAL referencing and fluoroscopy-guided, computer-navigated and robotic-guided systems. The target for cup positioning was defined as 40° of inclination and 20° of anteversion. Computer-navigated and robotic-assisted THAs were associated with a significantly greater number of acetabular components within Lewinneck's safe zone but when adjusted for Callanan's safe zones robotic-assisted acetabular navigation improved accuracy significantly when compared to all modalities. As with other studies looking at computer-assisted surgery, there was significantly less variance in the computer-navigated and robotic-assisted groups when compared with all other methods. It is important to note that there was one case of robotic-assisted failure requiring conversion to the conventional method of cup placement in the study. This is a reminder of the importance of surgery being supervised or performed by an experienced surgeon in order to use good judgement when required. Further theoretical limitations on employing such equipment include set-up and running costs, restriction of implant choice for robotic system compatibility, exposure to radiation via CT and operative time.

6.7 Conclusion

The literature has proven improvement in the accuracy of component alignment with navigation (in the reduction of variance), but this has not yet conclusively shown to translate to better functional results and improved survival of the implants. Long-term studies with large numbers will be needed to demonstrate this before navigation and its limitations become more readily accepted.

References

1. Khan W. Computer navigation in hip arthroplasty. In *The Hip Joint*, Iyer KM, ed. (Pan Stanford), 2016.
2. de Palma L, Procaccini R, Soccetti A, Marinelli M. Hospital cost of treating early dislocation following hip arthroplasty. *Hip Int*, 2012; **22**:62–67.
3. Najarian BC, Kilgore JE, Markel DC. Evaluation of component positioning in primary total hip arthroplasty using an imageless navigation device compared with traditional methods. *J Arthroplasty*, 2009; **24**:15–21.
4. Beaumont E, Beaumont P, Odermat D, Fontaine I, Jansen H, Prince F. Clinical validation of computer-assisted navigation in total hip arthroplasty. *Adv Orthop*, 2011; **2011**:171783.
5. Honl M, Schwieger K, Salineros M, Jacobs J, Morlock M, Wimmer M. Orientation of the acetabular component. A comparison of five navigation systems with conventional surgical technique. *J Bone Joint Surg Br*, 2006; **88**:1401–1405.
6. Dheerendra S, Khan W, Saeed MZ, Goddard N. Recent developments in total hip replacements: cementation, articulation, minimal-invasion and navigation. *J Perioper Pract*, 2010; **20**:133–138.
7. Punwar S, Khan WS, Longo UG. The use of computer navigation in hip arthroplasty: literature review and evidence today. *Ortop Traumatol Rehabil*, 2011; **13**:431–438.
8. Manzotti A, Cerveri P, De ME, Pullen C, Confalonieri N. Does computer-assisted surgery benefit leg length restoration in total hip replacement? Navigation versus conventional freehand. *Int Orthop*, 2011; **35**:19–24.
9. McCollum DE, Gray WJ. Dislocation after total hip arthroplasty. Causes and prevention. *Clin Orthop Relat Res*, 1990; 159–170.

10. DiGioia AM, III, Jaramaz B, Plakseychuk AY, Moody JE, Jr., Nikou C, Labarca RS, Levison TJ, Picard F. Comparison of a mechanical acetabular alignment guide with computer placement of the socket. *J Arthroplasty*, 2002; **17**:359–364.

11. Rittmeister M, Callitsis C. Factors influencing cup orientation in 500 consecutive total hip replacements. *Clin Orthop Relat Res*, 2006; **445**:192–196.

12. Callanan MC, Jarrett B, Bragdon CR, Zurakowski D, Rubash HE, Freiberg AA, Malchau H. The John Charnley Award: risk factors for cup malpositioning: quality improvement through a joint registry at a tertiary hospital. *Clin Orthop Relat Res*, 2011; **469**:319–329.

13. Archbold HA, Mockford B, Molloy D, McConway J, Ogonda L, Beverland D. The transverse acetabular ligament: an aid to orientation of the acetabular component during primary total hip replacement: a preliminary study of 1000 cases investigating postoperative stability. *J Bone Joint Surg Br*, 2006; **88**:883–886.

14. Epstein NJ, Woolson ST, Giori NJ. Acetabular component positioning using the transverse acetabular ligament: can you find it and does it help? *Clin Orthop Relat Res*, 2011; **469**:412–416.

15. Archbold HA, Slomczykowski M, Crone M, Eckman K, Jaramaz B, Beverland DE. The relationship of the orientation of the transverse acetabular ligament and acetabular labrum to the suggested safe zones of cup positioning in total hip arthroplasty. *Hip Int*, 2008; **18**:1–6.

16. Picardo NE, Khan W, Johnstone D. Computer-assisted navigation in high tibial osteotomy: a systematic review of the literature. *Open Orthop J*, 2012; **6**:305–312.

17. Wong JM, Khan WS, Saksena J. The role of navigation in total knee replacement surgery. *J Perioper Pract*, 2013; **23**:202–207.

18. Gandhi R, Marchie A, Farrokhyar F, Mahomed N. Computer navigation in total hip replacement: a meta-analysis. *Int Orthop*, 2009; **33**:593–597.

19. Sariali E, Boukhelifa N, Catonne Y, Pascal MH. Comparison of three-dimensional planning-assisted and conventional acetabular cup positioning in total hip arthroplasty: a randomized controlled trial. *J Bone Joint Surg Am*, 2016; **98**:108–116.

20. Parratte S, Argenson JN, Validation and usefulness of a computer-assisted cup-positioning system in total hip arthroplasty. A prospective, randomized, controlled study. *J Bone Joint Surg Am*, 2007; **89**:494–499.

21. Sugano N, Takao M, Sakai T, Nishii T, Miki H. Does CT-based navigation improve the long-term survival in ceramic-on-ceramic THA? *Clin Orthop Relat Res*, 2012; **470**:3054–3059.
22. Parratte S, Ollivier M, Lunebourg A, Flecher X, Argenson JN. No benefit after THA performed with computer-assisted cup placement: 10-year results of a randomized controlled study. *Clin Orthop Relat Res*, 2016; **474**:2085–2093.
23. Langton DJ, Jameson SS, Joyce TJ, Webb J, Nargol AV. The effect of component size and orientation on the concentrations of metal ions after resurfacing arthroplasty of the hip. *J Bone Joint Surg Br*, 2008; **90**:1143–1151.
24. De HR, Campbell PA, Su EP, De Smet KA. Revision of metal-on-metal resurfacing arthroplasty of the hip: the influence of malpositioning of the components. *J Bone Joint Surg Br*, 2008; **90**:1158–1163.
25. Hartmann A, Hannemann F, Lutzner J, Seidler A, Drexler H, Gunther KP, Schmitt J. Metal ion concentrations in body fluids after implantation of hip replacements with metal-on-metal bearing--systematic review of clinical and epidemiological studies. *PLoS One*, 2013; **8**:e70359.
26. Amstutz HC, Campbell PA, Le Duff MJ. Fracture of the neck of the femur after surface arthroplasty of the hip. *J Bone Joint Surg Am*, 2004; **86-A**:1874–1877.
27. Shimmin AJ, Bare J, Back DL. Complications associated with hip resurfacing arthroplasty. *Orthop Clin North Am*, 2005; **36**:187–193, ix.
28. Morison Z, Olsen M, Higgins GA, Zdero R, Schemitsch EH. The biomechanical effect of notch size, notch location, and femur orientation on hip resurfacing failure. *IEEE Trans Biomed Eng*, 2013; **60**:2214–2221.
29. Marker DR, Seyler TM, Jinnah RH, Delanois RE, Ulrich SD, Mont MA. Femoral neck fractures after metal-on-metal total hip resurfacing: a prospective cohort study. *J Arthroplasty*, 2007; **22**:66–71.
30. Liu H, Li L, Gao W, Wang M, Ni C. Computer navigation vs conventional mechanical jig technique in hip resurfacing arthroplasty: a meta-analysis based on 7 studies. *J Arthroplasty*, 2013; **28**:98–102.
31. Richolt JA, Effenberger H, Rittmeister ME. How does soft tissue distribution affect anteversion accuracy of the palpation procedure in image-free acetabular cup navigation? An ultrasonographic assessment. *Comput Aided Surg*, 2005; **10**:87–92.
32. Wolf A, DiGioia AM, III, Mor AB, Jaramaz B. Cup alignment error model for total hip arthroplasty. *Clin Orthop Relat Res*, 2005; 132–137.

33. Dorr LD, Hishiki Y, Wan Z, Newton D, Yun A. Development of imageless computer navigation for acetabular component position in total hip replacement. *Iowa Orthop J*, 2005; **25**:1–9.
34. Barbier O, Skalli W, Mainard L, Mainard D. The reliability of the anterior pelvic plane for computer navigated acetabular component placement during total hip arthroplasty: prospective study with the EOS imaging system. *Orthop Traumatol Surg Res*, 2014; **100**:S287–S291.
35. Domb BG, El Bitar YF, Sadik AY, Stake CE, Botser IB. Comparison of robotic-assisted and conventional acetabular cup placement in THA: a matched-pair controlled study. *Clin Orthop Relat Res*, 2014; **472**:329–336.
36. Domb BG, Redmond JM, Louis SS, Alden KJ, Daley RJ, LaReau JM, Petrakos AE, Gui C, Suarez-Ahedo C. Accuracy of component positioning in 1980 total hip arthroplasties: a comparative analysis by surgical technique and mode of guidance. *J Arthroplasty*, 2015; **30**:2208–2218.

Chapter 7

Advances in the Treatment of Meralgia Paresthetica in Surgery of the Hip Joint in Adults

K. Mohan Iyer
Formerly Royal Free Hampstead NHS Trust, Royal Free Hospital, Pond Street, London NW3 2QG, United Kingdom
kmiyer28@hotmail.com

Meralgia paresthetica (also called Bernhardt–Roth syndrome) is a distinctive condition, more common in men than in women, characterised by paraesthesiae and often burning pain over the anterolateral aspect of the thigh. Impaired or altered sensation is found in the same area, but there is no motor weakness or wasting and knee jerk is preserved, distinguishing it from radiculopathy. Martin Bernhardt, a German neuropathologist, first depicted the condition in 1878 [1, 2]). Vladimir Karlovich Roth, a Russian neurologist, coined the term 'meralgia paresthetica' when he noticed this in cavalry men who wore their belts too tightly, causing compression of the emerging femoral cutaneous nerve [3]. Meralgia paresthetica, or lateral femoral cutaneous neuropathy, is a neurological disorder due to the entrapment of the lateral femoral cutaneous nerve, with

---

The Hip Joint in Adults: Advances and Developments
Edited by K. Mohan Iyer
Copyright © 2018 Pan Stanford Publishing Pte. Ltd.
ISBN 978-981-4774-72-7 (Hardcover), 978-1-351-26244-6 (eBook)
www.panstanford.com

subsequent numbness and/or pain on the anterolateral aspect of the thigh. The lateral femoral cutaneous nerve (L2–L3) passes under the inguinal ligament to innervate the anterolateral skin of the thigh. Meralgia paresthetica is caused when one of the large sensory nerves (lateral femoral cutaneous nerve) to one of the legs is being compressed. This pain may sometimes be confused with hip pain, but it may be possible to reproduce the patient's symptoms by local percussion over the trigger area.

This compression results in a terrible burning sensation felt in the outer thigh. Common causes are repetitive motion of the legs, recent injuries to the hip, wearing of tight clothing and weight gain. Diagnosis includes complete medical history; a comprehensive clinical exam, including neurological exams; imaging studies such as X-rays and/or magnetic resonance imaging (MRI), electrodiagnostic studies electromyography [EMG] and diagnostic nerve block.

Meralgia paresthetica is an uncommon, but known, complication of total hip arthroplasty (THA) [4]. Neurologic complications after THA are relatively uncommon, with a reported overall incidence of 1% to 2%. Femoral acetabular impingement syndrome is to be considered as one of the possible spontaneous causes of meralgia paresthetica [5]. Injury to the lateral femoral cutaneous nerve (LFCN) is a known complication of anterior approaches to the hip and pelvis. No study has quantified its incidence in anterior arthroplasty procedures [4]. The anterior approach follows a well-delineated internervous plane and provides direct access to the hip with minimal muscle retraction. However, the LFCN and its branches are within the field of dissection and at risk of injury. Hip resurfacing was identified as a risk factor for LFCN neuropraxia. Intraoperative modifications to avoid injury include lateralising the skin incision, strict subfascial dissection within TFL and avoidance of medial subcutaneous fat pad dissection, and rigorous retraction of the rectus femoris during component implantation. Nerve compression injuries as a result of positioning have also been reported, usually as a result of inadequate padding of bony prominences such as the elbow or the knee intraoperatively. Peripheral nerve injuries can also result from excessively long surgeries or periods of hypoxia intraoperatively.

7.1 Aetiology and Risk Factors

The lateral femoral cutaneous nerve is the nerve that provides sensation to the outer thigh surface. The compression of this nerve causes meralgia paresthetica. As the lateral femoral cutaneous nerve is a sensory nerve, the ability to use leg muscles does not get affected. In many individuals, this nerve travels through the groin to the upper thigh without any hindrance; but in meralgia paresthetica, the lateral femoral cutaneous nerve becomes pinched/compressed/trapped, usually under the inguinal ligament. Conditions that increase pressure on the groin are the common causes of this compression, such as tight clothing, excess weight/obesity, pregnancy and the presence of scar tissue near the inguinal ligament. It could be due to injury or a previous surgery or standing, walking or cycling for prolonged periods of time; and injury to the nerve can occur in a motor vehicle accident or in diabetes, causing meralgia paresthetica.

The laparoscopic-assisted single-port approach, also known as the all-in-one appendectomy, has gained recent popularity (see case report [6]). We describe a child who suffered meralgia paresthetica (a neuropathy in the distribution of the lateral femoral cutaneous nerve) after a laparoscopic-assisted single-port appendectomy, perhaps secondary to mobilisation of the cecum [6].

7.2 Signs and Symptoms

- Symptoms commonly occur on one side of the body and may increase in intensity after walking or standing.
- There is upper thigh numbness or numbness and tingling in the lateral (outer) part of the thigh.
- There is a burning sensation or pain in or on the outer surface of the thigh.
- There is dull pain in the groin area or across the buttocks, although this is not that common.

The most common point for possible entrapment is as the nerve passes between the two slips of the inguinal ligament's lateral attachment to the anterior-superior iliac spine (ASIS), where it exits the pelvis. The nerve is tightly bordered by the tendinous fibres of the inguinal ligament at this point and makes a right-handed bend to change direction from a horizontal course in the pelvis to a more vertical course in the lateral and anterolateral thigh. The lower slip of the inguinal ligament also gives origin to some sartorius fibres. The nerve may pass in front of or through the sartorius into the thigh.

7.3 Differential Diagnosis

It is imperative to differentiate the symptoms of meralgia paresthetica from other causes of pain and nerve discomfort that can have similar clinical presentations. Included in the differential are (i) spinal nerve radiculopathy at L1–L3, (ii) malignancy or metastasis to the iliac crest, (iii) uterine fibroids or pelvic mass that compresses the nerve, (iv) avulsion fracture of the ASIS and (v) chronic appendicitis.

The lateral femoral cutaneous nerve is purely a sensory nerve and doesn't affect your ability to use your leg muscles. Meralgia paresthetica is a condition that causes numbness, pain or a burning feeling in your outer thigh. You might also hear it called Bernhardt–Roth syndrome. It happens when there's too much pressure on or damage to one of the nerves in your leg.

7.4 Treatment for Meralgia Paresthetica

- Conservative treatment is effective for most of the patients and helps in getting rid of the pain within a few weeks to months. Conservative measures include:
    - Wearing loose clothing
    - Losing excess weight
- Lifestyle changes such as avoiding tight clothing, avoiding standing or walking for prolonged periods of time, maintaining a healthy weight and losing excess weight help in preventing and relieving meralgia paresthetica.
- Over-the-counter pain killers, such as acetaminophen (Tylenol), ibuprofen (Advil, Motrin, etc.) and aspirin help in relieving pain.

- Tricyclic antidepressants also help in pain relief. Side effects include a dry mouth, drowsiness, constipation and impaired sexual functioning.
- Antiseizure medications, such as gabapentin (Neurontin) and pregabalin (Lyrica), also help in alleviating symptoms of pain. Side effects include nausea, dizziness, constipation, drowsiness and light-headedness.
- Corticosteroid injections are given in severe cases, where the symptoms persist despite conservative treatment. These injections help in reducing pain and inflammation. Side effects of corticosteroid injections include pain and whitening of the skin around the site of injection, infection of the joint and nerve damage.
- Rarely surgery is required in severe cases where the patient has persistent symptoms for a long time. Surgery is done in order to decompress the nerve.

References

1. Bernhardt M. *Neuropathologische Beobachtungen*. Deutsches Archiv für klinische Medicin. Leipzig 187822362–393.393.
2. Bernhardt M. *Ueber isoliert im Gebiete des N. cutaneus femoris externus vorkommende Paräesthesien*. Neurologisches Centralblatt 189514242–244.244.
3. Roth WK. *Meralgia Paraesthetica* (Berlin: Karger), 1895.
4. Goulding K, Beaulé PE, Kim PR, Fazekas A. Incidence of lateral femoral cutaneous nerve neuropraxia after anterior approach hip arthroplasty. *Clin Orthop Relat Res*, 2010; **468**(9):2397–2404.
5. Ahmed A. Meralgia paresthetica and femoral acetabular impingement: a possible association. *J Clin Med Res*, 2010; **2**(6):274–276.
6. Ng AM, Fike FB, Schneider SJ, Hong AR, Dolgin SE. Meralgia paresthetica after "all-in-one" appendectomy, *J Paediatr Surg Case Rep*, 2015; **3**(11):521–522.

Chapter 8

Advances in Surgery for Bursitis of the Hip Joint in Adults

K. Mohan Iyer
Formerly Royal Free Hampstead NHS Trust, Royal Free Hospital, Pond Street, London NW3 2QG, United Kingdom
kmiyer28@hotmail.com

Bursitis is the inflammation or irritation of the bursa. Numerous bursae have been described in the hip region, but only three have practical significance, namely bursa overlying the greater trochanter (Fig. 8.1), the ischio-gluteal bursa and the psoas bursae.

Bursitis are often the result of high-risk activities, which include gardening, raking, carpentry, shovelling, painting, scrubbing, tennis, golf, skiing, throwing and pitching. The incidence of trochanteric bursitis (TB) is the highest in middle-aged to elderly people. This is a common cause of hip pain [1]. Bursitis is most often caused by repetitive, minor impact on the area or by a sudden, more serious injury.

The Hip Joint in Adults: Advances and Developments
Edited by K. Mohan Iyer
Copyright © 2018 Pan Stanford Publishing Pte. Ltd.
ISBN 978-981-4774-72-7 (Hardcover), 978-1-351-26244-6 (eBook)
www.panstanford.com

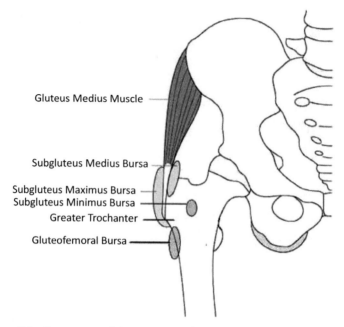

Figure 8.1 Bursae around the greater trochanter of the hip joint. Reproduced with the kind permission of Barak Haviv, Orthopedic Department, Tel-Aviv University, Israel.

8.1 Causes and Symptoms of Trochanteric Bursitis

Some of the typical causes of TB are:
- Injury to the point of the hip.
- Play or work activities that cause overuse of or injury to the joint areas. Such activities might include running up the stairs, climbing and standing for long periods of time.
- Incorrect posture. This condition can be caused by scoliosis, arthritis of the lumbar (lower) spine and other spine problems.
- Stress on the soft tissues.
- Other diseases or conditions. These may include rheumatoid arthritis, gout, psoriasis, thyroid disease or an unusual drug reaction. In rare cases, bursitis can result from infection.

- Previous surgery around the hip or prosthetic implants in the hip.
- Hip bone spurs or calcium deposits in the tendons that are attached to the trochanter. If bursitis persists and is left untreated, calcium deposits can form within the bursae. These calcium deposits limit the range of motion and can lead to a permanently stiff joint.

TB typically causes the following symptoms:
- Pain on the outside of the hip and thigh or in the buttock
- Pain when lying on the affected side
- Pain when you press in on the outside of the hip
- Pain that gets worse during activities such as getting up from a deep chair or getting out of a car
- Pain while walking up the stairs

TB is a self-limiting disorder in the majority of patients and typically responds to conservative measures. However, multiple courses of nonoperative treatment or surgical intervention may be necessary in refractory cases. It is a common problem seen by sports medicine practitioners, affecting as many as 5.6 patients per 1000 adults. Sometimes described as the 'greater trochanteric pain syndrome', or GTPS, it is characterised by chronic lateral hip pain exacerbated by active abduction, passive adduction and direct palpation. The incidence of TB is highest in middle-aged to elderly adults. This is a common cause of hip pain.

8.2   Investigations

- Fluid removal
- Aspiration, which may be sometimes necessary to remove fluid from the bursa to check for possible *Staphylococcus epidermis* (or *S. aureus*) bacterial infection
- X-rays
- Diagnostic ultrasound
- Magnetic resonance imaging (MRI) and computed tomography (CT) scans

Arthroscopic bursectomy appears to be an effective option for recalcitrant TB, which gives good pain relief and improved function compared with the preoperative status. Improvements in a patient's status are usually evident by one to three months after surgery and appear to be lasting [2].

Greater TB, also known as GTPS, is one of the most common causes of hip pain. Studies have explored the use of the arthroscope in the treatment of these patients. The iliotibial band (ITB) is the main cause of pain, inflammation and trochanteric impingement, leading to the development of bursitis. ITB release involves two incisions – one 4 cm proximal to the greater trochanter along the anterior border of the ITB and the other 4 cm distal and along the posterior border. The 30° arthroscope is introduced through the inferior portal, and a cannula is introduced through the superior portal. A 5.5 mm arthroscopic shaver is inserted through the superior cannula to clear off the surface of the ITB so that it may be adequately visualised. A hooked electrocautery probe is then used to longitudinally incise the ITB until it no longer rubs, causing impingement over the greater trochanter [3].

GTPS is often a manifestation of underlying gluteal tendinopathy. Extracorporeal shockwave therapy is effective in numerous types of tendinopathies. Shockwave therapy is an effective treatment for GTPS [4].

GTPS is associated with excessive tension between the ITB and the greater trochanter. The ITB and fascia lata act as a lateral tension band to resist tensile strains on the concave aspect of the femur and are often implicated as the source of GTPS. They describe an endoscopic technique to release the ITB and remove the bursa and conclude that endoscopic bursectomy with cross incision of the ITB is a safe approach to treat patients with refractory GTPS [5].

The differential diagnosis for GTPS includes pathologies around the hip and lower back. Usually nonoperative treatment that includes modified activity, physiotherapy, local injections and shockwave therapy is helpful [6]. Pathologies of the abductor musculature are demonstrated well with MRI. Typical findings are thickening of the gluteal tendons, local oedema, high signal and discontinuity of the tendon at the insertion site. Ultrasound can also be used for identifying tendon ruptures although MRI is preferable because of higher specificity and sensitivity. Keyhole surgery for hip pathologies has developed significantly in the last decade.

8.3 Psoas Bursitis

This is a large bursa that lies between the iliopsoas muscle and the pelvis. Posteriorly and above is the iliopectineal eminence, while below is the capsule of the hip joint. It usually accompanies the femoral nerve and frequently communicates with the hip joint. Clinically there may be pain and tenderness in the medial part of Scarpa's triangle, while in its late stages when there is suppuration, fluctuation may be demonstrable. The swelling may be large enough to obliterate the normal inguinal groove, or it may compress the femoral nerve to give rise to referred pain down the leg, usually the knee, as in hip joint disease. Flexion of the hip elicits pain, while with extension the pain increases. The diagnosis of this condition from hip joint disease and from psoas abscess may be extremely difficult. Always remember the presence of an obturator hernia, before considering aspiration of the swelling in this region.

8.4 Treatment

The treatment in suppurative conditions is incision and drainage, while in chronic infections, it is complete excision. The bursa is best approached by a vertical incision lateral to the femoral artery, when the muscle of the iliotibial (IT) tract is retracted medially and the bursa is thus exposed. It must be remembered that this bursa often communicates with the joint cavity, and hence it may be important to drain the hip joint also in purulent infection of the bursa.

8.5 Snapping Hip

In certain conditions, an audible click may be distinctly heard, which may be due to certain intra-articular conditions or extra-articular conditions.

There are three types of snapping hip syndrome:

- **Internal snapping hip:** This first type of snapping hip occurs when a tendon slides over protruding bony structures at the front of the hip joint, creating tension and then releasing with a 'snap'.

- **External snapping hip:** The more common extra-articular cause is similar to the luxation of the peroneal tendons at the ankle. The snap may be heard and felt when the knee is flexed and the hip joint is forcibly rotated medially. This may be seen at times as a tight band that slips backwards and forward over the greater trochanter. This may occur both in children and adults and is due to friction between the anterior border of the gluteus maximus and the trochanter or between a facial band and the bony prominence. This phenomenon is also encountered in arthritis or in an effusion in the bursa between the gluteus maximus and the femur. In such cases, a radiograph may be taken to rule out an osteoma or an osteochondritis. A snapping hip may become habitual, causing considerable discomfort in highly nervous people. If operative treatment becomes necessary, then dividing the offending band or tendon or surgical excision of the bony prominence may be necessary in some cases. Should an osteoma or exostosis be present, then complete excision offers complete cure. This syndrome is also associated with a tight IT band and so is sometimes called IT band syndrome.
- **Intra-articular snapping hip:** This is the most common cause of intra-articular snapping hip and may result from an acetabular labral tear, an injury to the articular cartilage or loose bodies of material in the hip.

References

1. Lustenberger DP, Ng VY, Best TM, Ellis TJ. Efficacy of treatment of trochanteric bursitis: a systematic review. *Clin J Sport Med*, 2011; **21**(5):447–453.
2. Baker CL, Jr, Massie RV, Hurt WG, Savory CG. Arthroscopic bursectomy for recalcitrant trochanteric bursitis. *Arthroscopy*, 2007; **23**(8):827–832.
3. Jambou S, Michels F, Allard M, Bousquet V, Colombet P, de Lavigne C. Arthroscopic treatment of the iliotibial band syndrome. *Arthroscopy*, 2012; **28**(9):347.
4. Furia JP, Rompe JD, Maffulli N. Low-energy extracorporeal shock wave therapy as a treatment for greater trochanteric pain syndrome. *Am J Sports Med*, 2009; **37**(9):1806–1813.

5. Govaert LHM, van Dijk CN, Zeegers AVCM, Albers GHR. Endoscopic bursectomy and iliotibial tract release as a treatment for refractory greater trochanteric pain syndrome: a new endoscopic approach with early results. *Arthrosc Tech*, 2012; **1**(2):e161–e164.
6. Haviv B. Update on trochanteric bursitis of the hip. *OA Orthop*, 2013; **1**(1):10.

Chapter 9

Advances in Surgery of the Hip Joint in Rheumatoid Arthritis in Adults

K. Mohan Iyer
Formerly Royal Free Hampstead NHS Trust, Royal Free Hospital, Pond Street, London NW3 2QG, United Kingdom
kmiyer28@hotmail.com

There is no cure for rheumatoid arthritis (RA). The aetiology remains unknown, though there are several initiating factors seen with evidence of immune overactivity. However, certain aetiological factors may be involved, and the American Rheumatism Association has laid down certain criteria. The details are mentioned in my book entitled *General Principles of Orthopedic and Trauma* [1].

RA symptoms develop gradually, and it is not always possible to know when the disease first developed. Many people have symptoms that are present continuously, some have symptoms that completely resolve and others have alternating periods of bothersome symptoms and complete resolution. The onset, severity and specific symptoms of this condition can vary greatly from person to person.

But recent discoveries indicate that remission of symptoms is more likely when treatment begins early with strong medications

---

The Hip Joint in Adults: Advances and Developments
Edited by K. Mohan Iyer
Copyright © 2018 Pan Stanford Publishing Pte. Ltd.
ISBN 978-981-4774-72-7 (Hardcover), 978-1-351-26244-6 (eBook)
www.panstanford.com

known as disease-modifying antirheumatic drugs (DMARDs). There are several types of DMARDs:

- Conventional synthetic DMARDs, sometimes termed 'traditional DMARDs' or 'small-molecule DMARDs', are produced by traditional drug-manufacturing techniques.
- Biologic DMARDs, sometimes termed 'targeted biologic agents', are manufactured using molecular biology (recombinant DNA) techniques.
- Another DMARD, tofacitinib, is produced by traditional drug-manufacturing techniques and can be taken as a pill but has adverse effects similar to those of the biologic DMARDs and is sometimes referred to as a 'targeted synthetic DMARD'.

Hip RA can cause symptoms such as severe pain, stiffness and swelling. With RA hip pain, you may have discomfort and stiffness in the thigh and groin.

RA is an autoimmune disease, but the following may play a role: (i) genetics, (ii) environmental factors and (iii) hormones.

Other tests that may be helpful in diagnosing RA include magnetic resonance imaging (MRI), ultrasound and bone scan.

DMARDs can slow the progression of RA and save the joints and other tissues from permanent damage. Azathioprine (Imuran) is a DMARD.

DMARDs also include drugs known as biologic modifiers. These drugs can target parts of the immune system that trigger inflammation that causes joint and tissue damage. These types of drugs also increase the risk of infections. Biologic DMARDs are usually most effective when paired with a nonbiologic DMARDs, such as methotrexate. Abatacept (Orencia) is a biologic agent.

Some common complementary and alternative treatments that have shown promise for RA are:

- **Fish oil:** Fish oil supplements may reduce RA pain and stiffness.
- **Plant oils:** Seeds of evening primrose, borage and black currant contain a type of fatty acid that may help with RA pain and morning stiffness.
- **Tai chi:** This movement therapy involves gentle exercises and stretches combined with deep breathing. Tai chi may reduce RA pain.

Regular exercise is important for RA Exercise strengthens muscles that support joints. Exercise also helps you stay flexible. This is important for preventing painful falls.

Synovectomy is extremely useful in the knee joint because a major part of the synovium is readily available.

*For patients with earlier-stage RA of the hip, hip arthroscopy may help to ease pain. For a more severe form of the disease, total joint replacement may be recommended.

Juvenile rheumatoid arthritis: This form of disease must be kept in mind when this disease occurs in children around the age of two to four years. This differs in some important aspects from the adult disease, and some people still prefer to call it Still's disease on account of chronic inflammatory polyarthritis along with systemic symptoms.

Reference

1. Iyer KM. *General Principles of Orthopedic and Trauma* (Springer), 2013.

Chapter 10

Advances in Surgery of the Hip Joint in Tuberculosis Arthritis in Adults

K. Mohan Iyer
Formerly Royal Free Hampstead NHS Trust, Royal Free Hospital, Pond Street, London NW3 2QG, United Kingdom
kmiyer28@hotmail.com

The surgical options vary from excision arthroplasty to hip replacement. Total hip replacement (THR) in the active stage of the disease is yet another area of controversy. Tuberculosis (TB) of the hip is still a common condition in developing countries. Early presentations are pain around the hip and a limp. Later the patient presents with deformities, shortening of the limb and restriction of movements. The constitutional symptoms may or may not be present in all the cases. Diagnosis is mainly clinicoradiological; however, supportive blood investigations and imaging modalities like ultrasonography (USG) and magnetic resonance imaging (MRI) are helpful. Histological proof may not be necessary in all the cases in the endemic zones for TB. The management depends upon the stage of clinical presentation and the severity of destruction as visible radiologically. From conservative therapy in the form of

---

The Hip Joint in Adults: Advances and Developments
Edited by K. Mohan Iyer
Copyright © 2018 Pan Stanford Publishing Pte. Ltd.
ISBN 978-981-4774-72-7 (Hardcover), 978-1-351-26244-6 (eBook)
www.panstanford.com

antitubercular chemotherapy (ATT) and traction to debridement and joint replacement, a variety of surgical procedures have been described. On an average 2%–5% of the patients report back with reactivation of the disease within about 20 years after the apparent clinical healing of the first lesion [1].

Cementless stems and sockets were used in all patients. The average follow-up period was 4.8 years. The reactivation of the infection was not detected in all cases. The result was excellent in all patients according to the Harris Hip Score [2]. Total hip arthroplasty (THA) in the tuberculous hip is a safe procedure and produces superior functional results compared with resection arthroplasty or arthrodesis. The results of primary THA in selected patients is satisfactory as they rapidly recover from the disease [2].

With thorough debridement followed by a complete course of ATT, active tuberculous infection should not be considered a contraindication for THA [3].

THR in patients with active TB of the hip is a safe procedure, providing symptomatic relief and functional improvement if undertaken in association with extensive debridement and appropriate antituberculosis treatment [4].

Two-stage THA is an alternative option to treat patients with advanced active TB of the hip under some difficult conditions. A hip with sinus tracts or destroyed extensively, where thorough debridement in a single operation is difficult, may indicate two-stage THA [5].

With antituberculous medications, in addition to antibiotic therapy for superinfection patients, the two-stage THA protocol offers the greatest chance for eradication of the infection [6].

Despite the state of TB, cementless THA is an effective treatment for advanced TB of the hip [7, 8].

References

1. Saraf SK, Tuli SM. Tuberculosis of hip: a current concept review. *Indian J Orthop*, 2015; **49**(1):1–9.
2. Yoon TR, Rowe SM, Santosa SB, Jung ST, Seon JK. Immediate cementless total hip arthroplasty for the treatment of active tuberculosis. *J Arthroplasty*, 2005; **20**(7):923–926.

3. Wang Q, Shen H, Jiang Y, Wang Q, Chen Y, Shao J, Zhang X. Cementless total hip arthroplasty for the treatment of advanced tuberculosis of the hip. *Orthopedics*, 2011; **34**(2):90.
4. Oztürkmen Y, Karamehmetoğlu M, Leblebici C, Gökçe A, Caniklioğlu M. Cementless total hip arthroplasty for the management of tuberculosis coxitis. *Arch Orthop Trauma Surg*, 2010; **130**(2):197–203.
5. Kim SJ, Postigo R, Koo S, Kim JH. Total hip replacement for patients with active tuberculosis of the hip: a systematic review and pooled analysis. *Bone Joint J*, 2013; **95-B**(5):578–582.
6. Li L, Chou K, Deng J, Shen F, He Z, Gao S, Li Y, Lei G. Two-stage total hip arthroplasty for patients with advanced active tuberculosis of the hip. *J Orthop Surg Res*, 2016; **11**:38.
7. Li L, Chou K, Deng J, Shen F, He Z, Gao S, Li Y, Lei G. Two-stage total hip arthroplasty for patients with advanced active tuberculosis of the hip. *J Orthop Surg Res*, 2016; **11**:38.
8. Zeng M, Hu Y, Leng Y, Xie J, Wang L, Li M, Zhu J. Cementless total hip arthroplasty in advanced tuberculosis of the hip. *Int Orthop*, 2015; **39**(11):2103–2107.

Chapter 11

Advances in Fractures in the Neck of the Femur in Adults

Dayanand Manjunath
Bangalore Medical College and Research Institute, Bengaluru, India
drdayanand.m@gmail.com

11.1 Fracture of the Neck Femur

11.1.1 Introduction

11.1.1.1 Epidemiology

A fractured neck of the femur is increasingly common due to the aging population, more common in women than in men and more common in whites.

The mechanism of injury is usually high energy in young patients, which constitutes around 3%–5% of the total neck fractures, compared to low-energy falls in older patients due to osteoporosis, which is more common.

The Hip Joint in Adults: Advances and Developments
Edited by K. Mohan Iyer
Copyright © 2018 Pan Stanford Publishing Pte. Ltd.
ISBN 978-981-4774-72-7 (Hardcover), 978-1-351-26244-6 (eBook)
www.panstanford.com

11.1.1.2 Pathophysiology

The healing potential of a fractured neck is poor because it is intracapsular, is bathed in synovial fluid and lacks the periosteal layer, resulting in limited callus formation, thereby affecting healing.

Of the femoral shaft fractures 6%–9% are associated with femoral neck fractures (FNFs).

Mortality is approximately 25%–30% at one year (higher than vertebral compression fractures) mainly in the older population. Preinjury mobility is the most significant determinant for postoperative survival. Coexisting comorbidities increase the risk of mortality up to 45%.

11.1.2 Anatomy

11.1.2.1 Osteology

The normal neck–shaft angle is 130° ± 7° and normal anteversion 10° ± 7°.

11.1.2.2 Blood supply to the femoral head

The blood supply of the femoral head comes from three main sources [1, 2]: the medial femoral circumflex artery (MFCA), the lateral femoral circumflex artery (LFCA) and the obturator artery. In the adult, the obturator artery provides little and variable amount of blood supply to the femoral head via the ligamentous teres. The LFCA gives rise to the inferior metaphyseal artery by way of the ascending branch and provides the majority of the inferoanterior femoral head. The largest contributor to the femoral head, especially the superolateral aspect of the femoral head, is the MFCA. The lateral epiphyseal artery complex comes from the MFCA and courses along the posterosuperior aspect of the femoral neck before supplying the femoral head. It is important to know and understand that these terminal branches supplying the femoral head are intracapsular. Thus, disruption or distortion due to fracture displacement of terminal branches to the femoral head plays a significant role in the development of osteonecrosis. Variables that have been hypothesised in contributing to femoral head osteonecrosis include vascular damage from the initial FNF, the quality of reduction or fixation of the fracture (restoring flow to the distorted arteries) and the elevated intracapsular pressure.

11.1.3 Classification

11.1.3.1 Garden's classification for the fractured neck of a femur

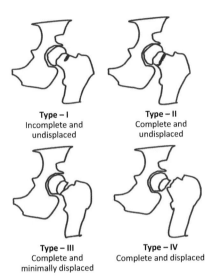

Type – I
Incomplete and undisplaced

Type – II
Complete and undisplaced

Type – III
Complete and minimally displaced

Type – IV
Complete and displaced

(Line drawing of Garden's classification of fracture of the neck of the femur; see Figure 6.5, Chapter 6, *The Hip Joint*[1])

11.1.3.2 Pauwels classification

(Based on a vertical orientation of the fracture line)

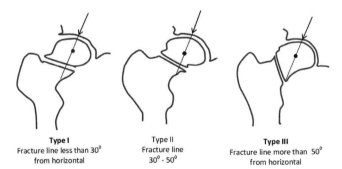

Type I
Fracture line less than 30° from horizontal

Type II
Fracture line 30° - 50°

Type III
Fracture line more than 50° from horizontal

---

[1] Iyer KM, Kempanna VK, Goyal S, Krishnan S, and Biring GS. Injuries around the hip joint, including periprosthetic fractures. In *The Hip Joint*, Iyer KM, ed. (Pan Stanford), 2016.

(Line drawing of Pauwels's classification of fracture of the neck of the femur: types, I, II, and III, depending upon the angle of the fracture line in relation to the horizontal line; see Figure 6.6, Chapter 6, *The Hip Joint*[2])

Type I	<30° from horizontal
Type II	30°–50° from horizontal
Type III	>50° from horizontal (most unstable with the highest risk of nonunion and avascular necrosis [AVN])

11.1.3.3 Anatomic classification

- Subcapital
- Transcervical
- Basal type

11.1.4 Presentation

- **Symptoms:** Patients with impacted and stress fractures might have a slight pain in the groin or pain referred along the medial side of the thigh and knee. Displaced fractures lead to pain in the entire hip region and the inability to move the limb.
- **Physical exam:** Impacted and stress fractures show no obvious clinical deformity. The patient may have minor discomfort with an active or passive hip range of motion and muscle spasms during extremes of motion pain with percussion over the greater trochanter. In displaced fractures, the leg will be in external rotation and abduction, with minimal shortening.

11.1.5 Imaging

- **Radiographs:** It is recommended to obtain anteroposterior (AP) pelvis and cross-table lateral, traction-internal rotation AP hip is best for defining the fracture type.
- **Computed tomography (CT):** This is helpful in determining the displacement and degree of comminution in some patients.
- **Magnetic resonance imaging (MRI):** This is helpful in ruling out occult and stress fractures, determining the access viability of the head in delayed presentation cases and ruling out AVN.

---
[2]Iyer KM, Kempanna VK, Goyal S, Krishnan S, and Biring GS. Injuries around the hip joint, including periprosthetic fractures. In *The Hip Joint*, Iyer KM, ed. (Pan Stanford), 2016.

11.1.6 Treatment

Treatment depends on the age of the patient, displacement and duration of the presentation.

Osteosynthesis is indicated for most patients <60 years of age and is considered a surgical emergency. It may be achieved using cannulated cancellous screws or dynamic hip screws (DHSs) [3–6, 17, 19].

There is no consensus regarding how quickly a FNF in a young individual must be reduced and fixed. As in any fracture, healing is dependent upon restoration of anatomic alignment, preservation of blood supply to both the bone and the surrounding tissues and stable fixation. Because the blood supply to the femoral head may be compromised by displacement or increased intracapsular pressures, some advocate early fixation of these fractures (within 6 to 12 hours). This allows early decompression of the capsule, reduction of the FNF and thereby the critical local vessels, and fracture stabilisation, all of which may decrease the rates of femoral head osteonecrosis. Haidukewych et al. [20] reported on his series of 73 FNFs in young patients and found that the rate of osteonecrosis was not statistically different between those treated within 24 hours from the time of diagnosis (24%) and those treated more than 24 hours from the time of diagnosis (20%).

11.1.6.1 Surgical considerations

Multiple studies have shown that intracapsular pressure after FNF is the lowest when the hip is flexed and externally rotated. Therefore, extension and internal rotation should be avoided prior to the time of decompression of the capsule. Because of this, most clinicians avoid skeletal or skin traction for these injuries. Without traction immobilisation, protection of the injured area from further injury due to instability is difficult. Fracture instability also causes pain.

11.1.6.2 Closed reduction manoeuvres

Leadbetter technique: The hip is flexed to 90°, with slight adduction, and traction is applied in line with the femur; while maintaining traction, the leg is internally rotated to 45° and slowly brought into slight abduction and full extension, while maintaining traction and internal rotation.

Whitman's method: The hip is extended and internally rotated, traction is applied and the limb is abducted.

Flynn method: The hip is slightly flexed and traction is applied in line of the femoral neck and extended.

Clinically the reduction is assessed using the **heel palm test**, where the surgeon holds both heels in his palms with both legs in abduction, and the internal rotation is released and the surgeon notes the amount of external rotation of both feet; if the fractured site has significantly more external rotation than the noninjured side, then reduction is probably not satisfactory; if the injured side stays in internal rotation, then the reduction is complete. Closed reduction is tried once or twice, and open reduction is indicated if closed reduction is not successful.

Assessment of reduction: On image intensification the outline of the femoral head and neck junction produces the image of an S or reversed S curve; if the outline reveals an unbroken C curve, then the fracture is not reduced.

Garden's alignment index: This is based on the angle of the compression trabeculae in the AP view relative to the longitudinal axis of the femoral shaft and the angle of the compression trabeculae in the lateral view relative to the femoral shaft; **acceptable reduction** lies within the range of 155°–180° in both views.

Open reduction is routinely done using the anterolateral approach.

11.1.6.3 The role of capsulotomy

Capsulotomy in FNFs remains a controversial issue, and the practice varies by trauma programme, region and country. There are both animal and clinical studies showing the benefit of capsulotomy. Animal studies have shown that increased hip intracapsular pressure results in a tamponade effect and may reduce blood flow to the femoral head. Clinical studies show that decompressing the intracapsular haematoma via capsulotomy or aspiration reduces the intracapsular pressures. This decrease in the intracapsular pressure results in improved blood flow to the femoral head and may reduce femoral head ischaemia. Most of these studies are small series, single centre and uncontrolled [20].

11.1.6.4 Fixation methods

Cannulated screw fixation is more commonly used in nondisplaced transcervical fractures (Garden's type I and II fracture patterns) in healthy young patients.

A minimum of three 6.5 mm cancellous screws are used. It is ensured that they are parallel and that the thread is in the head fragment and does not cross the fracture line (Fig. 11.2). A parallel configuration maintains uniform compression, obtained by ensuring as much screw spread as possible in the femoral neck. The starting point should be at or above the level of the lesser trochanter to avoid fracture and avoid multiple cortical perforations during the guide pin or screw placement to avoid the development of a lateral stress riser.

The pattern could be a triangle, an inverted triangle or a diamond shape, but it should be ensured that the inferior screw rests on the calcar (Fig. 11.1), the superior screws should buttress the superior margin and the posterior screw should buttress the posterior communition. Washers should be used to stop the screw heads from penetrating the bone of the greater trochanter.

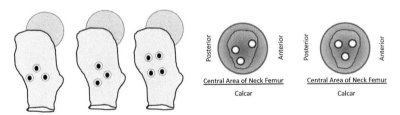

Figure 11.1 Cross-sectional view demonstrating the pattern of screw placement.

Sliding hip screws with antirotatory screws [6] are indicated in a basicervical fracture or a vertical fracture pattern in a young patient with osteoporotic bones (Fig. 11.3).

Biomechanically these screws are superior to cannulated screws, and the placement of an additional cannulated screw above the sliding hip screw is used to prevent rotation.

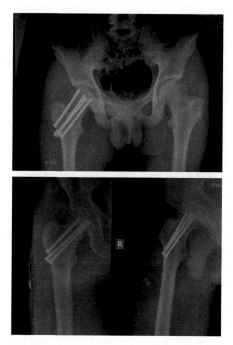

Figure 11.2 Radiograph of a fractured femur neck with cannulated cancellous screws.

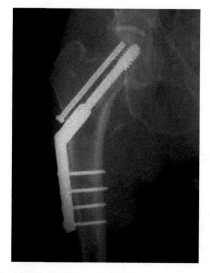

Figure 11.3 DHS with an antirotatory screw.

11.2 Fractured Femur Neck of More Than Three-Week Duration

When the fractured neck of the femur is of more than three weeks duration, internal fixation alone has a very high failure rate. Internal fixation has to be combined with some type of bone graft or osteotomy, particularly in young patients below the age of 55 years in whom it is desirable to preserve the joint. In patients above the age of 55 replacement arthroplasty is the preferred treatment if the patient can afford it and the lifestyle permits. Changes in the neck include progressive absorption of the neck of the femur, resulting in an increase in the gap between the fragments and a decrease in the size of the proximal fragment. The head of the femur may start showing signs of AVN.

A good CT scan or MRI of the pelvis can be extremely useful in accurately measuring the gap between the fragments and the size of the proximal fragment.

Osteosynthesis may be achieved using:
- Internal fixation with one screw and double fibular graft or two screws and one fibular graft [7, 8]. If the neck of the femur is narrow then one screw and one fibular graft may be given (Fig. 11.4).

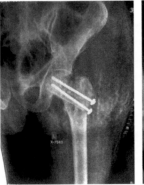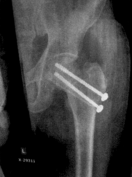

Figure 11.4   Fibular graft with a cancellous graft.

- Open reduction and bone muscle pedicle graft based on quadratus femoris [9] or sartorius or tensor fascia femoris (Fig. 11.5).

Figure 11.5 Quadratus muscle pedicle graft.

- Abduction osteotomy and osteosynthesis with a DHS or a 135°-angled blade plate [10, 11]. This procedure is particularly useful when the fracture is situated nearer the base and the length of the proximal fragment is 3.5 cm or more (Fig. 11.6).

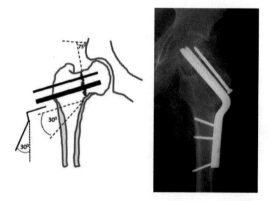

Figure 11.6 Valgus osteotomy and an angled blade plate.

11.3 Management of a Fractured Femur Neck in People >60 Years Old

Garden's type I and type II are good candidates for osteosynthesis by internal fixation with cannulated cancellous screws or DHSs with screws, whereas in Garden's type III and IV osteosynthesis is not recommended because of high failure rates, especially nonunion of around 36%.

Hence the treatment of choice is arthroplasty, which could be hemiarthroplasty or total hip arthroplasty (THA).

Recent randomised trials compared the outcomes of THA with those of unipolar and bipolar hemiarthroplasty in healthy [12–14] mobile patients with displaced FNFs. They noted overall better function in the THA group. A higher dislocation rate (6% versus 0%) and a lower overall revision rate (2.5% versus 14%; $P = .058$) were found in the THA group. Blomfeldt et al. [16] noted no differences in health related to quality-of-life outcomes and no difference between mortality of patients undergoing THA and those undergoing bipolar hemiarthroplasty. The authors noted higher Harris Hip Scores at 4 months and 12 months in patients undergoing THA.

11.3.1 Hemiarthroplasty

Hemiarthroplasty can be done with unipolar or bipolar prosthesis.

Bipolar prosthesis has the advantage of movement at two interfaces, thereby reducing the acetabular wear.

There are studies on earlier designs of bipolar prosthesis showing that the bipolar hemiarthroplasty functions as a unipolar a few months (3–12) after surgery, but recent studies that included modular bipolar prosthesis showed lesser acetabular erosion due to a better design, indicating real advantage in favour of the bipolar design, which is most probably due to the function of the dual-bearing system [12–14] (Fig. 11.9).

Hence monopolar is indicated in very elderly, fragile, household ambulatory whose life expectancy is less.

Most commonly used monopolar are Austin Moore's and Thompson's. Since Austin Moore's is uncemented (Fig. 11.7) and requires good calcar, which is not present in many patients, and there are more incidents of loosening, Thompson's cemented prosthesis is preferred (Fig. 11.8).

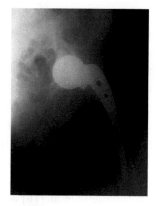

Figure 11.7 Austin Moore HA.

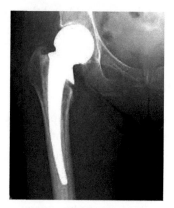

Figure 11.8 Thompson's HA.

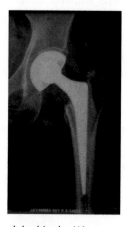

Figure 11.9 Cemented modular bipolar HA.

11.3.2 Complications of Hemiarthroplasty

These include dislocation, implant loosening, periprosthetic fractures and acetabular erosion.

Total hip replacement is usually indicated in pre-existing arthritis of the hip.

The latest meta-analysis comparing THA with hemiarthroplasty found that THA was preferable for healthy elderly patients with displaced FNFs, because of better functional outcomes and lower reoperation rates. Both reoperation rates and acetabular erosion rates are higher after hemiarthroplasty after more than four years. Some of the outcomes, such as EQ_{index}-5D, mobility and pain rate, were in favour of THA, while the operating time was in favour of THA and the risk of dislocation after THA was higher. No significant differences were found in other outcomes, including infection rate, general complication, one-year mortality, blood loss and length of postoperative hospital stay [15, 16, 18].

The anterolateral approach is recommended for THA and usage of a bigger femoral head to prevent dislocation.

THA could be cemented (Fig. 11.10) or uncemented (Fig. 11.11) depending on bone quality and Dorr classification.

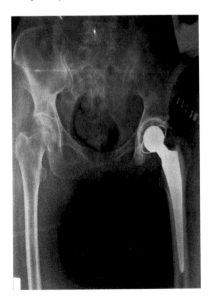

Figure 11.10 Cemented total hip arthroplasty.

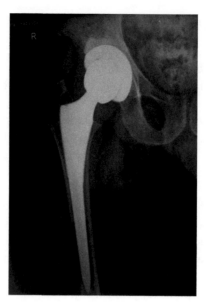

Figure 11.11 Uncemented total hip arthroplasty.

Some complications of THA are dislocation (10%), infection and implant loosening.

11.4 Complications

Osteonecrosis: The incidence of AVN is 10%–45%, and recent studies fail to demonstrate association between time to fracture reduction and subsequent AVN. There is an increased risk with an increase in the initial displacement of the fracture. AVN can still develop in nondisplaced injuries and nonanatomical reduction. Major symptoms are not always present when AVN develops, and management of AVN is THA.

- **Nonunion:** The incidence of nonunion is 5%–30%. Increased incidence is seen in displaced fractures and in older patients (>60 years). Nonanatomic varus malreduction most closely correlates to nonunion.

  Nonunion can be managed by valgus intertrochanteric osteotomy, which is indicated in younger patients as long as the neck is not severely collapsed and the head is viable. It turns a vertical fracture line into a horizontal fracture line and decreases shear forces across the fracture line and increases compressive force.

It can also be managed by free vascularised/nonvascularised fibula graft, as described earlier.

11.5 Other Methods of Osteosynthesis

Osteosynthesis can also be done using a muscle pedicle graft, thereby increasing the blood supply to the fracture site to promote healing. This includes quadratus femoris muscle pedicle graft (Myer's procedure) and tensor fascia lata muscle pedicle graft. The use of a muscle pedicle graft provides blood supply to the femoral head, structural bone graft to buttress the posterior femoral neck comminution and enhanced stability. The reported rate of nonunion is 10% and of femoral head osteonecrosis is 5%.

Arthroplasty is indicated in older patients or when the femoral head is not viable. It is also an option in younger patients with nonviable femoral heads as opposed to free vascularised fibular grafting (FVFG). Again hemiarthroplasty is preferred in very elderly patients and THA in young patients.

References

1. Sevitt S, Thompson RG. Distribution and anastomosis of arteries supplying the head of the femur. *J Bone Joint Surg Br*, 1966; **47**:560573.
2. Trueta J, Harrison MH. The normal vascular anatomy of human head of femur in adult man. *J Bone Joint Surg Br*, 1953; **35**:442–451.
3. Stankewich CJ, Chapman J, Muthusamy R, Quaid G, Schemitsch E, Tencer AF, et al. Relationship of mechanical factors to the strength of proximal femur fractures fixed with cancellous screws. *J Orthop Trauma*, 1996; **10**:248–257.
4. Bonnaire FA, Weber AT. Analysis of fracture gap changes, dynamic and static stability of different osteosynthetic procedures in the femoral neck. *Injury*, 2002; **33**:C24–C32.
5. Ly TV, Swiontkowski MF. Management of femoral neck fractures in young adults. *Indian J Orthop*, 2008; **42**:3–12.

6. Broos PL, Vercruysse R, Fourneau I, Driesen R, Stappaerts KH. Unstable femoral neck fractures in young adults: treatment with the AO 130-degree blade plate. *J Orthop Trauma*, 1998; **12**:235-239.
7. Nagi ON, Gautam VK, Marya SKS. Treatment of fracture neck of femur with a cancellous screw and fibular graft. *J Bone Joint Surg Br*, 1986; **63**:387-391.
8. Nagi ON, Dhillon MS, Goni VG. Open reduction, internal fixation and fibular autografting for neglected fracture of the femoral neck. *J Bone Joint Surg Br*, 1998; **80**:798-804.
9. Meyer MH, Harvey JP Jr, Moore TM. Treatment of displaced subcapital and transcervical fracture of the femoral neck by muscle pedicle bone graft and internal fixation. A preliminary report on one hundred and fifty cases. *J Bone Joint Surg Am*, 1973; **55**(2):257-274.
10. Gupta S, Kukreja S, Singh V. Valgus osteotomy and repositioning and fixation with a dynamic hip screw and a 135° single-angled barrel plate for un-united and neglected femoral neck fractures. *J Orthop Surg (Hong Kong)*, 2014; **22**:13-17.
11. Kalra M, Anand S. Valgus intertrochanteric osteotomy for neglected femoral neck fractures in young adults. *Int Orthop*, 2001; **25**:363-366.
12. Hedbeck CJ, Blomfeldt R, Lapidus G, Törnkvist H, Ponzer S, Tidermark J. Unipolar hemiarthroplasty versus bipolar hemiarthroplasty in the most elderly patients with displaced femoral neck fractures: a randomised, controlled trial. *Int Orthop*, 2011; **35**:1703-1711.
13. Cornell CN, Levine D, O'Doherty J, Lyden J. Unipolar versus bipolar hemiarthroplasty for the treatment of femoral neck fractures in the elderly. *Clin Orthop Relat Res*, 1998; **348**:67-71.
14. Raia FJ, Chapman CB, Herrera MF, Schweppe MW, Michelsen CB, Rosenwasser MP. Unipolar or bipolar hemiarthroplasty for femoral neck fractures in the elderly? *Clin Orthop Relat Res*, 2003; 259-265.
15. Baker RP, Squires B, Gargan MF, Bannister GC. Total hip arthroplasty and hemiarthroplasty in mobile, independent patients with a displaced intracapsular fracture of the femoral neck. A randomized, controlled trial. *J Bone Joint Surg Am*, **88**: 2583-2589.
16. Hedbeck C, Enocson A, Lapidus G, Blomfeldt R, Törnkvist H, Ponzer S, Tidermark J. Comparison of bipolar hemiarthroplasty with total hip arthroplasty for displaced femoral neck fractures. A concise four-year-follow-up of a randomized trial. *J Bone Joint Surg Am*, 2011; **93**(5):445-450.

17. Lu-yao GL, Keller RB, Littenberg B, Wennberg JE. Outcomes after displaced fractures of the femoral neck: a meta-analysis of one hundred and six published reports. *J Bone Joint Surg Am*, 1994; **76-A**:15–25.
18. Parker MJ, Gurusamy K. Internal fixation versus arthroplasty for intracapsular proximal femoral fractures in adults (Cochrane Review). In *The Cochrane Library*, Issue 4 (Chichester: Wiley), 2006.
19. Parker MJ, Stockton G, Gurusamy K. Internal fixation implants for intracapsular proximal femoral fractures in adults (Cochrane Review). In *The Cochrane Library*, Issue 4 (Chichester: Wiley), 2001.
20. Haidukewych GJ, Rothwell WS, Jacofsky DJ, Torchia ME, Berry DJ. Operative treatment of femoral neck fractures in patients between the ages of fifteen and fifty years. *J Bone Joint Surg Am*, 2004; **86**:1711–1716.

Chapter 12

Advances in Hip Arthroscopy

Prakash Chandran
Warrington and Halton Hospitals NHS Foundation Trust, United Kingdom
prachandran@gmail.com

12.1 Introduction

Hip arthroscopy is a less invasive alternative to various hip interventions that would otherwise require a major open procedure, including surgical dislocation of the hip. In addition, advancement in imaging and arthroscopic techniques allows surgeons to address intra-articular derangements that were previously undiagnosed or untreated. In 1802 Dr Phillipp Bozzini and in 1931 Dr Michael Burman demonstrated the arthroscopic technique on cadaveric hip joints, and its first clinical application was in 1939 by Dr Kenji Takagi for infection (suppurative and tubercular arthritis). He published it in the *Journal of Japanese Orthopedic Association* in 1939. In the last decade arthroscopic techniques have been used to deal with various hip pathologies with increasing success.

The Hip Joint in Adults: Advances and Developments
Edited by K. Mohan Iyer
Copyright © 2018 Pan Stanford Publishing Pte. Ltd.
ISBN 978-981-4774-72-7 (Hardcover), 978-1-351-26244-6 (eBook)
www.panstanford.com

Arthroscopic examination of the hip can be significantly challenging primarily because of the anatomical constraints. The femoral head is deeply recessed in the bony acetabulum. The tension in the fibrocapsular and muscular envelopes around the hip joint requires significant distraction force to allow safe instrumentation of the hip joint during arthroscopy. The relative proximity of the sciatic nerve, the lateral femoral cutaneous nerve and the remaining femoral neurovascular structures make portal placement more challenging. Technical advances in appropriate portal placement, needle positioning, distraction techniques and patient setup have all improved the accessibility of the hip joint.

Hip arthroscopy is now routinely used to deal with acetabular labral tears, femoroacetabular impingement (FAI), chondral injuries, loose bodies, joint infection, capsular laxity and injuries to the ligamentum teres. Extra-articular conditions and other less common indications for hip arthroscopy include management of internal and external snapping hip, synovial chondromatosis and other synovial abnormalities, crystalline hip arthropathy (gout and pseudo gout), management of posttraumatic intra-articular debris, osteonecrosis of the femoral head, management of mild to moderate hip osteoarthritis with mechanical symptoms and as an adjunct diagnostic and therapeutic tool in conjunction with open femoral and/or periacetabular osteotomy for dysplasia and complex hip deformities. With advances in techniques, reconstruction of the ligamentum teres, capsulorrhaphy in cases of instability and repair of injuries to the gluteal tendons are being studied. Occasionally, patients with longstanding, unresolved hip joint pain and positive physical findings may benefit from arthroscopic hip assessment [1, 2].

Hip arthroscopy is perceived not to be of much benefit in patients with hip fusion, advanced hip arthritis with obesity, stress fractures and severe dysplasia; and for some, heterotopic ossification, advanced osteoarthritis, protrusion and ankylosis are absolute contraindications. In the presence of infected open wounds, skin ulceration or cellulites in and around the hip area and acute osteomyelitis of the femur or acetabulum, hip arthroscopy is contraindicated to deal with other pathology (Table 12.1).

Table 12.1 Hip arthroscopy is perceived to be not of much benefit in some cases

Indications	Not of much benefit	Contraindicated
Labral tears FAI Chondral injuries Loose bodies Capsular laxity Ligamentum teres injuries Synovial chondromatosis Synovial abnormalities Crystalline hip arthropathy (gout and pseudo gout) Infection Management of posttraumatic intra-articular debris Osteonecrosis of the femoral head Mild to moderate hip osteoarthritis with mechanical symptoms Longstanding, unresolved hip joint pain with positive physical findings Internal and external snapping hip	Hip fusions Advanced hip arthritis Obesity Stress fractures Severe dysplasia Protrusion Heterotrophic ossification Ankylosis	Infected open wounds or cellulites in and around the hip area

12.2 Surgical Anatomy of the Hip

The hip is a ball-and-socket type of diarthrodial joint, formed by the union of the femoral head and the acetabulum. The hip joint can go through a triplanar motion, with the bony architecture providing the stability. The femoral head sits deep in the acetabular socket, and the acetabulum is further deepened by the fibrocartilaginous labrum. This inherent stability of the hip restricts the terminal range of movement. The capsule of the hip joint provides stability to the joint and is reinforced by the presence of the intrinsic capsular ligaments. Intra-articular structures such as the labrum and the ligamentum teres are also elements of the hip joint anatomy that add stability and restrict terminal mobility. The musculature of the hip joint (Table 12.2) contributes to both mobility and stability of the hip and can be compartmentalised regionally into the gluteal muscles and the muscles of the anterior, medial and posterior thigh. Innervation to the hip joint is provided by the nerves crossing the hip joint. The obturator nerve is considered the primary source of innervation to the hip; however, branches of the femoral and sciatic nerves also contribute to its sensory innervation. The primary source of blood supply to the hip joint is the medial femoral circumflex artery, with additional contributions from the femoral and gluteal vessels. Knowledge of the surgical anatomy and their surface markings and understanding the relationships between the bony, ligamentous, muscular and neurovascular structures are of great importance while considering arthroscopic surgical treatment of the hip joint pathologies.

Table 12.2 Hip muscles and movements

Hip movements	Muscles
Hip flexion	Iliopsoas, rectus femoris and sartorius
Hip extension	Gluteus maximus and hamstrings (semitendinosus, semimembranosus and the long head of biceps femoris)
Hip abduction	Gluteus medius and minimus The tensor fascia lata also helps with abduction in a flexed hip
Hip adduction	Adductor brevis, longus and magnus, and pectineus and gracilis
Hip external rotation	Obturator internus, obturator externus, superior and inferior gemellus, quadratus femoris and piriformis
Hip internal rotation	Secondary actions of the anterior fibres of the gluteus medius and minimus, tensor fascia lata, semimembranosus, semitendinosus, pectineus and posterior part of the adductor magnus

In the last decade, acetabular labral pathology has been gaining more importance. It is a fibrocartilaginous tissue attached to the rim of the acetabulum and continues as the transverse acetabular ligament, bridging the cotyloid fossa. Surrounding the periphery of the acetabular rim, the labrum increases the depth, surface area, volume, congruity and stability of the hip joint. The labrum has been shown to contribute an average of 22% to the articulating surface area and add 33% to the acetabular volume [3]. It provides a seal around the osseous acetabulum and femoral head. This fluid seal is one of the most important functions of the labrum, as it produces a negative intra-articular pressure, significantly increasing hip joint stability [4].

Microscopically the peripheral aspect of the acetabular labrum consists of dense connective tissue. The internal layer consists of type II collagen-positive fibrocartilage. Scanning electron microscopy reveals three distinct layers in the acetabular labrum: (i) the articular surface, covered by a meshwork of thin fibrils, (ii) beneath the superficial network, a layer of lamella-like collagen fibrils and (iii) the deeper layer wherein the majority of the collagen fibrils are oriented in a circular manner. The collagen fibres of the anterior labrum are arranged parallel to the labral-chondral junction, but at the posterior labrum they are aligned perpendicular to the junction. The orientation of the collagen fibres parallel to the labral-chondral junction in the anterior labrum may render it more prone to damage than the posterior labrum, where the collagen fibres are anchored in the acetabular cartilage. Histologically, the fibrocartilaginous labrum was contiguous with the acetabular articular cartilage through a 1–2 mm zone of transition. A consistent projection of bone extends from the bony acetabulum into the substance of the labrum that is attached via a zone of calcified cartilage with a well-defined tidemark [3, 5].

A biomechanical analysis of the labrum suggests that it gets predominately stressed when faced with a compressive load. Therefore, excision or removal of the labrum may alter physiological functions such as enhancing joint stability and load distribution. Sensory fibres, mechanoreceptors and free nerve fibres densely populate the acetabular labrum, capsule and transverse acetabular ligament, suggesting their potential roles as the source of hip pain. It is found that the anterior zone of the labrum contains the highest relative concentration of sensory fibres [6, 7].

The acetabular labrum receives its blood supply from radial branches of a periacetabular periosteal vascular ring that traverses the osseolabral junction on its capsular side and continues towards the labrum's free edge [8]. The distribution of vessels within the labrum is not homogeneous. Blood vessels can be detected in the peripheral one-third of the labrum. The internal part is relatively avascular. The vascular pattern identified should encourage surgeons to develop repair strategies for peripheral labral tears to maintain its functions in the hip [9].

Acetabular labrum tears have been implicated as a cause of hip pain in adult patients. Of the tears 74% are located in the anterosuperior quadrant (Fig. 12.7). Two distinct types of tears of the labrum were identified histologically. The first consisted of a detachment of the fibrocartilaginous labrum from the articular hyaline cartilage at the transition zone. The second consisted of one or more cleavage planes of variable depth within the substance of the labrum. Both types of labral tears were associated with increased microvessel formation seen within the tear. Labral tears occur early in the arthritic process of the hip and may be one of the causes of degenerative hip disease [6].

Under the influence of joint compression in a neutral hip position, the acetabular labrum continues to resist femoral head dislocation despite detachment from the acetabular rim. A radial tear in the acetabular labrum decreases adjacent labral strain, but removal of 2 cm or more of the acetabular labrum is needed before hip stability decreases [10]. Loss of labral function, either via tear or via debridement, may induce hip microinstability, subluxation or dislocation. Breakage of the labral fluid seal is the rationale behind loss of joint stability [9]. This may occur in hips with FAI or dysplasia during flexion and rotational manoeuvres. Although the majority of labral tears occur in the presence of osseous pathology, they may also occur without any obvious bony pathomorphology in patients performing certain sports or activities or in patients with iliopsoas impingement. Labral preservation, particularly with larger tears, may be important for maintaining hip stability [11]. Ligamentum teres connects the centre of the femoral head to the acetabular fossa. Lesions of the ligamentum teres are being increasingly recognised as a cause of persistent hip pain, particularly due to microinstability.

While understanding the surgical anatomy of the hip joint, the approach would be to differentiate the various structures, from the bony anatomy of the hip joint, including the proximal femur and pelvis and the lumbar and sacral spine; to cartilage structures, which include articular cartilage and the acetabular labrum, synovial tissue and capsule, including the ligaments, muscles, tendons and bursae; and the surrounding neurovascular structure. A systematic understanding of the surgical anatomy and their systematic evaluation are necessary to arrive at a diagnosis and formulate an effective management plan.

12.3 Imaging

Standard preoperative imaging for defining hip pathology includes plain radiographs, magnetic resonance (MR) arthrogram and computed tomography (CT) with 3D reformatting.

12.3.1 Radiography

Plain radiographic [11] examination and proper interpretation of images are the most basic and critical steps for the diagnosis of hip disorders. In plain radiography (X-ray), anteroposterior and lateral hip radiographs are usually taken; other views in the oblique plane and also in varying degrees of rotation of the hip can be performed as indicated. A standard anteroposterior hip radiograph includes images of both sides of the hip on the same film and projects towards the middle of the lines connecting the upper symphysis pubis and the anterior-superior iliac spine (ASIS); lower extremities should be internally rotated by 15°–20° to accommodate femoral anteversion, and the distance from the X-ray tube to the film should be 1.2 metres. There are multiple imaging techniques for lateral hip radiography, which includes the frog-leg lateral view, the Löwenstein view and the cross-table lateral view. Studies have observed that 87% of the patients with labral tears had an identifiable underlying structural abnormality that could be seen on the plain radiographs. A plain radiographic evaluation of the hip is dependent on patient positioning and subject to inconsistencies in the radiographic technique; image quality limits its reliability [4].

12.3.2 Evaluation of Images

To evaluate an image (Table 12.3) it is critical to confirm that the X-ray was properly taken and the patient was in an appropriate position. In particular, the tilt and rotation of the pelvis should be known precisely during evaluation of an anteroposterior hip radiograph. In a standard anteroposterior hip radiograph, the coccyx and symphysis pubis should be in a straight line and positioned in the middle line of the image and both sides of the iliac wings and obturator foramina should be symmetric, while the distance between the superior border of the pubic symphysis and the tip of the coccyx should be between 1 and 3 cm. In addition, the greater and lesser trochanters should be clearly distinguishable. Various lateral hip radiographs have specific advantages and limitations.

Table 12.3 Summary of radiographic measurements

Measurement in standard AP radiograph	Technique
Leg length	The height of the iliac crests/ tear drop/ischial tuberosity on both sides is compared with the prominent point on the greater or lesser trochanter.
Neck–shaft angle	This is the angle between the longitudinal axis of the femoral shaft and the axis of the femoral neck (normal range 125°–140°).
Acetabular coverage	This is the lateral CEA and femoral head extrusion index. The lateral CEA is the angle between the vertical line from the centre of the femoral head and the line connecting the femoral head centre to the lateral margin of the acetabulum (normal range 25°–40°, acetabular dysplasia if it is less than 20° and excessive coverage if it is more than 40°). The femoral head extrusion index indicates the percentage of the femoral head not covered by the acetabulum and is considered normal if it is less than 25%.

Measurement in standard AP radiograph	Technique
Acetabular depth	Coxa profunda: The acetabular fossa meets the ilioischial line. Protrusion acetabuli: The femoral head is medially displaced and overlaps the ilioischial line.
Acetabular inclination (acetabular roof angle of Tönnis)	The angle between the first line is drawn through the inferior aspect of the sclerotic acetabular sourcil parallel to the interteardrop line, and the second line connects the inferior and lateral aspects of the sclerotic acetabular sourcil. (Normal 0°–10°; if it exceeds 10°, there is a likelihood of hip instability, and if it is less than 0°, pincer-type FAI can occur.)
Acetabular version	This refers to the line along the anterior margin of the acetabulum and a line along the posterior margin of the acetabulum. Anteversion is the absence of an intersection between these lines. Retroversion is the presence of an intersection (cross-over or figure-of-eight sign). A deficient posterior wall (the centre of the femoral head is positioned lateral to the posterior margin of the acetabulum) and a prominent projection of the ischial spine into the pelvic cavity are additional signs of acetabular retroversion. Acetabular version can vary considerably in terms of tilt and rotation depending on the patient position.
Head sphericity	The femoral head deviates more than 2 mm from the reference circle (spherical).
Joint space width	Standing anteroposterior hip radiographs: This is the interbone distance between the highest margin of the femoral head and the lowest margin of the acetabulum.
Anterior CEA	A false-profile view, it is the angle between the vertical line from the centre of the femoral head and the posterior margin of the acetabulum is measured (an angle less than 20° means insufficient anterior coverage of the acetabulum).

Every radiograph provides important information critical for accurate diagnosis of hip disorders. Generally, anteroposterior and false-profile views provide information on the shape of the acetabulum, whereas other lateral images provide information on proximal femoral parts, including the femoral head. The following specific information can be obtained from anteroposterior hip radiographs: (i) leg length, (ii) neck–shaft angle, (iii) acetabular coverage – the lateral centre edge (CE) angle and femoral head extrusion index, (iv) acetabular depth, (v) acetabular inclination, (vi) acetabular version, (vii) head sphericity and (viii) joint space width. The lateral views help to evaluate the sphericity of the femoral head, joint congruency and the shape and offset of the head–neck junction, and in the false-profile view, the anterior coverage of the femoral head can be assessed.

In lateral hip radiographs, the shape and offset of the femoral head–neck junction as well as the offset alpha angle are assessed. Axial CT or magnetic resonance imaging (MRI) is used for more accuracy. A cam deformity is diagnosed if the alpha angle exceeds 50°–55°.

12.3.3 Computer Tomography Scan of the Hip and Pelvis

Both CT and MRI, with and without an intra-articular contrast medium, have become increasingly popular methods for screening for hip pathological anatomy. CT allows accurate evaluation of the bony anatomy of the hip and also the anatomic relationships between the femoral head and the acetabulum, independent of patient position. It has the advantage of quantifying femoral and acetabular version, and volumetric imaging techniques allow computers to generate a 3D model of the hip that can quantify femoral head coverage and asphericity. This will help with preoperative planning. In addition, finite element models from CT arthrography can provide insight into the relationship between abnormal anatomy and mechanical causes of cartilage and labrum degeneration in preosteoarthritic hips [6, 10, 12].

Bony structural abnormalities usually identified with CT scan include loose bodies. On the femur the CT scan helps measure the alpha angle, reduced or exaggerated femoral neck version angles, and coxa valga and coxa vara. On the acetabular side a CT scan helps in identifying anteversion/retroversion and inclination. A CT scan also helps in identifying a combination of these abnormalities. Ninety per cent of the patients with labral pathology show structural abnormalities that can be identified on the CT scans, and these structural abnormalities frequently occur in combination.

12.3.4 Ultrasound of the Hip

Ultrasonography is increasingly being used to evaluate pathologies of both intra-articular and extra-articular soft tissues in and around the hip joint, which includes muscles, tendons and bursae. It is gaining popularity due to its easy accessibility, availability, lack of radiation exposure and low costs. The distinct advantage of the ultrasound is that it allows for dynamic evaluation, where in the joint and soft tissues can be visualised while taking the joint through the motions. It is also a valuable tool in guiding infiltrations/intervention around the hip joint for both diagnostic and therapeutic purposes [12].

The high-frequency linear transducer, typically with frequencies at approximately 7–12 MHz, may have adequate penetration for hip examination and a lower-frequency sector transducer may be required if the patient is obese. Possible differential diagnoses would help focus the examination on the relative appropriate structures around the hip. For ultrasound examination, the hip area is divided into four quadrants: anterior, medial, lateral and posterior.

The ultrasound examination is performed sequentially starting – with the patient in a supine position – from the anterior quadrant and working around to lateral, posterior and medial sides. In the anterior aspect the Iliopsoas tendonopathy, bursitis and snapping iliopsoas tendon can be evaluated. On the lateral side evaluation focuses on Iliotibial band (ITB), tensor fascia lata, rectus femoris and gluteus medius tendon for hip abductor tendinopathy or tear. Posteriorly, hamstring tendons and muscles, ischiogluteal bursitis and piriformis and on the medial side the adductor longus and the gracilis can be visualised.

12.3.5 Magnetic Resonance

MRI is becoming increasingly essential in the work-up of bone and soft-tissue abnormalities of the hip. Intra-articular injection of a contrast medium is required to obtain a precise diagnosis of intra-articular pathology such as labral tears, cartilaginous lesions,

FAI and intra-articular foreign bodies and while evaluating for developmental dysplasia of the hip (DDH). Evaluation of the labrum and cartilage is better with a 3-T magnetic field. MR arthrography of the hip sequences includes fat-saturated T1-weighted 3 mm slices in axial oblique, coronal and sagittal planes; fat-saturated T2-weighted 4 mm slices in the coronal plane and fat-saturated 3D gradient echo T1-weighted 1 mm slices allowing for multiplanar and radial reconstructions. This helps in the detection of acetabular labrum lesions. As with all MR studies, absolute contraindications related to magnetic field exposure need to be respected [13–15].

12.3.5.1 Labral tears

In 1996, Czerny et al. [16] proposed an MRI classification (Table 12.4) of acetabular labrum lesions with reference to increasing stages of

Table 12.4 Czerny magnetic resonance arthrography; classification of labral tears

Stage	Description
Stage 0	The tear has homogeneous low signal intensity, triangular shape and continuous attachment to the lateral margins of the acetabulum without a notch or a sulcus. A recess between the joint capsule and the labrum consists of a linear collection of the contrast material extending between the cranial margin of the acetabular labrum and the joint capsule.
Stage 1A	The labral tear has an area of increased signal intensity in the centre that does not extend to the margins, a triangular shape and a continuous attachment to the lateral margin of the acetabulum without the sulcus. A normal labral recess is also present.
Stage 1B	The tear is similar to that in stage 1A but is thickened and no labral recess is present.
Stage 2A	The tear is an extension of contrast into the labrum without detachment from the acetabulum, is triangular and has a labral recess.
Stage 2B	The tear is like stage 2A but is thickened, and the labral recess is not present.
Stage 3A	The labrum is detached from the acetabulum but triangular in shape.
Stage 3B	The tear is like stage 3A but thickened.

severity and Blankenbaker [17] classified on the basis of the location of the tear (Table 12.5). Lage et al. [18] described an arthroscopic classification based on the morphology of labral tears (Table 12.6). However, further studies did not find a good correlation between these classifications and resorted to morphological description of the tear. A linear or curvilinear increased signal at the insertion of the labrum suggests a partial tear (Figs. 12.1 and 12.2), but this needs to be correlated clinically as it can sometimes be found in asymptomatic patients. The presence of a paralabral cyst suggests a complete labral tear [19].

Table 12.5 Blankenbaker magnetic resonance arthrography classification of labral tears

Type 1	Frayed: The labrum has irregular margins without a discrete tear.
Type 2	Flap tear: The contrast extends into or through the labral substance.
Type 3	Peripheral longitudinal: The contrast is partially or completely between the labral base, and there is acetabulum labral detachment.
Type 4	The tear is thickened and distorted and thus likely unstable.

Table 12.6 Lage et al.'s arthroscopic classification of labral tears

Radial flap	There is disruption of the free margin of the labrum with subsequent formation of a discrete flap.
Radial fibrillated	A shaving brush, hairy appearance occurs at the free margin.
Longitudinal peripheral	The tear is of variable length along the acetabular insertion of the labrum.
Unstable	There is subluxation of the labrum.

Figure 12.1 MRI arthrogram of the hip showing a labral tear and a cam lesion.

Figure 12.2 MRI arthrogram axial view showing a labral tear.

12.3.5.2 Femoroacetabular impingement

MR arthrography signs of cam-type FAI are a flattening or convexity of the anterosuperior part of the normally concave femoral head-neck junction [22] and an alpha angle exceeding 55° [24]. The alpha angle is measured on an axial-oblique image between the axis of the femoral neck and a line connecting the femoral head centre with the point of change in the radius of the curvature on the anterior part of the femoral head. A cam deformity is diagnosed if the alpha angle exceeds 50°–55° (Figs. 12.3 and 12.4).

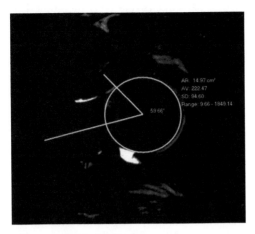

Figure 12.3 MRI measurement of the alpha angle.

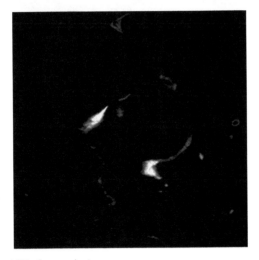

Figure 12.4 MRI of a cam lesion.

In more severe cases, there may be an associated subchondral oedema, subchondral microcyst and fibrocystic changes in the anterosuperior part of the femoral neck. As a consequence, there is an extensive anterosuperior labral injury [21, 23].

A pincer-type FAI causes direct contact between the anterior acetabular border and the femoral head–neck junction. In this type of FAI, the articular cartilage is relatively incomplete. In this type of FAI, there is a morphological anomaly of the acetabulum (Fig. 12.5), which causes an excessive coverage of the femoral head, such as coxa profunda, acetabular protrusion or acetabular retroversion [24–26]. In longstanding cases, lesions to the posteroinferior cartilage by a contrecoup mechanism may appear [20].

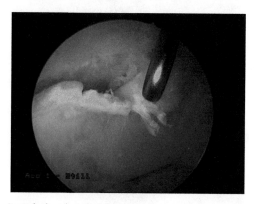

Figure 12.5 Acetabular chondral flap.

12.3.5.3 Abnormality of the cartilage

As the articular cartilage is so thin and due to the direct apposition of the femur and acetabulum, it is difficult to differentiate femoral from acetabular cartilage with standard MRI. The sensitivity and specificity of MR arthrography in the detection of chondral lesions using T1-weighted spin-echo sequence are estimated to be about 62%–79% and 77%–94% [65 ml, 66 ml]. Multidetector computed tomographic (MDCT) arthrography and 3D double-echo steady-state sequence optimised for cartilage may improve sensitivity, specificity and interobserver reproducibility [27]. Cartilages are categorised into normal, signal heterogeneity, fissuring, thinning to <50% of the normal thickness, thinning to >50% of the normal thickness and full-thickness cartilage loss. Cartilage lesions are also correlated with bone marrow oedema, subchondral cyst formation and subchondral osteosclerosis. A MR arthrogram can help assess the presence and severity of cartilage loss [28].

12.3.5.4   Intra-articular foreign bodies

MR and MDCT arthrography help differentiate intra-articular foreign bodies from osteophytes, synovial folds or hypertrophic synovitis. The intra-articular foreign bodies are seen surrounded by a contrast medium but only partially in the case of osteophytes and synovial.

12.3.5.5   Abnormality of the ligamentum teres

Abnormality of the ligamentum teres such as a complete or partial tear can be identified. A degenerated ligamentum teres can be seen in young patients often associated with avascular osteonecrosis of the femoral head or slipped capital femoral epiphysis (SCFE). Recently studies have demonstrated microinstability with a deficient ligamentum teres.

12.4   Physical Examination of the Hip and Pelvis

The history followed by the physical examination of the hip is the key for evaluation of patients presenting with hip pain. A comprehensive assessment of the hip and the surrounding structures includes assessment of osteochondral, capsulolabral, musculotendinous and neurovascular structures. The hip examination should be performed in a systematic and orderly approach, with clear understanding of the principles of each test used. The findings during the examination will provide direction towards the special tests and other areas to be examined in detail. The general principle of orthopaedic assessment 'look, feel, move' and special tests could be followed; also a hip examination should include assessment of the knee, lumbar spine and pelvis, including the opposite hip joint. For a sleeker and comprehensive examination, all the relevant inspection, palpation, movements and special tests could be performed in the following sequence. The most efficient order of the examination would be starting the assessment with tests in the standing position followed by tests in the sitting position, then the supine, then the lateral and finally the prone position.

A complete history should explore key symptoms in detail, including the characteristics of the symptom, and also look into the effects on the individual's function. A routine history taking will include the patient age, onset and duration of symptoms, mechanisms of injury/mode of onset, location of pain and other characters of pain. Typically a FAI will present with groin pain on sitting or deep flexion of the hip (squatting), pain during certain movements/positions (during sporting activities) and clicking/popping at the hip (e.g., during a golf swing). The pain is usually insidious onset and progressive and typically demonstrates the C sign (the patient will hold his or her hand in the shape of a C and place it above the greater trochanter, with the thumb positioned posterior to the trochanter and fingers extending into the groin). The pain will increase on flexion, adduction and internal rotation movements at the hip. The pain is typically intermittent and activity related. The nature of the pain can also give clues to the diagnosis; if the pain is sharp and acute, with associated symptoms of 'locking' or giving way, consider labral pathology. Pain associated with specific movements can be related to a single musculotendinous unit, tendinitis or bursitis.

Further history taking should consider other areas as a source of 'hip' pain. For example, in the lumbar spine, consider lumbar disc prolapse, osteoarthritis, ankylosing spondylitis and malignancy. Enquire about peripheral neurological symptoms; inguinal hernia, psoas pathology, abdominal wall or intra-abdominal and genitourinary pathology can all present with hip pain. Past history could include any previous hip problems or surgery in childhood, specifically DDH, Perthes' disease, transient synovitis, SCFE and hip dysplasia, infection, gout and malignancy.

A history of sports and recreational activities can help determine the type of injury and also guide the treatment, considering the patient's goals and expectations. Physical examination of the hip is summarised in Table 12.7.

It is important to be able to differentiate extra-articular from intra-articular pathology. A comprehensive assessment algorithm provides a rational approach to patient evaluation and improves clinical decision making and treatment plans.

Table 12.7 Comprehensive clinical examination of the hip

Standing	Inspection	
	From the front	Shoulders should be level. There will be deformity, skin changes, swelling, muscle wasting, scars and sinus and obvious leg length difference.
	From the sides	There will be attitude of the limb and deformity at the hip, knee, foot and ankle. In a lateral profile of the spine (lumbar lordosis, kyphosis), swelling, wasting, scars and sinus will be visible.
	From the back	Spine scoliosis, swelling, wasting, scars and sinus are present.
	Trendelenburg test	Feel for the ASIS bilaterally as the patient's palms rest on your forearms. As the patient stands on each leg in turn, the pelvis should remain level, and if the pelvis drops to the non-weight-bearing side, or the patient uses the contralateral upper limb for support on you, or the patient lurches towards the weight-bearing side, the test is positive in the standing limb. This identifies abductor weakness.
	Gait	
	Gait evaluation	Check for foot progression angle, pelvic rotation, stance phase and stride length, Trendelenburg, antalgic, short-limb gait, circumduction gait, etc.
Sitting	Inspection	
	Back	Check for structural and nonstructural scoliosis.

(Continued)

Physical Examination of the Hip and Pelvis | 159

Table 12.7 (Continued)

Supine	Inspection		
	Palpation	Tenderness	Check the ASIS, posterior superior iliac spine (PSIS), sacroiliac (SI) joint, pubic tubercle, ischial tuberosity, trochanteric bursa (bursitis), femoral head, capsule, adductor longus, hernia (cough impulse) and iliotibial band (ITB) for fascial defects.
	Movements (active and passive)	Flexion/Extension, fixed flexion, deformity	Check the range of flexion (100°–140°) and extension (10°–30°). Perform the Thomas test.
		Adduction/Abduction, fixed adduction and abduction, deformity	Check the range of adduction (20°–45°) and abduction (30°–60°) and assign the ASIS level (ASIS lower = abduction deformity; ASIS higher = adduction deformity). Square the pelvis to reveal the fixed adduction or fixed abduction deformity.
		Internal and external rotation, fixed internal and external rotation, deformity	Check the range of internal rotation (20°–45°) and external rotation (40°–60°). With the patient's hip in extension check the position of the patella. In the prone position the patient flexes the knee to 90°. Check with patient's hip and knee at 90° flexion.

Leg length measurements	Apparent length	This is the fixed midline point (e.g., xiphisternum) to the medial malleolus bilaterally.
	True length	It is called the Galeazzi test. Square the pelvis and measure from the ASIS to the medial malleolus on each side and compare, placing the unaffected limb in the same position as the affected side if there is a fixed flexion or adduction/abduction deformity. Take the segmental measurement of the femur and the tibia. Map Bryant's triangle.
Lateral position	Ober's test	The patient should be in a lateral position, resting on the unaffected hip. The affected hip is placed in abduction and extension and the left is unsupported. If the hip remains in abduction, the test is positive.
	ITB tightness	A set of passive adduction tests (similar to Ober's test) is performed with the leg in three positions: extension (tensor fascia lata contracture test), neutral (gluteus medius contracture test) and flexion (gluteus maximus contracture test).

(Continued)

Table 12.7 (Continued)

Special tests	Duncan–Ely test	Rectus femoris	The patient should be in a prone position, with the hip in extension, and should flex the knee. The ipsilateral thigh should remain horizontal, but a test is positive if the ipsilateral thigh and buttock rise off the couch.
	Impingement	Acetabular labrum, cam lesion, the dynamic internal rotatory impingement (DIRI) test, McCarthy test	This involves assessment of anterior femoroacetabular congruency. The patient is in a supine position. The anterior impingement test or apprehension test is performed by passive flexion, internal rotation and adduction of the hip. The posterior impingement test is performed with the supine patient's hips at the end of the couch and the contralateral hip and knee held flexed by the patient. A positive test is indicated by pain as the affected hip is moved into extension and external rotation.
		The dynamic external rotator impingement test (DEXRIT)	This test involves assessment of superolateral and posterior FAI.
	Hip telescoping test		The patient's hip and knee are at 90° flexion and elicit telescoping of the trochanter in relation to the pelvis. There may be nonunion of the femur, DDH and soft-tissue laxity.

FABER/Patrick test	Flexion, abduction, external rotation	Flexion, abduction and external rotation of the affected hip are involved. To assess the right hip, the right foot is placed over the left knee in the 'figure of-four' position and gentle downwards pressure is exerted on the right knee, increasing the degree of external rotation.
Craig's test	Femoral neck anteversion	The patient lies in a prone position. The affected hip is externally and internally rotated until the prominence of the greater tuberosity is felt laterally with the other hand. Anteversion is estimated as the angle between the leg and the vertical.
Piriformis test	Piriformis	The patient is in a lateral position, lying on the unaffected side. With the knee flexed and the hip flexed to 60° on the affected side, downwards pressure over the knee will elicit pain in the area of the Piriformis in a positive test.
Vascular system	Peripheral pulse	Check dorsalis pedis, posterior tibial popliteal pulses and capillary filling.
Neurologic systems	Neurological examination	This involves sensory examination, motor examination, and checking lower limb reflexes and nerve root tension signs.

12.5 Surgical Technique

An arthroscopic examination of the hip is significantly more challenging because of the anatomical constraints; the femoral head is deeply recessed in the bony acetabulum. The thick fibrocapsular and muscular envelopes around the hip joint increase the amount of force required for distraction of the hip during arthroscopy. The relative proximity of the sciatic nerve, the lateral femoral cutaneous nerve and remaining femoral neurovascular structures makes portal placement more challenging.

For an arthroscopic procedure of the hip, the surgical anatomy of the hip joint is divided into a central compartment, a peripheral compartment and extra-articular structures (Table 12.8). To visualise the central compartment, traction and distraction of the joint surfaces is required. A systematic sequential approach is followed to visualise the structures. Depending on personal preferences the central or peripheral compartment can be approached first. If the hip joint is tight, start with the peripheral compartment and then proceed to the central compartment. Flexible instruments allow for significantly improved access to most structures within the hip joint.

Table 12.8 Arthroscopic evaluation of the hip

Central compartment (with traction)	Peripheral compartment	Periarticular/ extra-articular structures
Anterior labrum	Peripheral femoral head and its articular cartilage	Gluteal musculature
Anterolateral labrum	Femoral neck	Iliotibial band
Posterior labrum	Joint capsule	Iliopsoas
Acetabular facets	Capsular portion of the labrum	Fascia lata
Acetabular articular cartilage	Medial synovial plica	Piriformis muscles
Femoral articular cartilage	Articular recess (medial, anterior, posterior recesses)	External rotators
Ligamentum teres	Zona orbicularis	Sciatic nerve
Transverse ligament	Dynamic evaluation of the hip articulation	Greater sciatic foramen exploration

12.5.1 Patient Position

Supine and lateral positions are widely used for hip arthroscopic procedures; the choice of the position depends on ones understanding of the orientation of structures and familiarity with the position. Both supine and lateral positions have shown to give good access to the hip joint and the extra-articular structures.

The supine position was described by Byrd in 1994 and has been popular. The patient is positioned on the traction table, with the leg positioned in adduction, with the hip flexed to 10° and the femur in a slight internal rotation. A well-padded perineal support is required to ensure good countertraction. The lateral position was described by Glick in 1987. Special table attachments allow for adequate pelvic stabilisation, and a well-padded perineal post provides the countertraction in a lateral position when traction is applied.

12.5.2 Portals

The standard portals commonly used are anterolateral, posterolateral, anterior and modified anterior. Other portals described are the proximal transgluteal portal; super-lateral modified anterior portal (MAP), proximal and distal; and anterodistal portal.

12.5.3 Theatre Layout and Equipment

The layout of the theatre equipment would depend on the surgeon's preference for supine or lateral positioning of the patient. The equipment should be positioned such that the monitor screens are within visual range and minimal movement of equipment is needed during the procedure.

A standard traction table or additional traction extension attachments are used to achieve traction. A well-padded perineal post (at least 9 cm of perineal padding) is used for distraction of the hip. The foot is wrapped in wool and placed in a foam boot and firmly secured in the distraction boot. While traction is applied, care must be taken not to crush the contralateral thigh between the post and the table when operating on a patient in a lateral position. The perineum is checked before and after traction has been applied. Traction is cautiously applied under image intensifier guidance, and the hip's distractibility is confirmed prior to draping. General anaesthesia or regional anaesthesia with good muscle relaxation is necessary for safe distraction of the hip. Hypotensive anaesthesia helps the surgeon to minimise pump pressure and decrease fluid extravasation.

In cases where longer traction times are expected (e.g., more than two hours of distraction time), as an alternative to traction, invasive distraction can be used. By avoiding traction on the entire lower limb, nerve injury and skin injury are minimised. This can remove the two-hour time barrier for the more complex cases and when advanced techniques are used, for example, in fully arthroscopic cartilage transplantation.

12.5.4 Portal Placement

Traction is reapplied to distract the joint. A fluoroscope is used to guide the initial portal placement. Some surgeons pass a spinal needle into the hip joint at this stage, and an air arthrogram is performed, which allows for loss of vacuum, and the joint can be distracted further. Distraction can also be improved by injecting normal saline into the joint under pressure. A 14-gauge needle and nitinol guide wire is used to create an anterolateral portal. The needle is advanced into the hip joint under image intensifier control, with the bevel surface facing the femoral head. Once in the joint the nitinol guide wire is inserted and the needle is removed, the portal is dilated and a 70° arthroscope is introduced. Under direct vision the MAP is placed at an approximately 30° angle anterior to the anterolateral portal using a 14-gauge spinal needle and nitinol wire. Depending on the location of the pathology and the interventions required, additional portals may be necessary. Care should be taken to avoid damage to the labrum and the femoral articular surface. An interportal capsulotomy is then created with a curved beaver blade; a satisfactory cuff of capsular tissue is left adjacent to the labrum and acetabular rim. A systematic sequential approach is followed to visualise and evaluate the structures and deal with the pathology. Capsulotomy is further extended as required, with the radiofrequency (RF) ablator aiming to produce an incision in the capsule that parallels the labral margin.

12.5.5 Central Compartment

The anterior, superior and posterior chondrolabral complex are assessed using an arthroscopic probe/hook. The acetabular articular surface is visualised and evaluated. The ligamentum teres is viewed and dynamically assessed whilst rotating the limb into the maximum internal and then external rotation. The femoral head surface is then assessed. Additional portals may be placed depending on the location pathology that needs addressing.

12.5.6 Peripheral Compartment

Multiple techniques are described to access the peripheral compartment. Some surgeons recommend a superolateral viewing portal. The anterior capsule is relaxed by releasing the traction and flexing the hip 20°–30°. A 14-gauge × 6-inch spinal needle is directed under image intensifier control to the superior head–neck junction. The bevel is rotated such that it is opposed to the bony surface. The spinal needle is then advanced, with the bevel sliding along the anterior surface of the femoral neck. The portal is then established in the usual manner. Care must be taken to prevent the guide wire penetrating the medial capsule. Also, care should be taken against breakage of the guide wire.

Alternatively some surgeons use a swing-over technique to move from the central compartment to the peripheral compartment. Here, the existing central compartment portals are used. The traction is released and the hip flexed during this process. This technique is especially useful in the management of superior and posterosuperior peripheral compartment pathology.

The peripheral compartment is systematically assessed and the image intensifier used to ensure a complete assessment of the peripheral compartment. Dynamic examination of the hip joint is performed to check for FAI and adequacy of surgery.

The choice of instruments available continues to increase. Hand instruments are still widely in use. The initial capsulotomy is done using a beaver blade; graspers are available for removing loose bodies. A wide range of suture-passing instruments are used for labral repair and capsular plication. A microfracture awl is used to treat a full-thickness acetabular chondral defect. Powered instruments such as soft-tissue shavers and a variety of bone burrs are used. RF probes are increasingly being used. Repair of labral tears is usually done with the help of bone anchors and knot or knotless systems.

12.6 Complications

Along with the anaesthetic risks, complications specifically associated with this procedure are traction injury and pressure necrosis, which can be significantly reduced by using a perineal post that is well padded and at least greater than 9 cm in diameter. Use of traction force should be kept to minimal to distract the joint adequately and also keep the traction time to the minimum. Testicular and labial injury, including vaginal tears, has been reported. Perineal nerve neuropraxia is the commonest nerve injury reported. Traction injuries to pudendal, femoral or sciatic nerves have also been reported. Most of the traction injuries mend fully within hours of surgery. Traction injuries are shown to be directly related to the traction force and duration of traction. Direct injury to the lateral femoral cutaneous nerve has been reported during anterior or midanterior portal placement, and it is important to have detailed knowledge of the surface anatomy. Other structures at risk from portal placement are the femoral neurovascular bundle, which is in close proximity to the anterior portal, and the sciatic nerve, which is close to the posterolateral portal.

Iatrogenic chondral and labral injuries are usually underreported. Adequate distraction, careful initial cannulation and a gentle technique can minimise articular scuffing, gouging and labral damage. To minimise iatrogenic injuries, some surgeons access the peripheral compartment first and place a guide wire into the central compartment under direct vision. Broken guide wires and instruments though infrequent can be a challenge to retrieve. Careful retraction of the guide wire and controlled cannula insertion minimises the risk of wire breakage.

Postoperative hip instability and subluxation have been reported, particularly following acetabular rim recession in dysplastic hips. Care must be taken not to recede the acetabular rim to the point of destabilising the hip. Femoral neck fractures have been associated following femoral cam excision. Intra-abdominal or retroperitoneal fluid extravasation can be life threatening and is a particular risk when the procedure is performed following an acetabular fracture.

Postoperative infection is rare; one has to be vigilant of aseptic techniques to prevent infection. Intraoperative bleeding can obscure the surgeon's view; it can be overcome by increase in fluid pressure and RF coagulation of bleeding points. Postoperative bleeding, though rare, has been reported following damage to branches of major vessels, particularly the branch of the superior gluteal artery. Other rare complications reported are osteonecrosis, impotence, trochanteric bursitis and chronic regional pain syndrome (CRPS).

Failure to improve the symptoms and on occasion make them worse should be discussed with the patient prior to undertaking the procedure. Failure to gain access to the hip may also be regarded as a complication since it is not always possible to enter the hip arthroscopically.

12.7   Specific Conditions

12.7.1   Acetabular Labral Tears

Independent of the zetiology, labral tears are more common in the anterosuperior quadrant (Fig. 12.6).

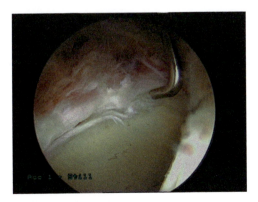

Figure 12.6   Acetabular labral tear.

At this location, the mechanical resistance of the labrum to traction (associated with instability) and compression (associated with FAI) is less compared to all the other regions. Functionally it acts as a sealant, enhances fluid lubrication and maintains synovial pressure and under negative pressure, it provides stability to the hip and prevents excessive contact pressure between the cartilages of the acetabulum and the femoral head. Acetabular labral tears are usually a consequence of bone deformity, FAI, trauma, degeneration, dysplasia, capsular laxity, instability and supraphysiological movements of the hip. It is believed that labral tears rarely occur in the absence of bony abnormalities. Labral lesions are associated with chondropathy, with delamination of the acetabular cartilage; the cartilage sheet appears like a wave when separated from the underlying bone, described as a wave sign (Fig. 12.7).

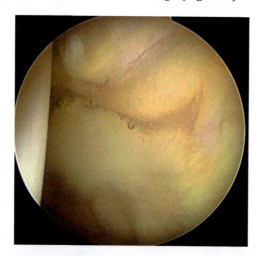

Figure 12.7 Acetabular cartilage delamination with a wave sign.

An unstable labrum can cause mechanical impingement, and with time, direct contact between the abnormal labrum and the femoral chondral surface leads to the formation of bony protrusions progressively evolving into osteoarthritis.

The goal of arthroscopic intervention would be to retain as much healthy labrum as possible and remove nonsalvageable, loose and devitalised tissue, mainly to relieve pain due to mechanical reasons. Stabilisation/repair or partial resection of the labrum, together with correction of the factor that triggered the lesion, is vital. It is believed that arthroscopic surgical management of the labral tear in conjunction with complete correction of any osseous pathomorphology could influence the progression of the osteoarthritis in the hip joint. Therefore, early restoration of the normal morphology of the hip joint is advised before the setting in of chondropathy and osteoarthritis. The labrum has the capacity to heal after refixation. At one year following surgery, labral fixation is shown to achieve better results than resection and has a lower reoperation rate. Studies following labral repairs have shown up to 88% good results and a mean Harris Hip Score (HHS) of 94.3%, and when the labrum was resected, 66.7% showed good results, with a mean HHS of 88.9%. Poor prognosis is associated with the presence of cartilage thinning or cartilage injuries of Outerbridge type IV or Tönnis type III or IV. If the joint space is less than 2 mm, progression to arthroplasty occurs in 80% of such cases within two years on an average.

Most of the vascular supply to the labrum comes from the capsular contribution [29]; the articular surface of the labrum has decreased vascularity and has limited synovial covering. The labrum is thinner in the anterior inferior section and is thicker and slightly rounded posteriorly. During repair the labral tear is identified and the margins defined. A motorised shaver is then used to debride and remove the torn portion of the labrum. A high-speed burr is used to decorticate the acetabular rim. This provides a bleeding surface for the graft to heal. The labrum is repaired by using a bioabsorbable suture anchor, which is placed on the rim of the acetabulum between the labrum and the capsule. Once the anchor is placed, the suture material is passed through the split in the labrum in a vertical mattress suture technique. The suture is tied down by using standard arthroscopic knot-tying techniques or knotless methods. For an intrasubstance split in the labrum, a bioabsorbable suture is passed around the split by using a suture lasso or similar suture-passing instrument. The suture is tied, thus reapproximating the split labral tissue. It is vital that the associated pathology be dealt with, to minimise recurrence and improve outcome. In cases where a significant portion of the labrum is deficient or removed, the labrum can be reconstructed with rectus auto graft, ITB and hamstrings auto or allograft.

12.7.2 Femoroacetabular Impingement

FAI is a purely mechanical disorder. The impingement of the head–neck junction occurs against the rim of the acetabulum in the position of flexion and internal rotation. As a consequence labral tears occur with associated injuries to the adjacent cartilage on the rim of the acetabulum, which predispose the patient towards arthritis. The estimated prevalence of asymptomatic FAI in the general population is 10%–15%. Ganz described two clinical types: the cam-type and the pincer- or tong-type lesions. In the cam type, the anterosuperior part of the anatomical neck of the femur is hypertrophied and convex in shape and impinges on the labrum and acetabulum; and in the pincer-type impingement, there is excessive coverage of the femoral head by the acetabulum. The cam type is more frequent compared to the pincer type. In more than 70% of the cases both acetabular and femoral lesions are found, and these are described as 'mixed' impingement. The deformity may also be secondary to previously undiagnosed silent slips of the femoral epiphysis or other congenital reasons, for example, acetabular retroversion. In some cases, the cartilage injury may occur in atypical locations, particularly seen in ballet dancers, where the impingement is due to overloading, above the physiological level, without secondary structural causes identified.

Early surgical intervention in cases of femoroacetabular deformities, before irreversible cartilage injuries occur, can possibly delay the evolution of the hip arthrosis. Conventionally, an open surgical dislocation technique for removal of excessive bone at the head–neck junction was performed and popularised by Ganz. The procedure is now effectively performed arthroscopically by using appropriate hip distraction techniques, portals and instruments. Arthroscopic management comes with the advantages of a smaller incision, shorter recovery time and potentially fewer complications than open surgery. However, many studies have claimed that the open procedures have results similar to those from arthroscopy.

Surgical treatment involves correction of the deformities on both sides of the joint, by means of osteochondroplasty. On accessing the peripheral compartment as described earlier, the head–neck recess is visualised and the location of the head–neck junction identified. The surgery involves sequential management of the pathology; the sequence may vary across surgeons. It must be ensured that all the underlaying pathology is addressed. Depending on the surgeon's preference the peripheral compartment or the central compartment can be accessed first. The author's choice would be to excise the pincer lesion and perform the labral repair/reconstruction first, followed by excision of the cam lesion (Fig. 12.8).

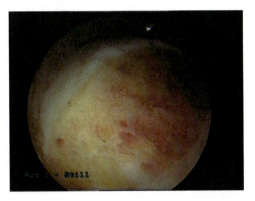

Figure 12.8 After excision of the cam lesion.

In cases of retroversion, anterior acetabular osteoplasty and in the case of protrusion circumferential, osteoplasty is performed. Acetabular retroversion may require periacetabular osteotomy.

The nonsalvageable labral tissue is debrided and the salvageable labral tissue should be preserved and repaired. If there is significant loss of labral tissue, reconstructed can be done with capsular tissue, gracilis, ITB or allograft. The capsule over the anterosuperior acetabulum is elevated from the bony rim by use of cautery. The acetabuloplasty is performed with a high-speed burr. The pincer lesion is defined and excised. The acetabular rim adjacent to the labral damage is decorticated to expose the bleeding bone for repair/reconstruction. Repair is performed as described above. A smaller labral deficiency can be reconstructed using a capsular autograft. The capsular tissue is a less robust graft source; it is locally available and can be used to fill small segmental defects. The key point for capsular autograft reconstruction is to ensure that the capsulotomy is performed close to the femoral head. This allows sufficient capsular tissue for reconstruction. Borderline dysplasia patients should undergo capsular closure to avoid instability; if found deficient reconstruction could be undertaken. Anchors are placed in the area of labral deficiency and are spaced 5 to 8 mm apart. Anchors are placed at the terminal ends of the labral deficiency to overlap capsular tissue on labral tissue; this may better healing and restore labral function. Excess capsular tissue is trimmed back so that the reconstructed tissue protrudes similarly to the adjacent healthy labral tissue. The repaired/reconstructed tissue should restore the suction seal of the labrum. The femoral head–neck recess is visualised, head and neck osteoplasty is performed to remove cam deformity and the aim is to recreate an alpha angle of 43° as measured on standard plain films to allow for FAI-free motion of the hip. In cases with acetabular dysplasia or borderline dysplasia (e.g., a lateral centre edge angle less than 25°) or in cases of ligamentous laxity, patients benefit from capsular closure or capsular plication to restore hip stability, and capsular tissue reconstruction in itself would be inadequate and is likely to leave these patients with persistent instability.

Studies comparing open and arthroscopic surgical techniques showed that there were no significant differences in the precision of osteochondroplasty of the femoral head in cases of cam-like impingement. Significant improvements in the Western Ontario and McMaster Universities (WOMAC) arthritis score and HHS score have been published. Conversion to arthroplasty has been reported from 3% to 9% at two years following the procedure. With minimum complications reported, 76% of the athletes treated for FAI had fully returned to their sports after an arthroscopic FAI surgery. The factors that predicted improvements were preoperative HHS >80, previous joint space larger than 2 mm and labral repair rather than resection during the arthroscopy procedure.

12.7.3 Articular Cartilage

Articular cartilage defects can occur on both the femoral head and the acetabular surface. Labral tears occurring at the watershed zone may destabilise the adjacent acetabular conditions. Arthroscopic observations support the concept that labral disruption, acetabular chondral lesions or both frequently are part of a continuum of degenerative joint disease. Chondral injuries may occur in association with a multitude of hip conditions, including labral tears, loose bodies, osteonecrosis, SCFE, dysplasia and degenerative arthritis. Cartilage integrity is evaluated, and various restoration techniques, like fibrin glue for cartilage delamination, microfracture in cartilage loss, stem cell therapy for cartilage lesions, autologous chondrocyte implantation, mosaic plasty, osteochondral autograft and osteochondral allograft are described. Although good results have been reported, most studies lack a control group, and the small number of patients limits these studies.

12.7.4 Hip Septic Arthritis

Arthroscopy has been successfully used to assess the hip joint for draining infective collection, washout of the joint and debridement and to obtain synovial tissue samples. It is minimally invasive compared to the conventional open arthrotomy. Numerous studies have looked at the outcome of arthroscopic washout in both children and adults and have reported a superior outcome by the arthroscopic method (90% of the results are good with the arthroscopic technique compared with the open technique, in which 70% of the results are good).

12.7.5 Pigmented Villonodular Synovitis and Synovial Chondromatosis

Pigmented villonodular synovitis presents as a focal or as a diffuse lesion. Both of these presentations can be addressed using arthroscopic excision. The prognosis is poorer in diffuse cases, with early progression to arthritis.

Synovial chondromatosis and loose bodies can be managed arthroscopically. Studies have observed that some of the patients recovered well and did not need further treatments, up to 20% of the patients needed further arthroscopy, 37% required an open surgery and 19% progressed on to have a total hip arthroplasty. Arthroscopy is an effective method for removing free chondromatous bodies. Due to difficulties in accessing posteromedial and posterolateral areas in the peripheral compartment, recurrences can be higher.

Irrigation and debridement of the inflammatory tissue in autoimmune diseases, ankylosing spondylitis, rheumatoid arthritis and psoriatic arthritis have shown to improve the range of motion and diminished synovitis on MRI; studies show 75% of the patients being satisfied with the outcome.

12.7.6 Arthroscopy in Hip Trauma

Hip arthroscopy can be useful in posttraumatic cases both in traumatic hip dislocation and in fractures. Most commonly arthroscopy can be used to deal with posttraumatic bony or cartilaginous free bodies in the joint; to assess the state of the articular cartilage, fractured acetabular border or femoral head, teres ligament injuries and joint instability; and in arthroscopic-assisted fracture fixation. Bleeding obscuring the arthroscopic view and fluid extravasation are the added risks. However, there is no report of increased complications due to extravasation of fluid when the pump pressure was maintained at 30 mm Hg.

12.7.7 Osteonecrosis of the Femoral Head

Hip arthroscopy can be used to evaluate the joint cartilage and its integrity in osteonecrosis. Studies have found damage to the femoral cartilage that had not been detected through MRI in up to 36% of the patients. Arthroscopy is usually performed together with decompression of the femoral head, in Ficat stages I or IIa of osteonecrosis. With minimal risk to the circulation of the femoral head, Ellenrieder et al. demonstrated that, in patients with Steinberg stages II and III, without head collapse or chondral lesions, decompression can be performed in association with grafting using autologous graft cylinders. In cases of collapse (Steinberg IV), reduction of the collapsed portion was performed with fluoroscopy control and articular surface checked with arthroscopy.

12.7.8 Painful Hip Arthroplasty

Evaluation of painful hip arthroplasty by means of arthroscopy is helpful for collection of synovial fluid and tissue samples for analysis and microbiological studies. Other sources of pain can also be evaluated, such as joint instability, aseptic loosening, impingement between components and adhesions. Extra-articular causes can be evaluated and addressed, for example, iliopsoas tendinitis and interposition of foreign bodies.

12.7.9 Hip Dysplasia

Arthroscopy in cases of dysplasia has limited indications and should be considered cautiously. Some studies report good results from arthroscopic procedures in borderline dysplastic hips. However, others studies have found acceleration of the degenerative process, which evolved into severe arthritis, lateral migration of the femoral head with instability and a persisting pain symptom following the arthroscopic procedure. Some authors consider the presence of dysplasia as a contraindication to performing hip arthroscopy.

Arthroscopy and labral repair together with or after periacetabular osteotomy has shown good results. The acetabular reorientation provides a better environment and altered stress, allowing the repaired labrum to heal.

12.7.10 Slipped Capital Femoral Epiphysis and Perthes Sequelae

SCFE and Perthes sequelae cause cam-like FAI. Careful evaluation of the deformities by radiographs and CT reconstructions is necessary to plan intervention. In cases of alteration of the femoral offset, that is, significant posterior slippage and realignment of the proximal femur by means of intra-articular or subtrochanteric osteotomy may be needed along with osteochondroplasty to correct the FAI. The arthroscopic procedure has shown not to affect the natural history of Perthes disease; however, it helps to improve the quality of life.

12.7.11 Teres Ligament Injuries and Capsule Repair in Cases of Instability

The function of the ligamentum teres is not entirely clear; it can be tensioned with flexion, abduction and external rotation or by extension in external rotation of the hip. Injuries to the ligamentum teres is divided into three groups: partial traumatic, total traumatic and degenerative rupture. After the initial description by Philippon et al., reconstruction of the ligamentum teres is reported in a selected group of patients with complaints of instability and supraphysiological movements. The capsule-ligament stabilisers of the hip are being continuously studied, and their role has still not been fully defined. Techniques for capsule repair have been described; however, the long-term outcome remains unknown [30].

12.7.12 Extra-articular Pathological Conditions around the Hip

The commonest indications for extra-articular arthroscopy are trochanteric bursitis, external snapping and tendinopathy of the gluteus minimus and gluteus maximus, which together cover the concept of the painful syndrome of the greater trochanter, internal snapping and piriform syndrome (deep gluteal pain).

12.7.13 External Snapping

External snapping is produced due to friction between the greater trochanter and the thickened posterior portion of the ITB or the thickened anterior fibres of the gluteus maximus during flexion or extension. It may or may not be painful. The primary management relies on physiotherapy and stretching. Release of the structures can be performed as an open or an arthroscopic procedure to diminish tension in the ITB. The techniques described also include half-releasing the gluteal tendon at its femoral insertion, on the linea aspera, to release the tension.

12.7.14 Trochanteric Bursitis and Injuries of the Gluteal Muscles

As the aetiology is varied, there are many therapeutic options for trochanteric bursitis. The management relies principally on physiotherapy and stretching. In cases that are refractory to conservative treatment, surgical interventions by means of arthroscopic debridement and release of the tight ITB have shown to be safe and efficient.

Incomplete or complete tears of the gluteal muscle insertions have been shown to be associated with chronic trochanteric bursitis with a positive Trendelenburg sign. These injuries are often under diagnosed. There is paucity of literature on the long-term prognosis of these tears with or without interventions.

12.7.15 Internal Snapping

Internal snapping generally occurs when the tendon of the iliopsoas rubs against the iliopectineal eminence or the femoral head. Arthroscopically a tenotomy can be performed at the level of the lesser trochanter or through the anterior capsular region of the hip. Studies have shown no significant difference between the outcomes of these two techniques. However, the arthroscopic release had a lower complication rate and less postoperative pain than with the open technique. Iliopsoas pressure symptoms and pain have been reported after hip arthroplasty; arthroscopic release of the Iliopsoas tendon has shown to improve the symptoms.

12.7.16 Piriformis Syndrome

Also known as deep gluteal pain, piriformis syndrome is a pathological condition diagnosed by ruling out other causes and is hypothesised to be due to muscle spasms or compression by piriform muscle, fibrous bands, vascular malformations and adherences to the obturator muscles and the quadratus femoris muscle. It manifests as pain in the gluteal region, with or without accompanying sciatic pain. It worsens with local deep pressure and generally continues for years.

The management is mainly conservative. Surgery is indicated in refractory cases. Classically open decompression is described. Proximity to the sciatic nerve and risk of potential nerve injury warrants extreme caution. Endoscopic exploration of the sciatic nerve accompanied by tenotomy of the piriformis and neurolysis of the sciatic nerve, with intraoperative neural monitoring (evoked potential and electroneuromyography), achieved significant improvement after surgery, without recurrences and without neurological injuries. The sciatic pain, which occurred in a sitting position, disappeared in 83% of the cases.

12.8 Rehabilitation

The postoperative recovery period is mainly focused on early physical therapy and a structured, progressive rehabilitation programme. The rehabilitation depends on the operative findings, and the procedure performed should be tailored to the individual needs. Postoperative rehabilitation progresses through the following stages: (i) regaining of joint motion, (ii) strengthening of muscles, (iii) development of proprioception, (iv) advanced strengthening and (v) development of agility. The final stages of rehabilitation should be sport specific.

In general, patients following FAI surgery/labral reconstruction toe-touch weight to partial weight bearing is maintained for six weeks. A range of motion is allowed within a pain-free range during this period. The patient subsequently progresses to full weight bearing and full range of motion. Therapy is sequentially advanced as shown in the Table 12.9. When the range of motion and strength are satisfactory, progression is made to sport-specific training.

Table 12.9 Rehabilitation programme

Week 1	• Perform isometrics – gluteal, quads, trans abs, hip abduction. • Do passive stretching, piriformis stretch (side lying), quads stretch (prone) and adductor stretch (sitting). • Use a stationary bike, high seat, no resistance.
Week 2	• Perform Week 1 exercises. • Do quadruped rocking. • Do standing hip IR exercises.
Weeks 3 and 4	• Undergo gait re-education. • Do ROM exercises (continued from Week 1 and 2 exercises). • Do calf, hamstring and ITB stretching. • Use the bike, no resistance, and increase the time. • Use the leg-press, the cross-trainer and the Swiss ball. • Work on core stability. • Undergo hydrotherapy.
Weeks 5 and 6	• Continue previous exercises. • Use the bike – with resistance. • Perform balance work – for example, use the wobble board. • Work on core stability – progress as able. • Do lunges, lateral side steps and knee bends; jog/walk.
Week 7+	• Increase hydrotherapy exercises (squats, step-ups/step-downs, ¼–½ lunges). • Run – progress from a straight line to multidirectional. • Do sport-specific exercise.

12.9 Outcome Assessment

For assessing the outcome following hip arthroscopy, numerous patient-reported outcome measures are in use. The commonly used outcome measures look at either the general well-being of the person, for example, SF12, or specific improvement with the hip symptoms and function, for example, the Hip Outcome Score (HOS). The most recommended tools for assessment of young adults with hip pain and undergoing nonsurgical treatment or hip arthroscopy are the Hip and Groin Outcome Score (HAGOS), the HOS, the International Hip Outcome Tool-12 (IHOT-12) and the IHOT-33. These patient-reported outcomes are validated and contain adequate measurement qualities. They are all checked for test-retest reliability, construct validity, responsiveness and interpretability [30].

Various scoring systems are available for quantification of hip pain, hip function and severity of symptoms. The HHS, the modified HHS, the Merle d'Aubigné score, the Non-arthritic Hip Score, the musculoskeletal function assessment (MFA) and the WOMAC are in use. A visual analogue score is also useful for the quantification of pain at rest and pain with activities.

References

1. Kelly BT, Buly RL. Hip arthroscopy update. *HSS J*, 2005; **1**(1): 40–48.
2. Edwards DJ, Lomas D, Villar RN. Diagnosis of the painful hip by magnetic resonance imaging and arthroscopy. *J Bone Joint Surg Br*, 1995; **77**(3):374–376.
3. Seldes RM, Tan V, Hunt J, Katz M, Winiarsky R, Fitzgerald RH Jr. Anatomy, histologic features, and vascularity of the adult acetabular labrum. *Clin Orthop Relat Res*, 2001; (382):232–240.
4. Philippon MJ, Nepple JJ, Campbell KJ, Dornan GJ, Jansson KS, LaPrade RF, Wijdicks CA. The hip fluid seal--Part I: the effect of an acetabular labral tear, repair, resection, -and reconstruction on hip fluid pressurization. *Knee Surg Sports Traumatol Arthrosc*, 2014; **22**(4):722–729.
5. Cashin M, Uhthoff H, O'Neill M, Beaulé PE. Embryology of the acetabular labral-chondral complex. *J Bone Joint Surg Br*, 2008; **90**(8):1019–1024.
6. Kılıçarslan K, Kılıçarslan A, Demirkale İ, Aytekin MN, Aksekili MA, Uğurlu M. Immunohistochemical analysis of mechanoreceptors in

transverse acetabular ligament and labrum: a prospective analysis of 35 cases. *Acta Orthop Traumatol Turc*, 2015; **49**(4):394–398.

7. Gerhardt M, Johnson K, Atkinson R, Snow B, Shaw C, Brown A, Vangsness CT Jr. Characterisation and classification of the neural anatomy in the human hip joint. *Hip Int*, 2012; **22**(1):75–81.

8. Kalhor M, Horowitz K, Beck M, Nazparvar B, Ganz R. Vascular supply to the acetabular labrum. *J Bone Joint Surg Am*, 2010; **92**(15):2570–2575.

9. Petersen W, Petersen F, Tillmann B. Structure and vascularization of the acetabular labrum with regard to the pathogenesis and healing of labral lesions. *Arch Orthop Trauma Surg*, 2003; **123**(6):283–288.

10. Smith MV, Panchal HB, Ruberte Thiele RA, Sekiya JK. Effect of acetabular labrum tears on hip stability and labral strain in a joint compression model. *Am J Sports Med*, 2011; **39** Suppl:103S–110S.

11. Lim S-J, Park Y-S. Plain radiography of the hip: a review of radiographic techniques and image features, *Hip Pelvis*, 2015; **27**(3):125–134.

12. Lin Y-T, Wang T-G. Ultrasonographic examination of the adult hip. *J Med Ultrasound*, 2012; **20**(4):201–209.

13. Aubry S, Bélanger D, Giguère C, Lavigne M. Magnetic resonance arthrography of the hip: technique and spectrum of findings in younger patients. *Insights Imaging*, 2010; **1**(2):72–82.

14. Plotz GM, Brossmann J, von Knoch M, Muhle C, Heller M, Hassenpflug J. Magnetic resonance arthrography of the acetabular labrum: value of radial reconstructions. *Arch Orthop Trauma Surg*, 2001; **121**(8):450–457.

15. Kubo T, Horii M, Harada Y, Noguchi Y, Yutani Y, Ohashi H, Hachiya Y, Miyaoka H, Naruse S, Hirasawa Y. Radial-sequence magnetic resonance imaging in evaluation of acetabular labrum. *J Orthop Sci*, 1999; **4**(5):328–332.

16. Czerny C, Hofmann S, Neuhold A, Tschauner C, Engel A, Recht MP, Kramer J. Lesions of the acetabular labrum: accuracy of MR imaging and MR arthrography in detection and staging. *Radiology*, 1996; **200**(1):225–230.

17. Blankenbaker DG, DeSmet AA, Keene JS, Fine JP. Classification and localization of acetabular labral tears. *Skeletal Radiol*, 2007; **36**(5):391–397.

18. Lage LA, Patel JV, Villar RN. The acetabular labral tear: an arthroscopic classification. *Arthroscopy*, 1996; **12**(3):269–272.

19. Magee T, Hinson G. Association of paralabral cysts with acetabular disorders. *AJR Am J Roentgenol*, 2000; **174**(5):1381–1384.

20. Bredella MA, Stoller DW. MR imaging of femoroacetabular impingement. *Magn Reson Imaging Clin N Am*, 2005; **13**(4):653–664.
21. Ganz R, Parvizi J, Beck M, Leunig M, Notzli H, Siebenrock KA. Femoroacetabular impingement: a cause for osteoarthritis of the hip. *Clin Orthop Relat Res*, 2003; (417):112–120.
22. Tanzer M, Noiseux N. Osseous abnormalities and early osteoarthritis: the role of hip impingement. *Clin Orthop Relat Res*, 2004; **429**:170–177.
23. Kassarjian A, Yoon LS, Belzile E, Connolly SA, Millis MB, Palmer WE. Triad of MR arthrographic findings in patients with cam type femoroacetabular impingement. *Radiology*, 2005; **236**(2):588–592.
24. Notzli HP, Wyss TF, Stoecklin CH, Schmid MR, Treiber K, Hodler J. The contour of the femoral head-neck junction as a predictor for the risk of anterior impingement. *J Bone Joint Surg Br*, 2002; **84**(4):556–560.
25. Beck M, Kalhor M, Leunig M, Ganz R. Hip morphology influences the pattern of damage to the acetabular cartilage: femoroacetabular impingement as a cause of early osteoarthritis of the hip. *J Bone Joint Surg Br*, 2005; **87**(7):1012–1018.
26. Siebenrock KA, Schoeniger R, Ganz R. Anterior femoroacetabular impingement due to acetabular retroversion. Treatment with periacetabular osteotomy. *J Bone Joint Surg Am*, 2003; **85-A**(2):278–286.
27. Nishii T, Tanaka H, Nakanishi K, Sugano N, Miki H, Yoshikawa H. Fat-suppressed 3D spoiled gradient-echo MRI and MDCT arthrography of articular cartilage in patients with hip dysplasia. *AJR Am J Roentgenol*, 2005; **185**(2):379–385.
28. Neumann G, Mendicuti AD, Zou KH, Minas T, Coblyn J, Winalski CS, Lang P. Prevalence of labral tears and cartilage loss in patients with mechanical symptoms of the hip: evaluation using MR arthrography. *Osteoarthritis Cartilage*, 2007; **15**(8):909–917.
29. Beaulé PE, Zaragoza E, Copelan N. Magnetic resonance imaging with gadolinium arthrography to assess acetabular cartilage delamination. A report of four cases. *J Bone Joint Surg Am*, 2004; **86-A**(10):2294–2298.
30. Cerezal L, Arnaiz J, Canga A, Piedra T, Altónaga JR, Munafo R, Pérez-Carro L. Emerging topics on the hip: ligamentum teres and hip microinstability. *Eur J Radiol*, 2012; **81**(12):3745–3754.

Chapter 13

Mesenchymal Stem Cell Treatment of Cartilage Lesions in the Hip

George Hourston, Stephen McDonnell and Wasim Khan

Division of Trauma & Orthopaedics, Addenbrooke's Hospital, University of Cambridge, Cambridge CB2 0QQ, United Kingdom
wasimkhan@doctors.org.uk

13.1 Introduction

There has been a great deal of excitement in the literature over the last few decades over the emergence of novel therapies and techniques that have the potential to cure cartilage disease. Our understanding of the cartilage has been hampered by inconsistency of terminology in the literature, and we have still not settled on a ubiquitous classification system for chondral lesions in the hip joint. This discussion is broached first. Once this classification has been achieved, we must then look to why our current treatment strategies and techniques, such as bone marrow stimulation, offer such inconsequential outcomes, and the limitations of these therapies are covered here. With interest in regenerative medicine and the applications of stem cells soaring, there has been an inevitable

---

The Hip Joint in Adults: Advances and Developments
Edited by K. Mohan Iyer
Copyright © 2018 Pan Stanford Publishing Pte. Ltd.
ISBN 978-981-4774-72-7 (Hardcover), 978-1-351-26244-6 (eBook)
www.panstanford.com

overlap of these fields with the treatment of musculoskeletal pathology. The history of mesenchymal stem cells (MSCs) and their utility in animal models and clinical trials will therefore be discussed at length. Finally, this chapter will touch upon some of the future prospects that biological therapies may present in our search of a cure for cartilage disease.

13.2 Cartilage Lesions

Chondral lesions of the hip and early osteoarthritis (OA) present challenges in terms of diagnosis and treatment. OA is a degenerative joint disease characterised by cartilage erosion that, particularly in ageing Western populations, represents a growing burden on healthcare services [1]. OA is insidious in its development and can be very debilitating for patients. One contributor to hip pain and OA in later life is damage to the cartilage. Chondral lesions are frequently associated with femoroacetabular impingement (FAI) [2, 3], labral tears, dislocation of the hip, osteonecrosis, slipped capital femoral epiphysis and dysplasia [4]. FAI is characterised by excessive abutment of the femoral head-neck junction against the acetabular rim, leading to intra-articular cartilage delamination and labral detachment [5, 6]. In one study, the frequency of chondral lesions in hip arthroscopy for FAI was over two-thirds [6]. Chondral lesions are therefore an increasingly recognised and important factor contributing towards the growing number of patients presenting with hip pain.

Classifying chondral lesions such as these has been a point of debate for over half a century. It is important that a classification system be used when formulating an algorithmic therapeutic strategy in order that patients receive the most effective treatment for their complaint. The most commonly used chondral lesion classification system was supplied by Outerbridge in 1961 [7]. Outerbridge described macroscopic changes of chondromalacia of the patella seen at meniscectomy and classified these changes according to four grades: grade 1 involves softening and swelling of the cartilage, grade 2 involves some fragmentation and fissuring of an area 'half an inch or less in diameter', grade 3 is equivalent to grade 2 only over a larger area and grade 4 involves cartilage eroded to the bone [7]. For decades, however, an adequate classification system for chondral lesions in the hip joint remained elusive. Another attempt came from Beck et al., who inspected acetabular and femoral cartilage for signs of degeneration in patients undergoing surgical treatment for anterior FAI, as shown in Table 13.1 [8]. Konan et al. developed a classification system for acetabular chondral lesions identified at arthroscopy in patients with FAI [9]. This system combines an anatomical division of the acetabulum into six zones by Ilizaliturri et al. [10] and an assessment of the chondrolabral junction and articular cartilage into grades 0 to 4, with 0 being normal cartilage; grade 1 classified as loss of fixation to the subchondral bone; the presence of a cleavage tear indicating grade 2; delamination of the articular cartilage classified as grade 3; grade 4 involving exposed bone in the acetabulum and finally grades 1, 3 and 4 being further grouped as A, B or C on the basis of the distance of the lesion from the acetabular rim to the cotyloid fossa [9].

Table 13.1 Classification of cartilage damage [8]

Stage	Description	Criteria
0	Normal	Macroscopically sound cartilage
1	Malacia	Roughening of surface, fibrillation
2	Pitting malacia	Roughening, partial thinning and full-thickness defects or deep fissuring to the bone
3	Debonding	Loss of fixation to the subchondral bone, macroscopically sound cartilage, carpet phenomenon
4	Cleavage	Loss of fixation to the subchondral bone, frayed edges, thinning of the cartilage
5	Defect	Full-thickness defect

13.3 Traditional Therapeutic Strategies and Limitations

Repair of articular hyaline cartilage lesions such as those described above is difficult because this tissue is avascular, aneural and alymphatic and exists in the unforgiving biomechanical environment of diarthrodial joints [11]. The definitive treatment for full-thickness cartilage defects and the ensuing end-stage OA is arthroplastic prosthetic replacement. However, this is unsatisfactory in younger patients since these prostheses have finite lifespans [12]. Historically, cartilage surgery has involved pain relief and bone marrow stimulation techniques, such as abrasion arthroplasty, drilling and microfracture [13]. Pain relief can be achieved by arthroscopic lavage to remove any excess fluids and loose bodies, including detached cartilage, and debridement of unviable roughened cartilage, and this has been shown to be a safe procedure and is clinically indicated in patients with mechanical symptoms and persistent pain [4, 14]. Thermal chondroplasty has been shown to have improved efficacy over mechanical debridement in laboratory and animal studies [15–17]. Therefore, this technique, which utilises radiofrequency energy, has been recommended for Konan grades 1 and 2 chondral lesions [18].

Bone marrow stimulation techniques include microfracture, a procedure developed by Steadman in Colorado [19], which involves drilling into the subchondral bone, leading to the formation of a blood clot in the defect site that mobilises bone marrow elements and promotes a degree of repair [20]. Microfracture has been recommended for delamination or full-thickness chondral lesions, represented by Konan's classification system as a grade 3 or 4 [18]. Fibrin adhesive has also been used for chondral delamination, as reported by Tzaveas and Villar, in a series of 19 patients treated successfully for pain and dysfunction [21]. Despite some temporary symptomatic improvement, microfracture has failed to yield durable and long-lasting effects because it typically promotes the development of a scar-type fibrocartilage tissue [20]. It has been proposed that this technique is more suited to younger patients since a thorough rehabilitation regime is required [22].

Related to microfracture, another therapeutic strategy that employs material from the affected joint is autologous chondrocyte implantation (ACI). ACI is a two-part procedure involving the isolation and expansion of chondrocytes from an intact non-weight-bearing region of the joint and then transplantation of these cells into the osteochondral defect in the weight-bearing part of the joint, which are then covered with a periosteal flap [23, 24]. ACI has been shown to produce more hyaline cartilage, containing type II collagen fibres, which are critical for the macromolecular structure of the extracellular matrix, which gives articular cartilage its unique biomechanical properties [24]. Alternatively, the implanted chondrocytes can be seeded onto a scaffold, a biodegradable type I/III collagen membrane, and this technique is commonly referred to as matrix-induced autologous chondrocyte implantation (MACI) [25, 26]. MACI is thought to offer improved surgical application and clinical outcomes when compared with ACI [26, 27]. These procedures involving chondrocyte implantation do offer an improvement over microfracture in reconstituting articular cartilage, but there are several problems. The chondrocytes used have a limited expansion capacity, and furthermore, ACI/MACI requires two surgeries: an initial arthroscopy to gather the chondrocytes and a cell implantation step, often an open surgery, exposing the patient to a risk of infection and prolonging his or her recovery [28]. It is also expensive owing to the laboratory costs of in vivo culturing of the cells as well as the extra theatre time [28]. These disadvantages have caused scarce adoption of ACI among hip surgeons, despite frequent use in the knee [28].

In the hip, debridement and microfracture have remained the treatment of choice, although this has seen improvement in the same way as ACI. Again, the use of a type I/III collagen matrix has enhanced the microfracture technique, resulting in a novel approach known as autologous matrix-induced chondrogenesis (AMIC). The matrix used in AMIC, Chondro-Gide (Geistlich Pharma AG, Wolhusen, Switzerland), helps to stabilise the clot and provides a platform for structure in the scar tissue formed [29]. In 2012, AMIC was performed fully arthroscopically in the hip by a group in Italy, in a single-step procedure, thereby avoiding several of the limitations and risks associated with chondrocyte implantation [28]. Long-term clinical data are not yet available. This novel technique has been repeated with some success in the ankle [30] and shoulder [31].

The elements of bone marrow liberated through microfracture and AMIC are growth factors, healing proteins and, importantly, a progenitor population of cells with properties of self-renewal and multilineage differentiation potential, which have come to be known as MSCs. This cell population does not suffer the same limited expansion capacity and inefficiency of redifferentiation that plagues the dedifferentiated chondrocytes used in ACI [32]. Harnessing the capabilities of MSCs has therefore been thought to be of great importance in cartilage repair and thus the treatment of OA among other joint conditions. The pursuit for effective biological therapies such as stem cells is explored in the following section.

13.4 Mesenchymal Stem Cells: A History

To give a thorough account of the history of the discovery of MSCs one must consider the earlier discovery of the haematopoietic stem cell (HSC). The existence of the HSC was first proposed by a Russian biologist called Maximov in 1909. Maximov postulated that the diverse range of cells formed in adult human bone marrow and the different blood lineages and stages of differentiation that they give rise to through the process of haematopoiesis all derive from the common precursor at the top of the cellular hierarchy, the HSC [33]. The existence of a single subpopulation of cells with a multilineage differentiation potential was confirmed in 1961, when Till and McCulloch studied HSCs in clonal in vivo repopulation assays [23, 24]. On the basis of these assays, the main defining principle of a stem cell came to be known, the ability to self-renew. This was illustrated by the fact that HSCs could fully reconstitute haematopoiesis in secondary as well as primary recipients in animal models [34, 35]. Subsequently, bone marrow transplantation was adopted for the treatment of lymphoproliferative disease, which involved chemotherapy and radiotherapy to ablate the tumour and transplantation of healthy HSCs from tissue-compatible donors [36, 37].

It was in this period, in the 1970s, that Dr Alexander Friedenstein noted during his studies in the Soviet Union that bone marrow seeded in glass flasks and maintained in basic culture media produced colonies of nonhaematopoietic cells that were transplantable [38]. Friedenstein found that colonies he transplanted were capable of self-renewal and differentiating into mature nonhaematopoietic lineages in recipient adult mammals, including bone and cartilage, and realised that he may have found another subset of true stem cells [39]. The significance of this work was not realised in the West until the 1990s, when Arnold Caplan proposed a concept similar to haematopoiesis but for mesenchymal tissues, and in fact he was the first to coin the term 'mesenchymal stem cell' [40]. He proposed that the MSC was the top of a cellular hierarchy with a differentiation potential limited to mesodermal tissues via more mature cells such as osteoprogenitors giving rise to bone or chondroprogenitors giving rise to cartilage [41, 42].

Clearly, the high proliferative potential of culture-expanded MSCs offered conceivable therapeutic strategies. A biotechnology company called Osiris Therapeutics was the first to explore this idea in 1999 [42, 43]. Their article 'Science' demonstrated novel in vitro functional assays for MSCs that still form the basis for their characterisation today. It also proposed a set of markers to characterise MSCs following culture [42, 43]. This group, among others, also showed that culture-expanded MSCs have an immunomodulatory effect in the bone marrow microenvironment [44], and this was further demonstrated in human cells the following year [45]. This may prove to be a major hurdle in the widespread clinical application of MSCs since the immune system has a well-documented antitumour role [46].

The plasticity of MSCs in vitro was so broad that interest in these cells, and the experimental therapies explored, swelled even into the fields of nonmesenchyme liver and neuronal tissue regeneration [47, 48]. With regard to the skeletal system, MSCs were first found in the synovial membrane in 2001 [49]. That year, a similar population of fibroblast-like cells with mesodermal lineage differentiation potential was identified in adipose tissue [50]. Indeed, MSCs were being isolated from many different sources of tissue with varying proliferative capacities and differentiation potentials. DiGirolamo et al. also described the senescence of MSCs after prolonged culture in vitro [42, 51]. Senescence in the samples this group worked with resulted in a reduced ability to differentiate into certain cell types such as adipocytes [51]. Whether these aged cells could still be truly called MSCs was up for debate. All of these findings culminated in a pressing need for standardisation in the field. This was met in 2006 by the publication of a position statement on the defining criteria for MSCs by the International Society for Cell Therapy (ISCT):

First, MSC must be plastic-adherent when maintained in standard culture conditions. Second, MSC must express CD105, CD73 and CD90, and lack expression of CD45, CD34, CD14 or CD11b, CD79alpha or CD19 and HLA-DR surface molecules. Third, MSC must differentiate to osteoblasts, adipocytes and chondroblasts in vitro [52].

The ISCT subcommittee decided to use the term 'multipotent mesenchymal stromal cells' rather than 'mesenchymal stem cells' and this difference in nomenclature alone highlights the variability demonstrated by the literature in the field at the time. Although this definition has broadly been used ever since, it was not perfect. One consideration not met by these criteria is how proliferative MSCs ought to be. Some researchers consider that true MSCs can only be determined once all clones display multipotency for the mesenchymal lineages after around 20–30 divisions (or 'population doublings') in culture [42, 53, 54]. The ISCT conceded that these criteria would require modification in time, and over a decade later, in 2017, still, a marker expression profile for MSCs, and clarification on their 'stemness' and heterogeneity, has not yet been universally stipulated.

MSCs indeed seem to be a heterogeneous population of cells, such that their differentiation propensity depends on their tissue of residence. Other musculoskeletal MSCs that are compatible with the aforementioned ISCT definition have been found in bone [55], synovial fluid [53], joint fat pad [56], tendon [57], periosteum [54] and juvenile cartilage [58], as highlighted by Jones et al. [42]. However, periosteal MSCs have been shown to be more osteogenic than synovial MSCs [59], and synovial MSCs have been shown to be more chondrogenic than most other MSCs [60]. This poses a clear problem in the search for a unifying definition of the MSC, and more significantly, in selecting the appropriate MSC for clinical regeneration of specific musculoskeletal tissues [42]. Elucidating the character and mechanisms of these MSCs will be essential before their clinical potential can be realised.

13.5 Mesenchymal Stem Cells: Clinical Potential

Cartilage repair using MSCs has unwittingly been at the centre of chondral defect treatment since the emergence of bone marrow stimulation techniques such as microfracture in 2001 [19]. Despite this, it has been a long-held view that cartilage itself lacks any reparative capacity. The discovery of MSCs in the superficial layer of cartilage by Dowthwaite et al. disputes this adage and explains to some extent the occasional occurrence of spontaneous cartilage repair [58]. Indeed, cartilage regeneration has been noted following joint realignment osteotomy procedures [61] and in joint 'unloading' or distraction procedures [62, 63]. The theory of joint unloading is based on the idea that increased mechanical force, 'loading,' through joints is a major factor in the progression of OA [62]. Perhaps this unloading process mobilises or induces proliferation of MSCs [64]. As alluded to earlier, chondrogenesis was found to be more effective in synovium-derived MSCs than other tissue types [60]. There are also MSCs in the synovial fluid, and these are thought to originate from synovium [65]. Synovial fluid MSCs have been identified both in normal joints and in a greater number in the early stages of OA [65]. Synovial fluid MSCs have also been shown to increase in number following ligamentous injury [66]. It is therefore possible that synovial fluid MSCs are responsible for the physiological repair of cartilage damaged through 'wear and tear' or trauma, via the synovial lining to the denuded areas of the articular surface.

Therapeutically exploiting these synovial fluid MSCs has therefore been considered. In situ cartilage regeneration may be possible by augmenting the endogenous mechanisms of synovial fluid MSCs. This may involve enhancing their trafficking to chondral lesions. Yet undetermined substances in the synovial fluid may be responsible for the migration of MSCs from the synovium [67] or bone marrow [68]. A more detailed chemical profiling of synovial fluid will be required to determine whether this is true and before this can be clinically manipulated [64]. It seems clear, however, that upregulating this migration event could hold therapeutic value. Chemotactic homing of endogenous MSCs has also been explored and used to induce chondrogenesis with the appropriate growth factors in vitro [69]. Bioactive polymer scaffolds enhanced with control-released chemokines, such as stromal-derived factor-1 (SDF-1), have helped to specify this homing of MSCs to the chondral lesion [69, 70]. More in-depth investigation into the nature of the MSC role in the synovium and the physiological response to 'wear and tear' or injury will be needed to realise the full potential of these mechanisms in clinical practice [64].

13.6 Animal Models of Chondral Lesions

Novel therapeutic strategies have begun to be developed utilising culture-expanded MSCs in animal models of bone defects [71] and in tendon repair both in animal models [72, 73] as well as in veterinary applications, such as race horses [74]. It appears as though cartilage and osteochondral lesions offer more subtle challenges in animal models for tissue engineering. Osteochondral repair, for example, must appreciate both the bone and cartilage tissues and also the bone-cartilage interface [75]. This has proven difficult to model in a way that mimics clinical situations. A recent study from the Mayo Clinic demonstrated that cartilage regrowth can be difficult to achieve in an animal model of an osteochondral defect [76]. This study found that a durable bilayer implant composed of trabecular metal with autologous periosteum on top was excellent in reconstituting osseous defects, but despite the stem cell population of the periosteum and the fact that it acts as a biological scaffold, neocartilage generation was not satisfactory [76]. Pure cartilage tissue engineering in animal models has struggled in a similar way. These models must not have defects that penetrate into the subchondral bone, they must lack angiogenesis and they must provide mechanical stimulation by being in weight-bearing regions of joints in order to be relevant to clinical practice. Therefore, murine and lapine models are useful pilot and proof-of-concept studies, but their small size limits their translational value [77]. Canine, caprine and ovine models are better, but the joint size and cartilage thickness remain significantly smaller than human equivalents [77]. The average human medial femoral condyle cartilage thickness is 2.35 mm [78]. The average cartilage thickness in dogs used in such models of cartilage repair is 0.95 mm, and the commonest defect diameter is 4 mm, and these are significantly less than in humans [78]. Goats and sheep are reasonable models owing to the anatomy of their joints, but commonly these models have a significant degree of subchondral bone involvement, and since 95% of human cartilage defects do not involve subchondral bone [79], their utility is also limited [78]. Porcine models represent a further improvement, with an average medial femoral condyle cartilage thickness of 1.5 mm, but are not commonly used due to handling difficulties and logistical requirements [77, 78].

Equine models remain the most attractive owing to the size of horses and clinical cartilage problems similar to those of humans [77, 78]. However, they are not perfect since they are so large that they are also expensive to house and they are not typically bred for biomedical research. Despite this, and largely due to the race industry, there has been much investigation into cartilage repair in horses and horse MSCs have indeed been shown to have a good chondrogenic potential [80]. ACI has been investigated in equine models, too, even with genetic modification of chondrocytes infected with adenovirus vector encoding bone morphogenetic protein-7 to enhance the hyaline-like repair tissue [77, 81, 82]. More crucially, chondrogenesis has been enhanced in the early stages of articular defect repair in young, mature horses when injection of a self-polymerising autogenous fibrin vehicle containing MSCs was compared with autogenous fibrin alone in control joints [83]. However, long-term assessment revealed no significant benefit of the MSC treatment [83]. Evidently, much research has been done into cartilage repair in horses, and even into cartilage tissue regeneration. However, the major benefit of using horses is the availability of a 'ready-made' model consistent with OA, particularly among retired athletic horses [84], which cannot be easily achieved in any other animal model. For instance, one study investigating miniature pigs showed that partial-thickness cartilage defects over a large area can be resurfaced efficiently with hyaline-like type II collagen containing cartilage formed by transgene-activated periosteal cells [85]. However, superficial zones of cartilage tended to dedifferentiate, and the long-term stability of good articular cartilage seemed to depend on physicobiochemical factors that the porcine model could not effectively recapitulate [85].

There have been several studies using animal models to investigate the efficacy of exogenous MSC transplantation in cartilage repair [86–89]. These have shown variable degrees of success in promoting chondrogenesis in arthritic and prearthritic joints. However, for the reasons stated above, these studies struggle to replicate the clinical situation. Furthermore, due to the ease of accessing the joint and of performing 'relook' arthroscopies, these studies almost ubiquitously focus on the knee joint. More animal model work is needed in the field of hip cartilage regeneration in chondral lesions, including but not limited to OA, before we have the evidence base to progress to large clinical trials.

13.7 Clinical Trials

Several clinical studies investigating tissue regeneration approaches to chondral repair have been carried out, although these are largely in small patient groups and restricted to the knee [90–93]. It is therefore difficult to draw conclusions from such studies about the hip and how effective these therapeutic strategies will be for the population as a whole. In orthopaedics, the widespread application of MSCs in humans is hampered by our sheer lack of knowledge as to the physiology of these cells and also by concerns such as tumourigenicity [94]. Cancer formation is a possible outcome of stem cell transplantation, particularly with the aforementioned immunomodulatory effects of MSCs and the apparent knock-on effect on coagulation [95, 96]. The safety and efficacy of MSCs must therefore be tested in clinical trials before MSC-based therapies become widely available. There are many open clinical trials listed in clinicaltrials.gov addressing OA with MSC-based therapies. Those investigating the hip joint are shown in Table 13.2 [97].

13.8 Future Prospects

The source of cells used in biological therapy needs careful consideration if regenerative medicine is to become a viable treatment for much of the population. A theoretically unlimited supply of specialised cells for cartilage repair could come from human embryonic stem cells (hESCs) and induced pluripotent stem cells (iPSCs) [98]. Ethical and political controversies present significant challenges both to research and therapy using these cells [99]. iPSCs have been explored with great enthusiasm since they were developed in 2006 and could be preferable to hESCs since they are so easily retrieved from patients.

Table 13.2 Ongoing clinical trials using MSCs for the treatment of OA of the hip [97]

Trial	Sponsor	Type/Phase	Intervention/Treatment
Safety & Effectiveness of Autologous Regenerative Cell Therapy on Pain & Inflammation of Osteoarthritis of the Hip; NCT02844764	VivaTech International, Inc.	Interventional, Phase 2	StroMed + platelet-rich plasma (PRP)
Overview: This is a prospective clinical study of 50 patients in the United States to determine safety and the effects on pain and inflammation with the use of autologous cell therapy in hip osteoarthritis due to degeneration or chronic injury. Patients will be treated with autologous adipose cells and PRP that will be directly injected in the hip joint. The outcome measures will include laboratory inflammation markers, patient-reported outcome measures – including the Hip Dysfunction and Osteoarthritis Outcome Score (HOOS) questionnaire and SF-36 forms (a quality-of-life measure) – assessment of safety and efficacy of treatment, any reduction in patient medication, any delays in pending hip replacement therapy and MRI scans. The study will be followed up by a larger sample of 4000 patients.			
Outcomes Data of Adipose Stem Cells to Treat Osteoarthritis; NCT02241408	StemGenex	Observational	Intra-articular injection of autologous stromal vascular fraction (SVF)
Overview: This is a prospective cohort clinical study with an estimated enrolment of 50 patients in the United States to determine the effect of autologous adipose-derived stromal vascular fraction (AD-SVF) infusion on pain and functionality in joints of patients with osteoarthritis. Outcomes will include patient reported outcome measures, including the HOOS LK 2.0 questionnaire over the course of 12 months, as well as change from baseline in function in daily living and in sport and recreation.			
Autologous Adipose-Derived Stromal Cells Delivered Intra-articularly in Patients with Osteoarthritis; NCT01739504	Ageless Regenerative Institute	Interventional, Phase 1/Phase 2	Procedure: Liposuction with local anaesthesia; biological: intra-articular infusion of AD-SVF
Overview: This is a prospective clinical study with an estimated enrolment of 100 patients in the United States to determine the safety and efficacy of AD-SVF infusion into joints with osteoarthritis. The patient's AD-SVF will be collected from his or her own body fat. Liposuction will be performed under a local anaesthetic to collect the adipose tissue specimen. The adipose tissue will then be transferred to the laboratory for separation of the AD-SVF. Additionally, each patient's peripheral blood will be collected for isolation of PRP. This PRP will then be combined with the AD-SVF for intra-articular administration of the affected joint. The outcome measures will include change from baseline visual analogue scale (VAS), change from baseline quality-of-life score, reduction in analgesia and number of adverse events reported and secondarily, changes in imaging (X-ray, sonogram or MRI) of the affected joint compared to the baseline.			

(Continued)

Table 13.2 (Continued)

Trial	Sponsor	Type/Phase	Intervention/Treatment
Injections of FloGraft Therapy, Autologous Stem Cells, or Platelet Rich Plasma for the Treatment of Degenerative Joint Pain; NCT01978639	Arizona Pain Specialists	Prospective cohort; recruiting	Intra-articular injection of autologous bone marrow MSCs
Overview: This is an open-label, nonrandomised prospective cohort study with an estimated enrolment of 300 patients in the United States assessing the efficacy of three potentially regenerative treatments for degenerative conditions of the joints of the lower back (facet and sacroiliac), upper extremities (e.g., shoulder) and lower extremities (e.g. hip and knee). These three treatments are a single injection of Applied Biologics' FloGraftTM (a cryopreserved, injectable amniotic fluid-derived allograft), a single injection of autologous bone marrow–derived stem cells or a single injection of PRP. Patients will be assessed prior to treatment for their level of pain, quality of life (using SF-36 forms) and analgesia usage. Patients will be followed up at 4 weeks, 8 weeks, 12 weeks and 24 weeks after treatment. At each of these follow-up visits, pain, quality of life and analgesia usage will be assessed. The investigators hypothesise that all three treatment groups will experience reduced pain, improved quality of life and reduced pain medication usage at follow-up.			
Safety and Clinical Outcomes Study: SVF Deployment for Orthopedic, Neurologic, Urologic, and Cardio-pulmonary Conditions; NCT01953523	Cell Surgical Network Inc.	Phase 1	Intra-articular injection of autologous SVF
Overview: This is a prospective clinical study with an estimated enrolment of 3000 patients in the United States to determine the safety and efficacy of AD-SVF infusion into several organ systems, including joints with osteoarthritis. The SVF will be obtained by lipoharvesting, procurement and lipotransfer as a same-day operative procedure. The primary outcome measures will include the number of participants with adverse events related to either SVF deployment or the lipoharvesting procedure. Patient-reported outcome measures will include several questionnaires for the different organ systems involved. Particularly the HOOS will be taken every three months for the first three years.			

iPSCs were developed by Yamanaka, who found that by introducing four factors, Oct3/4, Sox2, c-Myc and Klf4, to adult mouse fibroblasts under embryonic stem cell culture conditions, these adult cells can develop pluripotency [100]. The same group repeated this the following year using human fibroblasts to demonstrate the possibility of producing human iPSCs (hiPSCs) [101]. hiPSCs can be amplified into an inexhaustible homogeneous population of almost any cell, having been derived from only a small skin biopsy. Their therapeutic potential was realised when in 2014, a Japanese woman was the first ever recipient of her own iPSCs, which had been differentiated into retinal tissue [102]. These cells could be a boundless source of highly conserved orthopaedic cell-based implants, such as articular cartilage [98]. In addition, the differentiated cells produced from iPSCs tend to exhibit younger traits and so these implants would be fast growing and long lasting [98]. iPSCs can be induced to differentiate into mesenchymal lineages using cytokines and altering gene expression [98, 103–107]. Generating MSCs from iPSCs has been well documented, but the protocols that exist have limitations such as the issues of utilising embryoid bodies, which can result in unpredictable differentiation into undesired cell lines [108], or of coculture with primary cells, which can risk a reduction in the purity of the cell population. Coculture of hESCs with human articular chondrocytes has been shown to improve chondrogenic differentiation [109]. However, this technique is problematic because the therapeutic cells need to be separated from the feeder cells. Impurity is a major consideration in iPSC-based therapy since contamination of an otherwise uniformly differentiated cell population required for clinical use by undifferentiated iPSCs poses a significant threat of teratocarcinoma formation [110]. Despite these risks, it seems highly profitable in therapeutic terms to pursue hiPSCs as a source of tissue for cartilage repair since retrieving the cells is straightforward and they have a potential to make robust biological implants with a long shelf life.

13.9 Concluding Remarks

Chondral lesions of the hip have not been investigated sufficiently to provide us with a robust pathway of treatment. Repair in a tissue devoid of blood supply is challenging, and we still have far to go before we are capable of reproducing articular cartilage that is a biomechanical match for our own healthy tissue. Traditional therapies have given patients an improved quality of life but have failed in becoming a curative option. Chondrocyte-based therapies have further improved outcomes, and this has facilitated minimally invasive options but at a seemingly limited restoration capacity. Surgical skills in hip arthroscopy are also developing, and this is sure to improve cartilage restoration techniques regardless of the cells being used. Perhaps the future of cartilage disease lies with regenerative medicine. The rate of MSC characterisation and the rate at which the biotechnology industry is developing scaffolds and other biomaterial delivery methods make the regenerative field an exciting one. This is only intensified by the potential of pluripotent stem cells as an infinite source of therapeutic MSCs. However, the ethical and political as well as scientific barriers to the clinical application of these cells present a major challenge. Surely, with adequate collaboration between stem cell scientists and orthopaedic surgeons and with support of the industry, MSC-based regenerative therapies could become a reality in our lifetime.

References

1. Zhang Y, Jordan JM. Epidemiology of osteoarthritis. *Clin Geriatr Med*, 2010; **26**(3):355–369.
2. Ganz R, Parvizi J, Beck M, Leunig M, Nötzli H, Siebenrock KA. Femoroacetabular impingement: a cause for osteoarthritis of the hip. *Clin Orthop*, 2003; (417):112–120.
3. Zhang C, Li L, Forster BB, Kopec JA, Ratzlaff C, Halai L, et al. Femoroacetabular impingement and osteoarthritis of the hip. *Can Fam Physician*, 2015; **61**(12):1055–1060.
4. Khanduja V, Villar RN. Arthroscopic surgery of the hip. *Bone Joint J*, 2006; **88-B**(12):1557–1566.
5. Beck M, Kalhor M, Leunig M, Ganz R. Hip morphology influences the pattern of damage to the acetabular cartilage. *Bone Joint J*, 2005; **87-B**(7):1012–1018.
6. Nepple JJ, Carlisle JC, Nunley RM, Clohisy JC. Clinical and radiographic predictors of intra-articular hip disease in arthroscopy. *Am J Sports Med*, 2011; **39**(2):296–303.

7. Outerbridge RE. The etiology of chondromalacia patellae. *Bone Joint J*, 1961; **43-B**(4):752–757.
8. Beck M, Leunig M, Parvizi J, Boutier V, Wyss D, Ganz R. Anterior femoroacetabular impingement: part II. Midterm results of surgical treatment. *Clin Orthop Relat Res*, 2004; (418):67–73.
9. Konan S, Rayan F, Meermans G, Witt J, Haddad FS. Validation of the classification system for acetabular chondral lesions identified at arthroscopy in patients with femoroacetabular impingement. *J Bone Joint Surg Br*, 2011; **93-B**(3):332–336.
10. Ilizaliturri Jr VM, Byrd JWT, Sampson TG, Guanche CA, Philippon MJ, Kelly BT, et al. A geographic zone method to describe intra-articular pathology in hip arthroscopy: cadaveric study and preliminary report. *Arthrosc J Arthrosc Relat Surg*, 2008; **24**(5):534–539.
11. Sophia Fox AJ, Bedi A, Rodeo SA. The basic science of articular cartilage. *Sports Health*, 2009; **1**(6):461–468.
12. Liu X-W, Zi Y, Xiang L-B, Wang Y. Total hip arthroplasty: areview of advances, advantages and limitations. *Int J Clin Exp Med*, 2015; **8**(1):27–36.
13. Gilbert JE. Current treatment options for the restoration of articular cartilage. *Am J Knee Surg*, 1998; **11**(1):42–46.
14. Diulus CA, Krebs VE, Hanna G, Barsoum WK. Hip arthroscopy: technique and indications. *J Arthroplasty*, 2006; **21**(4 Suppl):68–73.
15. Kaplan LD, Chu CR, Bradley JP, Fu FH, Studer RK. Recovery of chondrocyte metabolic activity after thermal exposure. *Am J Sports Med*, 2003; **31**(3):392–398.
16. Lotto ML, Wright EJ, Appleby D, Zelicof SB, Lemos MJ, Lubowitz JH. Ex vivo comparison of mechanical versus thermal chondroplasty: assessment of tissue effect at the surgical endpoint. *Arthrosc J Arthrosc Relat Surg Off Publ Arthrosc Assoc N Am Int Arthrosc Assoc*, 2008; **24**(4):410–415.
17. Edwards RB, Lu Y, Cole BJ, Muir P, Markel MD. Comparison of radiofrequency treatment and mechanical debridement of fibrillated cartilage in an equine model. *Vet Comp Orthop Traumatol*, 2008; **21**(1):41–48.
18. Mardones R, Larrain C. Cartilage restoration technique of the hip. *J Hip Preserv Surg*, 2016; **3**(1):30–36.
19. Steadman JR, Rodkey WG, Rodrigo JJ. Microfracture: surgical technique and rehabilitation to treat chondral defects. *Clin Orthop*, 2001; (391 Suppl):S362–S369.

20. Minas T, Nehrer S. Current concepts in the treatment of articular cartilage defects. *Orthopedics*, 1997; **20**(6):525–538.
21. Tzaveas AP, Villar RN. Arthroscopic repair of acetabular chondral delamination with fibrin adhesive. *Hip Int*, 2010; **20**(1):115–119.
22. Hurst JM, Steadman JR, O'Brien L, Rodkey WG, Briggs KK. Rehabilitation following microfracture for chondral injury in the knee. *Clin Sports Med*, 2010; **29**(2):257–265.
23. Grande DA, Pitman MI, Peterson L, Menche D, Klein M. The repair of experimentally produced defects in rabbit articular cartilage by autologous chondrocyte transplantation. *J Orthop Res*, 1989; **7**(2):208–218.
24. Brittberg M, Lindahl A, Nilsson A, Ohlsson C, Isaksson O, Peterson L. Treatment of deep cartilage defects in the knee with autologous chondrocyte transplantation. *N Engl J Med*, 1994; **331**(14):889–895.
25. Jacobi M, Villa V, Magnussen RA, Neyret P. MACI: a new era? *Sports Med Arthrosc Rehabil Ther Technol*, 2011; **3**:10.
26. Dunkin BS, Lattermann C. New and emerging techniques in cartilage repair: MACI. *Oper Tech Sports Med*, 2013; **21**(2):100–107.
27. Bartlett W, Gooding CR, Carrington RWJ, Skinner JA, Briggs TWR, Bentley G. Autologous chondrocyte implantation at the knee using a bilayer collagen membrane with bone graft. *Bone Joint J*, 2005; **87-B**(3):330–332.
28. Fontana A. A novel technique for treating cartilage defects in the hip: a fully arthroscopic approach to using autologous matrix-induced chondrogenesis. *Arthrosc Tech*, 2012; **1**(1):e63–e68.
29. Benthien JP, Behrens P. The treatment of chondral and osteochondral defects of the knee with autologous matrix-induced chondrogenesis (AMIC): method description and recent developments. *Knee Surg Sports Traumatol Arthrosc*, 2011; **19**(8):1316–1319.
30. Usuelli FG, de Girolamo L, Grassi M, D'Ambrosi R, Montrasio UA, Boga M. All-arthroscopic autologous matrix-induced chondrogenesis for the treatment of osteochondral lesions of the talus. *Arthrosc Tech*, 2015; **4**(3):e255–e259.
31. Cuéllar A, Ruiz-Ibán MÁ, Cuéllar R. The use of all-arthroscopic autologous matrix-induced chondrogenesis for the management of humeral and glenoid chondral defects in the shoulder. *Arthrosc Tech*, 2016; **5**(2):e223– e227.

32. Wakitani S, Goto T, Pineda SJ, Young RG, Mansour JM, Caplan AI, et al. Mesenchymal cell-based repair of large, full-thickness defects of articular cartilage. *J Bone Joint Surg Am*, 1994; **76**(4):579–592.

33. Doulatov S, Notta F, Laurenti E, Dick JE. Hematopoiesis: a human perspective. *Cell Stem Cell*, 2012; **10**(2):120–136.

34. Till JE, McCulloch EA. A direct measurement of the radiation sensitivity of normal mouse bone marrow cells. *Radiat Res*, 1961; **14**(2):213–222.

35. Becker AJ, McCulloch EA, Till JE. Cytological demonstration of the clonal nature of spleen colonies derived from transplanted mouse marrow cells. *Nature*, 1963; **197**:452–454.

36. Thomas ED, Storb R, Clift RA, Fefer A, Johnson FL, Neiman PE, et al. Bone-marrow transplantation. *N Engl J Med*, 1975; **292**(16):832–843.

37. Appelbaum FR. Hematopoietic-cell transplantation at 50. *N Engl J Med*, 2007; **357**(15):1472–1475.

38. Friedenstein AJ, Gorskaja JF, Kulagina NN. Fibroblast precursors in normal and irradiated mouse hematopoietic organs. *Exp Hematol*, 1976; **4**(5):267–274.

39. Friedenstein AJ. Precursor cells of mechanocytes. *Int Rev Cytol*, 1976; **47**:327–359.

40. Caplan AI. Mesenchymal stem cells. *J Orthop Res*, 1991; **9**(5):641–650.

41. Caplan AI. The mesengenic process. *Clin Plast Surg*, 1994; **21**(3):429–435.

42. Jones EA, Yang X, Giannoudis P, McGonagle D. Chapter 2: Mesenchymal stem cells: discovery in bone marrow and beyond. In *Mesenchymal Stem Cells and Skeletal Regeneration* [Internet], (Boston: Academic Press), 2013; [cited 2017 Feb 23], pp. 7–13. Available from: http://www.sciencedirect.com/science/article/pii/B9780124079151000027

43. Pittenger MF, Mackay AM, Beck SC, Jaiswal RK, Douglas R, Mosca JD, et al. Multilineage potential of adult human mesenchymal stem cells. *Science*, 1999; **284**(5411):143–147.

44. Bartholomew A, Sturgeon C, Siatskas M, Ferrer K, McIntosh K, Patil S, et al. Mesenchymal stem cells suppress lymphocyte proliferation in vitro and prolong skin graft survival in vivo. *Exp Hematol*, 2002; **30**(1):42–48.

45. Le Blanc K, Tammik L, Sundberg B, Haynesworth SE, Ringdén O. Mesenchymal stem cells inhibit and stimulate mixed lymphocyte cultures and mitogenic responses independently of the major histocompatibility complex. *Scand J Immunol*, 2003; **57**(1):11–20.

46. Disis ML. Immune regulation of cancer. *J Clin Oncol*, 2010; **28**(29):4531–4538.
47. Min AD, Theise ND. Prospects for cell-based therapies for liver disease. *Panminerva Med*, 2004; **46**(1):43–48.
48. Barry FP. Biology and clinical applications of mesenchymal stem cells. *Birth Defects Res Part C Embryo Today Rev*, 2003; **69**(3):250–256.
49. De Bari C, Dell'Accio F, Tylzanowski P, Luyten FP. Multipotent mesenchymal stem cells from adult human synovial membrane. *Arthritis Rheum*, 2001; **44**(8):1928–1942.
50. Zuk PA, Zhu M, Mizuno H, Huang J, Futrell JW, Katz AJ, et al. Multilineage cells from human adipose tissue: implications for cell-based therapies. *Tissue Eng*, 2001; **7**(2):211–228.
51. DiGirolamo CM, Stokes D, Colter D, Phinney DG, Class R, Prockop DJ. Propagation and senescence of human marrow stromal cells in culture: a simple colony-forming assay identifies samples with the greatest potential to propagate and differentiate. *Br J Haematol*, 1999; **107**(2):275–281.
52. Dominici M, Le Blanc K, Mueller I, Slaper-Cortenbach I, Marini F, Krause D, et al. Minimal criteria for defining multipotent mesenchymal stromal cells. The International Society for Cellular Therapy position statement. *Cytotherapy*, 2006; **8**(4):315–317.
53. Jones EA, English A, Henshaw K, Kinsey SE, Markham AF, Emery P, et al. Enumeration and phenotypic characterization of synovial fluid multipotential mesenchymal progenitor cells in inflammatory and degenerative arthritis. *Arthritis Rheum*, 2004; **50**(3):817–827.
54. De Bari C, Dell'Accio F, Vanlauwe J, Eyckmans J, Khan IM, Archer CW, et al. Mesenchymal multipotency of adult human periosteal cells demonstrated by single-cell lineage analysis. *Arthritis Rheum*, 2006; **54**(4):1209–1221.
55. Tuli R, Tuli S, Nandi S, Wang ML, Alexander PG, Haleem-Smith H, et al. Characterization of multipotential mesenchymal progenitor cells derived from human trabecular bone. *Stem Cells*, 2003; **21**(6):681–693.
56. Wickham MQ, Erickson GR, Gimble JM, Vail TP, Guilak F. Multipotent stromal cells derived from the infrapatellar fat pad of the knee. *Clin Orthop*, 2003; (412):196–212.
57. Salingcarnboriboon R, Yoshitake H, Tsuji K, Obinata M, Amagasa T, Nifuji A, et al. Establishment of tendon-derived cell lines exhibiting

pluripotent mesenchymal stem cell-like property. *Exp Cell Res*, 2003; **287**(2):289–300.

58. Dowthwaite GP, Bishop JC, Redman SN, Khan IM, Rooney P, Evans DJR, et al. The surface of articular cartilage contains a progenitor cell population. *J Cell Sci*, 2004; **117**(Pt 6):889–897.

59. De Bari C, Dell'Accio F, Karystinou A, Guillot PV, Fisk NM, Jones EA, et al. A biomarker-based mathematical model to predict bone-forming potency of human synovial and periosteal mesenchymal stem cells. *Arthritis Rheum*, 2008; **58**(1):240–250.

60. Sakaguchi Y, Sekiya I, Yagishita K, Muneta T. Comparison of human stem cells derived from various mesenchymal tissues: superiority of synovium as a cell source. *Arthritis Rheum*, 2005; **52**(8):2521–2529.

61. Koshino T, Wada S, Ara Y, Saito T. Regeneration of degenerated articular cartilage after high tibial valgus osteotomy for medial compartmental osteoarthritis of the knee. *Knee*, 2003; **10**(3):229–236.

62. Lafeber FPJG, Intema F, Van Roermund PM, Marijnissen ACA. Unloading joints to treat osteoarthritis, including joint distraction. *Curr Opin Rheumatol*, 2006; **18**(5):519–525.

63. Intema F, Van Roermund PM, Marijnissen ACA, Cotofana S, Eckstein F, Castelein RM, et al. Tissue structure modification in knee osteoarthritis by use of joint distraction: an open 1-year pilot study. *Ann Rheum Dis*, 2011; **70**(8):1441–1446.

64. Jones EA, Yang X, Giannoudis P, McGonagle D. Chapter 4: Challenges for cartilage regeneration. In *Mesenchymal Stem Cells and Skeletal Regeneration* [Internet], (Boston: Academic Press), 2013; [cited 2017 Feb 23], pp. 21–26. Available from: http://www.sciencedirect.com/science/article/pii/B9780124079151000040

65. Jones EA, Crawford A, English A, Henshaw K, Mundy J, Corscadden D, et al. Synovial fluid mesenchymal stem cells in health and early osteoarthritis: detection and functional evaluation at the single-cell level. *Arthritis Rheum*, 2008; **58**(6):1731–1740.

66. Morito T, Muneta T, Hara K, Ju Y-J, Mochizuki T, Makino H, et al. Synovial fluid-derived mesenchymal stem cells increase after intra-articular ligament injury in humans. *Rheumatol Oxf Engl*, 2008; **47**(8):1137–1143.

67. Zhang S, Muneta T, Morito T, Mochizuki T, Sekiya I. Autologous synovial fluid enhances migration of mesenchymal stem cells from synovium of osteoarthritis patients in tissue culture system. *J Orthop Res*, 2008; **26**(10):1413–1418.

68. Endres M, Neumann K, Häupl T, Erggelet C, Ringe J, Sittinger M, et al. Synovial fluid recruits human mesenchymal progenitors from subchondral spongious bone marrow. *J Orthop Res*, 2007; **25**(10):1299–1307.

69. Mendelson A, Frank E, Allred C, Jones E, Chen M, Zhao W, et al. Chondrogenesis by chemotactic homing of synovium, bone marrow, and adipose stem cells in vitro. *FASEB J*, 2011; **25**(10):3496–3504.

70. Schantz J-T, Chim H, Whiteman M. Cell guidance in tissue engineering: SDF-1 mediates site-directed homing of mesenchymal stem cells within three-dimensional polycaprolactone scaffolds. *Tissue Eng*, 2007; **13**(11):2615–2624.

71. Bruder SP, Kraus KH, Goldberg VM, Kadiyala S. The effect of implants loaded with autologous mesenchymal stem cells on the healing of canine segmental bone defects. *J Bone Joint Surg Am*, 1998; **80**(7):985–996.

72. Young RG, Butler DL, Weber W, Caplan AI, Gordon SL, Fink DJ. Use of mesenchymal stem cells in a collagen matrix for achilles tendon repair. *J Orthop Res*, 1998; **16**(4):406–413.

73. Awad HA, Butler DL, Boivin GP, Smith FNL, Malaviya P, Huibregtse B, et al. Autologous mesenchymal stem cell-mediated repair of tendon. *Tissue Eng*, 1999; **5**(3):267–277.

74. Alves AGL, Stewart AA, Dudhia J, Kasashima Y, Goodship AE, Smith RKW. Cell-based therapies for tendon and ligament injuries. *Vet Clin North Am Equine Pract*, 2011; **27**(2):315–333.

75. Nukavarapu SP, Dorcemus DL. Osteochondral tissue engineering: current strategies and challenges. *Biotechnol Adv*, 2013; **31**(5):706–721.

76. Mrosek EH, Chung H-W, Fitzsimmons JS, O'Driscoll SW, Reinholz GG, Schagemann JC. Porous tantalum biocomposites for osteochondral defect repair. *Bone Joint Res*, 2016; **5**(9):403–411.

77. Chu CR, Szczodry M, Bruno S. Animal models for cartilage regeneration and repair. *Tissue Eng Part B Rev*, 2010; **16**(1):105–115.

78. Ahern BJ, Parvizi J, Boston R, Schaer TP. Preclinical animal models in single site cartilage defect testing: a systematic review. *Osteoarthritis Cartilage*, 2009; **17**(6):705–713.

79. Hjelle K, Solheim E, Strand T, Muri R, Brittberg M. Articular cartilage defects in 1,000 knee arthroscopies. *Arthrosc J Arthrosc Relat Surg*, 2002; **18**(7):730–734.

80. Vidal MA, Robinson SO, Lopez MJ, Paulsen DB, Borkhsenious O, Johnson JR, et al. Comparison of chondrogenic potential in equine mesenchymal stromal cells derived from adipose tissue and bone marrow. *Vet Surg*, 2008; **37**(8):713–724.

81. Litzke LE, Wagner E, Baumgaertner W, Hetzel U, Josimović-Alasević O, Libera J. Repair of extensive articular cartilage defects in horses by autologous chondrocyte transplantation. *Ann Biomed Eng*, 2004; **32**(1):57–69.

82. Hidaka C, Goodrich LR, Chen C-T, Warren RF, Crystal RG, Nixon AJ. Acceleration of cartilage repair by genetically modified chondrocytes over expressing bone morphogenetic protein-7. *J Orthop Res*, 2003; **21**(4):573–583.

83. Wilke MM, Nydam DV, Nixon AJ. Enhanced early chondrogenesis in articular defects following arthroscopic mesenchymal stem cell implantation in an equine model. *J Orthop Res*, 2007; **25**(7):913–925.

84. McIlwraith CW, Frisbie DD, Kawcak CE. The horse as a model of naturally occurring osteoarthritis. *Bone Joint Res*, 2012; **1**(11):297–309.

85. Gelse K, Mühle C, Franke O, Park J, Jehle M, Durst K, et al. Cell-based resurfacing of large cartilage defects: long-term evaluation of grafts from autologous transgene-activated periosteal cells in a porcine model of osteoarthritis. *Arthritis Rheum*, 2008; **58**(2):475–488.

86. Murphy JM, Fink DJ, Hunziker EB, Barry FP. Stem cell therapy in a caprine model of osteoarthritis. *Arthritis Rheum*, 2003; **48**(12):3464–3474.

87. Lee KBL, Hui JHP, Song IC, Ardany L, Lee EH. Injectable mesenchymal stem cell therapy for large cartilage defects--a porcine model. *Stem Cells Dayt Ohio*, 2007; **25**(11):2964–2971.

88. Diekman BO, Wu C-L, Louer CR, Furman BD, Huebner JL, Kraus VB, et al. Intra-articular delivery of purified mesenchymal stem cells from C57BL/6 or MRL/MpJ superhealer mice prevents posttraumatic arthritis. *Cell Transplant*, 2013; **22**(8):1395–1408.

89. Mak J, Jablonski CL, Leonard CA, Dunn JF, Raharjo E, Matyas JR, et al. Intra-articular injection of synovial mesenchymal stem cells improves cartilage repair in a mouse injury model. *Sci Rep*, 2016; **6**:23076.

90. Centeno CJ, Busse D, Kisiday J, Keohan C, Freeman M, Karli D. Increased knee cartilage volume in degenerative joint disease using percutaneously implanted, autologous mesenchymal stem cells. *Pain Physician*, 2008; **11**(3):343–353.

91. Nejadnik H, Hui JH, Feng Choong EP, Tai B-C, Lee EH. Autologous bone marrow–derived mesenchymal stem cells versus autologous chondrocyte implantation: an observational cohort study. *Am J Sports Med*, 2010; **38**(6):1110–1116.

92. Kasemkijwattana C, Hongeng S, Kesprayura S, Rungsinaporn V, Chaipinyo K, Chansiri K. Autologous bone marrow mesenchymal stem cells implantation for cartilage defects: two cases report. *J Med Assoc Thail Chotmaihet Thangphaet*, 2011; **94**(3):395–400.

93. Davatchi F, Abdollahi BS, Mohyeddin M, Shahram F, Nikbin B. Mesenchymal stem cell therapy for knee osteoarthritis. Preliminary report of four patients. *Int J Rheum Dis*, 2011; **14**(2):211–215.

94. Barkholt L, Flory E, Jekerle V, Lucas-Samuel S, Ahnert P, Bisset L, et al. Risk of tumorigenicity in mesenchymal stromal cell–based therapies—bridging scientific observations and regulatory viewpoints. *Cytotherapy*, 2013; **15**(7):753–759.

95. Wang B, Wu S-M, Wang T, Liu K, Zhang G, Zhang X-Q, et al. Pre-treatment with bone marrow-derived mesenchymal stem cells inhibits systemic intravascular coagulation and attenuates organ dysfunction in lipopolysaccharide-induced disseminated intravascular coagulation rat model. *Chin Med J (Engl)*, 2012; **125**(10):1753–1759.

96. Falanga A, Marchetti M, Vignoli A. Coagulation and cancer: biological and clinical aspects. *J Thromb Haemost*, 2013; **11**(2):223–233.

97. Search of: 'osteoarthritis' AND 'hip' | Open Studies: list results; ClinicalTrials.gov [Internet], [cited 2017 Jan 15]. Available from: https://clinicaltrials.gov/ct2/results/displayOpt?flds=a&flds=b&flds =i&flds =f&flds=c&flds=j&flds=k&submit_fld_opt=on&term=%22oste oarthritis%22+AND+%22hip%22&recr=Open&show_flds=Y

98. Lietman SA. Induced pluripotent stem cells in cartilage repair. *World J Orthop*, 2016; **7**(3):149–55.

99. King NM, Perrin J. Ethical issues in stem cell research and therapy. *Stem Cell Res Ther*, 2014; **5**(4):85.

100. Takahashi K, Yamanaka S. Induction of pluripotent stem cells from mouse embryonic and adult fibroblast cultures by defined factors. *Cell*, 2006; **126**(4):663–676.

101. Takahashi K, Tanabe K, Ohnuki M, Narita M, Ichisaka T, Tomoda K, et al. Induction of pluripotent stem cells from adult human fibroblasts by defined factors. *Cell*, 2007; **131**(5):861–872.

102. Cyranoski D. Japanese woman is first recipient of next-generation stem cells. *Nature* [Internet], 2014 [cited 2015 Apr 12]. Available from: http://www.nature.com/doifinder/10.1038/nature.2014.15915

103. Noguchi M, Hosoda K, Nakane M, Mori E, Nakao K, Taura D, et al. In vitro characterization and engraftment of adipocytes derived from human induced pluripotent stem cells and embryonic stem cells. *Stem Cells Dev*, 2013; **22**(21):2895–2905.

104. Guzzo RM, Scanlon V, Sanjay A, Xu R-H, Drissi H. Establishment of human cell type-specific iPS cells with enhanced chondrogenic potential. *Stem Cell Rev*, 2014; **10**(6):820–829.

105. Okada M, Ikegawa S, Morioka M, Yamashita A, Saito A, Sawai H, et al. Modeling type II collagenopathy skeletal dysplasia by directed conversion and induced pluripotent stem cells. *Hum Mol Genet*, 2015; **24**(2):299–313.

106. Outani H, Okada M, Yamashita A, Nakagawa K, Yoshikawa H, Tsumaki N. Direct induction of chondrogenic cells from human dermal fibroblast culture by defined factors. *PLoS One* [Internet], 2013 [cited 2017 Jan 18]; **8**(10). Available from: http://www.ncbi.nlm.nih.gov/pmc/articles/PMC3797820/

107. Uto S, Nishizawa S, Takasawa Y, Asawa Y, Fujihara Y, Takato T, et al. Bone and cartilage repair by transplantation of induced pluripotent stem cells in murine joint defect model. *Biomed Res Tokyo Jpn*, 2013; **34**(6):281–288.

108. Nejadnik H, Diecke S, Lenkov OD, Chapelin F, Donig J, Tong X, et al. Improved approach for chondrogenic differentiation of human induced pluripotent stem cells. *Stem Cell Rev*, 2015; **11**(2):242–253.

109. Bigdeli N, Karlsson C, Strehl R, Concaro S, Hyllner J, Lindahl A. Coculture of human embryonic stem cells and human articular chondrocytes results in significantly altered phenotype and improved chondrogenic differentiation. *Stem Cells*, 2009; **27**(8):1812–1821.

110. Knoepfler PS. Deconstructing stem cell tumorigenicity: a roadmap to safe regenerative medicine. *Stem Cells Dayt Ohio*, 2009; **27**(5):1050–1056.

Chapter 14

Advances in Short-Stem Total Hip Arthroplasty

Karl Philipp Kutzner
Department of Orthopaedic Surgery and Traumatology,
St. Josefs Hospital Wiesbaden, Germany
kkutzner@joho.de

14.1 Background

Total hip arthroplasty (THA) is one of the most successful procedures of the last century providing excellent long-term results [1]. However, worldwide, increasingly, young and active patients with osteoarthritis are treated with THA, thus making it more demanding in terms of postoperative clinical function and physical activity. In Europe already over 20% of all patients treated with THA are under the age of 60 years [2]. The request for surgical procedures and implants allowing an active, high-quality daily life is constantly advancing. Consequently minimally invasive techniques have been developed, allowing muscle- and soft-tissue sparing implantation. In modern THA, however, not only the choice of approach determines the postoperative outcome, but also the type of implant. Choosing

---
The Hip Joint in Adults: Advances and Developments
Edited by K. Mohan Iyer
Copyright © 2018 Pan Stanford Publishing Pte. Ltd.
ISBN 978-981-4774-72-7 (Hardcover), 978-1-351-26244-6 (eBook)
www.panstanford.com

the adequate stem contributes highly to being able to optimally use minimally invasive techniques [3, 4].

Short stems have already been developed decades ago, in order to ensure bone- and soft-tissue sparing implantation. Since already the first short femoral implants, like the MAYO stem (Zimmer) and the collum femoris–preserving (C.F.P.) stem (Link), provided encouraging midterm results, in recent years numerous innovations and modifications have emerged in the market. However, at the same time, some short-stem designs have already been withdrawn from the market due to different reasons. The concept of short-stem THA, especially in Europe, since then has become increasingly important, and implantation figures constantly go up year by year. However, there is a large variety of different models of short stems, differing in design and function [5] (Fig. 14.1).

Figure 14.1 To date, a large variety of different short-stem designs is available in the market [5].

For most modern short-stem designs to date only short- and midterm results have been published [6–10]. At present there are almost no data on the long-term outcomes of new-generation short stems. A major concern in reducing the length and diaphyseal fixation of the femoral stem is still the concomitant reduction of implant stability and the increase of interface micromotion, which by interfering with osteointegration might increase the risk of implant loosening in the long term [11].

14.2 Classification of Short Stems

Because of the heterogeneity of different short-stem designs, a universally applicable classification seems necessary [5, 12]. However, despite some good approaches, to date no widely accepted classification could prevail. Feyen and Shimmin [13] proposed a classification in 2015 focusing on the length of the stem. Some of the short stems are designed similar to conventional straight-stem implants. However, they have just been shortened (ESKA, ESKA Implants; Microplasty, Biomet; TRI-LOCK, DePuy-Synthes; Fitmore, Zimmer; and AMIS, Medacta) (Fig. 14.2).

Figure 14.2 Examples of shortened conventional stems (Fitmore, Zimmer, 2007; short-stem ESKA, ESKA Implants, 2009).

Shortening of a conventional stem might change its biomechanical properties and doesn't correspond to the philosophy of modern short stems of the latest generation.

A more expedient classification has been proposed by Jerosch in 2012 [14] and has been adjusted by Falez et al. in 2015 [5]. It uses the corresponding level of resection of the femoral neck. It differentiates neck-retaining, partially neck retaining and neck-resecting short stems (Fig. 14.3a–c).

This takes into account distinctions in terms of biomechanics and implantation techniques [15].

Neck-retaining implants best allow sparing of bone (see Fig. 14.3a).

However, a superior quality of bone is also essential for these kinds of implants. The reconstruction of individual anatomy and biomechanics is very limited. Due to high revision rates, these stems did not prove to be reliable and mostly have been withdrawn from the market (Cut, ESKA Implants; Silent, DePuy-Synthes) [16].

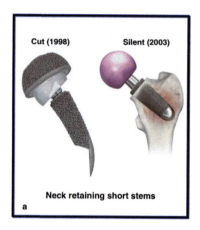

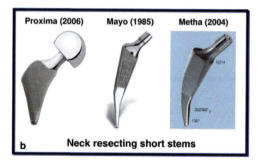

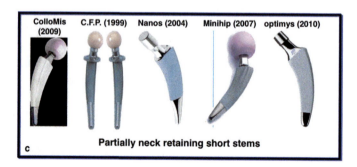

Figure 14.3 (a–c) Classification of short stems [14]. (a) Examples of neck-retaining short stems (Cut, ESKA Implants, 1998; Silent, Depuy-Synthes, 2003); (b) examples of neck-resecting short stems (Proxima, Depuy-Synthes, 2006; MAYO, Zimmer, 1985; Metha, Aesculap, 2004); and (c) examples of partially neck retaining short stems (ColloMis, Lima, 2009; C.F.P., Link, 1999; Nanos, Smith & Nephew, 2004; MiniHip, Corin, 2007; optimys, Mathys Ltd Bettlach, 2010).

Neck-resecting short stems largely correspond to the philosophy of past conventional straight stems and require a similar quality of bone (see Fig. 14.3b).

To reconstruct hip geometry in terms of offset and leg length, a high amount of modularity and different sizes are needed and often accuracy is poor [17]. The MAYO stem (Zimmer) is the one providing most long-term results but has been already been withdrawn from the market due to the advancing progress. A similar implant, the Metha stem (Aesculap) started off with problems regarding fretting and corrosion of the modular necks [18]. The monoblock version, providing different caput-collum-diaphyseal (CCD) angles, still is broadly used today. The trochanter-filling Proxima stem (DePuy-Synthes) did not prevail and has already been withdrawn from the market due to numerous disadvantages. The neck-resecting Fitmore stem inherits a special status. Given its calcar-orientated design, it focuses on a precise reconstruction of the anatomical parameters [19]. This is achieved by a high amount of variants, however, using a standardised level of neck resection [14].

The third group in the classification by Jerosch consists of the partially neck retaining short stems (see Fig. 14.3c).

Almost all new-generation short-stem designs derive from this group. They enable a great variability using a special calcar-guided implantation technique. An individualised level of resection of the femoral neck plays the most important role [20]. The level of resection is individually chosen according to the preoperative planning in order to reconstruct offset and leg length [21]. The positioning of the implants thus can be varied [22]. Early representatives of these partially neck retaining implants are the Pipino stem (Link) and the C.F.P. (Link). The newest generation of modern short stems consists exclusively of calcar-guided short stems, such as the Nanos stem (Smith & Nephew), the Minihip stem (Corin) and the optimys stem (Mathys Ltd Bettlach) (Fig. 14.3c).

14.3   Philosophy of Modern Short-Stem THA

In summary, the development of modern short stems in THA aims at a precise reconstruction of the individual anatomic hip geometry, bone- and soft-tissue sparing implantation and a physiological loading in the metaphysis in order to conserve proximal bone stock in the long term.

14.3.1 Individualised Positioning: Reconstruction of the Anatomy

Conventional straight-stem designs, due to their diaphyseal anchorage, often lack accuracy in reconstructing offset and leg length (Fig. 14.4).

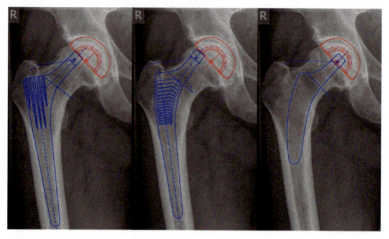

Figure 14.4 Conventional straight-stem designs, due to their diaphyseal anchorage, often lack accuracy in reconstructing offset and leg length (left and middle panels). The optimys short stem (Mathys Ltd Bettlach) is able to reconstruct the offset even in pronounced varus anatomies (right panel).

Also, neck-resecting short stems, such as the MAYO stem and the Metha stem, have been shown to offer minor properties in reconstructing varus anatomies [17] (Fig. 14.5).

The heart of the design in modern calcar-guided short stems consists of its shape being adapted to the medial anatomical calcar curve. The positioning thus is done according to the individual anatomy alongside the calcar curve [20]. This feature differs compared to straight stems and many other short-stem designs. These short stems follow a valgus anatomy into a valgus positioning and given a varus neck, the stem aligns itself into a varus position [22]. This is accomplished by intraoperatively choosing an adjusted level of resection according to the preoperative planning. A high resection of the femoral neck leads to a varus position with corresponding high offset, a low resection results in a valgus alignment and corresponding low offset [23] (Fig. 14.6).

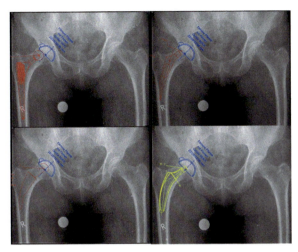

Figure 14.5 Neck-resecting short stems, such as the MAYO stem and the Metha stem (upper-right and lower-left X-rays), have been shown to offer minor properties in reconstructing varus anatomies [17]. Using calcar-guided short stems, varus anatomies can be managed well (lower-right X-ray).

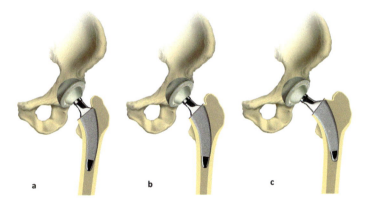

Figure 14.6 Individually adjusted level of resection according to the preoperative planning. A low resection of the femoral neck results in a valgus alignment and corresponding low offset, and a high resection leads to a varus position with corresponding high offset [23]. Courtesy of Mathys Ltd Bettlach.

The individual anatomy of the proximal femur thus can be reconstructed in a broad bandwidth, and offset and leg length can be restored [21]. Additionally, regarding the second plane, natural anteversion of the femoral neck can only be preserved by a short femoral implant, without needing to apply high antetorsion, again leading to restoration of offset (Fig. 14.7).

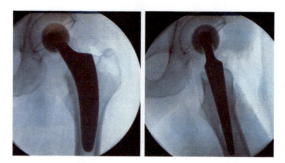

Figure 14.7 (Left) Anteroposterior radiograph; (right) axial view. Natural anteversion of the femoral neck can only be preserved by a short stem, without needing to apply high antetorsion.

14.3.2 Bone- and Soft-Tissue Sparing Implantation

The positioning of the rounded short stem alongside the calcar curve leads to another attribute of these implants. Unlike in conventional straight-stem THA, given the short and rounded design the insertion of the instruments as well as the implantation can be done in the 'round-the-corner' technique, sparing the greater trochanter region completely [20] (Figs. 14.8 and 14.9).

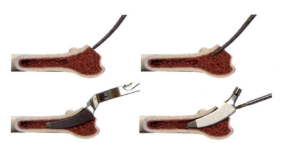

Figure 14.8 A calcar-guided short stem is implanted in the 'round-the-corner' technique, sparing the greater trochanter region completely [20]. Courtesy of Mathys Ltd Bettlach.

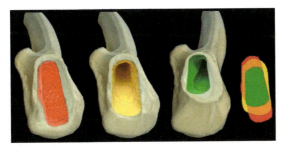

Figure 14.9 Short stems allow a bone sparing implantation (green) compared to conventional straight-stem designs (orange and yellow). Courtesy of Mathys Ltd Bettlach.

This is convenient, not only in terms of the incidence of possible fractures to the trochanter, but also in terms of reducing damage to muscle- and soft-tissue inserting at the piriformis fossa and the greater trochanter, such as the crucial gluteal muscles [15]. Consequently the gluteals can be spared completely, which is considered to have obvious impact on postoperative joint function. The usage of minimally invasive approaches, without transection or damage to the muscles, thus, is clearly facilitated using this technique.

14.3.3 Metaphyseal Anchorage

Unlike conventional straight stems with classical diaphyseal anchorage, short stems focus on anchoring in the metaphysis. However, in certain alignments also a three-point anchoring or a pronounced diaphyseal anchorage is possible. Metaphyseal anchoring stems are designed conical in three planes, allowing force transmission in all directions. The conical design aims at sufficient wedging of the stem in the metaphyseal femoral bone, leading to high primary stability. Postoperative subsidence thus should be prevented, and rotational stability should be ensured [24, 25]. This is particularly important when striving for immediate postoperative full weight bearing, given young and active patients [26]. The pronounced metaphyseal anchorage of short stems also aims at physiological loading of the proximal femoral bone. Thus, bony alterations such as stress shielding and the formation of osteolyses are supposed to be minimised [27]. Preservation of bone stock thus is beneficial in the case of revision surgery. Also typical signs of diaphyseal stress, such as cortical hypertrophy, which is commonly seen in several conventional straight-stem designs, are expected to be reduced and thigh pain thus avoided. Almost all of the newest-generation short stems present with a polished tip, in order to reduce peak stresses and prevent distal ingrowth (Fig. 14.3c).

14.4 Osteointegration and Migration Pattern

One of the most important attributes of femoral implants is the early achievement of secondary stability by attaining osteointegration into the femoral bone. In the case of delayed osteointegration, persistent micromovements are to be suspected of causing aseptic loosening and implant failure in the long term [28]. To achieve early stable osteointegration, the implant's surface has to guarantee fast ingrowth after implantation. Many modern short stems thus provide a two-layer coating, for example, such as a rough titan-plasma spray coating to increase the surface area, accompanied by a calcium phosphate coating, inducing bone synthesis on the interface (Fig. 14.10).

Figure 14.10 Two-layer coating of the optimys short stem. Courtesy of Mathys Ltd Bettlach.

Reducing the length and diaphyseal fixation of the femoral stem in short-stem THA are still causing concerns and might lead to a concomitant reduction of implant stability and the increase of interface micromotion, which by interfering with osteointegration might increase the risk of aseptic loosening [11]. The issue of long-term stability in short-stem THA is crucial regarding long-term outcomes and revision rates. The consequent follow-up of mid- and long-term results of new-generation short stems, using clinical and radiological studies, as well as dual-energy X-ray absorptiometry (DEXA) and radiostereometric analyses (RSA) is of fundamental importance.

In a biomechanical study Bieger et al. [25] conducted a comparing investigation of the common straight stem (CBC, Mathys Ltd Bettlach) versus a new-generation short stem (optimys, Mathys Ltd Bettlach), regarding primary stability and micromovements. The analyses resulted in equivalent axial stability and even fewer micromovements for the short stem compared to the straight stem. Studies investigating the migration pattern of modern short stems using EBRA-FCA (Ein-Bild-Roentgen-Analyse-femoral component analysis) suggest an initially pronounced settlement into the metaphyseal bone after full weight bearing, with subsequent stabilisation [26, 29–32]. To date, revision rates are very low, but long-term outcomes are still lacking [33].

14.5 Indications and Contraindications

Indications for short-stem THA have been constantly expanding in the last few years. Originally short stems were developed and thus have been primarily used in young, active osteoarthritis patients with high demands regarding postoperative function. To date, this group of patients still is mainly to be treated with short-stem THA. However, elderly patients and patients with limited bone quality are also increasingly indicated for short-stem THA with encouraging early results [34]. To date, there is no clear evidence on indications and contraindications due to the limited amount of data. In terms of anatomical conditions, such as primary or secondary deformities of the femur or dysplasia, calcar-guided short-stem THA offers a broad 'safe zone', allowing the implantation in almost any patient with adequate bone quality. However, conventional stem designs, which have been proven over decades, cannot be replaced completely.

In our experience, rheumatoid arthritis does not present as a contraindication as results in those patients are very encouraging. Regarding aseptic necrosis of the femoral head recent investigations could demonstrate good short- to midterm outcomes [35, 36]. However, an MRI should be conducted preoperatively to exclude the possibility of the osteonecrosis exceeding the femoral neck. Age limits are not provided by the companies, and investigations proved encouraging results in elderly patients [34]. Properties such as quality of bone, level of activity and accompanying illnesses should be considered preferably. In severe osteoporosis a cemented implant should be chosen [37]. As known for conventional straight stems, naturally also in short-stem THA the rate of complications, such as delayed wound healing and infection, is increased markedly in obese patients [38]. Additionally recent studies indicate that subsidence might be increased in patients in the obese category, especially given immediate full weight bearing postoperatively [26].

14.6 Conclusions

Modern short-stem THA offers numerous distinct advantages compared to conventional straight-stem designs. Besides an improved potential to reconstruct the individual anatomical hip geometry and more physiological loading in the metaphyseal bone, preventing resorption, particularly a simplified and bone- and soft-tissue sparing implantation technique represents a true accomplishment. Results of clinical and radiological short- and midterm follow-up are encouraging, and until today no disadvantages compared to standard implants are striking. Long-term results, however, are lacking. If the upcoming years will confirm encouraging results and additionally the registry data will show comparable revision rates, in the future worldwide there will be no getting around short stems in modern THA.

References

1. Learmonth ID, Young C, Rorabeck C. The operation of the century: total hip replacement. *Lancet*, 2007; **370**:1508–1519. doi: 10.1016/S0140-6736(07)60457-7

2. Jerosch J. *Kurzschaftendoprothesen: Wo liegen die Unterschiede?* (Köln: Deutscher Ärzteverlag), 2012.

3. Pfeil J. *Minimally Invasive Surgery in Total Hip Arthroplasty* [Englisch] (Auflage: Springer), 2010.

4. Kutzner KP, Hechtner M, Pfeil D, et al. Incidence of heterotopic ossification in minimally invasive short-stem THA using the modified anterolateral approach. *Hip Int*, 2017; **27**(2):162–168. doi: 10.5301/hipint.5000448

5. Falez F, Casella F, Papalia M. Current concepts, classification, and results in short stem hip arthroplasty. *Orthopedics*, 2015; **38**:S6–S13. doi: 10.3928/01477447-20150215-50

6. Kovacevic MP, Pfeil J, Kutzner KP. Implantation of a new short stem in simultaneous bilateral hip arthroplasty – a prospective study on clinical and radiographic data of 54 consecutive patients. *OUP*, 2014; **10**:456–461. doi: 10.3238/oup.2014.0456-0461

7. Gustke K. Short stems for total hip arthroplasty: initial experience with the Fitmore stem. *J Bone Joint Surg Br*, 2012; **94**:47–51. doi: 10.1302/0301-620X.94B11.30677

8. Anderl C. 2-Jahres-Ergebnisse mit dem Optimys-Kurzschaft über den direkten anterolateralen Zugang. *Orthopädie Rheumatol*, 2015; **2015**(05):32–34.

9. Ettinger M, Ettinger P, Lerch M, et al. The NANOS short stem in total hip arthroplasty: a mid term follow-up. *Hip Int*, 2011; **21**:583–586. doi: 10.5301/HIP.2011.8658

10. von Lewinski G, Floerkemeier T. 10-year experience with short stem total hip arthroplasty. *Orthopedics*, 2015; **38**:S51– S56. doi: 10.3928/01477447-20150215-57

11. Cinotti G, Della Rocca A, Sessa P, et al. Thigh pain, subsidence and survival using a short cementless femoral stem with pure metaphyseal fixation at minimum 9-year follow-up. *Orthop Traumatol Surg Res*, 2013; **99**:30–36. doi: 10.1016/j.otsr.2012.09.016

12. Khanuja HS, Banerjee S, Jain D, et al. Short bone-conserving stems in cementless hip arthroplasty. *J Bone Joint Surg Am*, 2014; **96**:1742–1752. doi: 10.2106/JBJS.M.00780

13. Feyen H, Shimmin AJ. Is the length of the femoral component important in primary total hip replacement? *Bone Joint J*, 2014; **96-B**:442–448. doi: 10.1302/0301-620X.96B4.33036

14. Jerosch J. Kurzschaft ist nicht gleich Kurzschaft: eine Klassifikation der Kurzschaftprothesen. *OUP*, 2012; **1**(7–8):304–312.

15. Mai S, Pfeil J, Siebert W, Kutzner KP. Kalkar-geführte Kurzschäfte in der Hüftendoprothetik: eine Übersicht. *OUP*, 2016; **5**:342–347. doi: 10.3238/oup.2016.0342-0347

16. Nieuwenhuijse MJ, Valstar ER, Nelissen RGHH. 5-year clinical and radiostereometric analysis (RSA) follow-up of 39 CUT femoral neck total hip prostheses in young osteoarthritis patients. *Acta Orthop*, 2012; **83**:334–341. doi: 10.3109/17453674.2012.702392

17. Höhle P, Schröder SM, Pfeil J. Comparison between preoperative digital planning and postoperative outcomes in 197 hip endoprosthesis cases using short stem prostheses. *Clin Biomech (Bristol, Avon)*, 2015; **30**(1):46–52. doi: 10.1016/j.clinbiomech.2014.11.005

18. Ceretti M, Falez F. Modular titanium alloy neck failure in total hip replacement: analysis of a relapse case. *SICOT-J*, 2016; **2**:20. doi: 10.1051/sicotj/2016009

19. Jerosch J, Funken S. Change of offset after implantation of hip alloarthroplasties. *Unfallchirurg*, 2004; **107**:475–482. doi: 10.1007/s00113-004-0758-2

20. Kutzner KP, Donner S, Schneider M, et al. One-stage bilateral implantation of a calcar-guided short-stem in total hip arthroplasty. *Oper Orthop Traumatol*, 2017; **29**(2):180–192. doi: 10.1007/s00064-016-0481-5

21. Kutzner KP, Kovacevic MP, Roeder C, et al. Reconstruction of femoroacetabular offsets using a short-stem. *Int Orthop*, 2015; **39**:1269–1275. doi: 10.1007/s00264-014-2632-3

22. Kutzner KP, Freitag T, Donner S, Kovacevic MP, Bieger R. Outcome of extensive varus and valgus stem alignment in short-stem THA: clinical and radiological analysis using EBRA-FCA. *Arch Orthop Trauma Surg*, 2017; **137**:431–439. doi: 10.1007/s00402-017-2640-z

23. Jerosch J, Grasselli C, Kothny PC, et al. [Reproduction of the anatomy (offset, CCD, leg length) with a modern short stem hip design--a radiological study]. *Z Orthop Unfall*, 2012; **150**:20–26. doi: 10.1055/s-0030-1270965

24. Bieger R, Ignatius A, Decking R, et al. Primary stability and strain distribution of cementless hip stems as a function of implant design. *Clin Biomech (Bristol, Avon)*, 2012; **27**:158–164. doi: 10.1016/j.clinbiomech.2011.08.004

25. Bieger R, Ignatius A, Reichel H, Dürselen L. Biomechanics of a short stem: in vitro primary stability and stress shielding of a conservative cementless hip stem. *J Orthop Res*, 2013; **31**:1180–1186. doi: 10.1002/jor.22349

26. Kutzner KP, Kovacevic MP, Freitag T, et al. Influence of patient-related characteristics on early migration in calcar-guided short-stem total hip arthroplasty: a 2-year migration analysis using EBRA-FCA. *J Orthop Surg Res*, 2016; **11**:29. doi: 10.1186/s13018-016-0363-4

27. Kutzner KP, Pfeil D, Kovacevic MP, et al. Radiographic alterations in short-stem total hip arthroplasty: a 2-year follow-up study of 216 cases. *Hip Int*, 2016; **26**:278–283. doi: 10.5301/hipint.5000339

28. Krismer M, Biedermann R, Stöckl B, et al. The prediction of failure of the stem in THR by measurement of early migration using EBRA-FCA. Einzel-Bild-Roentgen-Analyse-femoral component analysis. *J Bone Joint Surg Br*, 1999; **81**:273–280.

29. Kutzner K, Freitag T, Kovacevis M-P, et al. One-stage bilateral versus unilateral short-stem total hip arthroplasty: comparison of migration patterns using "Ein-Bild-Roentgen-Analysis Femoral-Component-Analysis". *Int Orthop*, 2017; **41**:61–66. doi: 10.1007/s00264-016-3184-5

30. Freitag T, Kappe T, Fuchs M, et al. Migration pattern of a femoral short-stem prosthesis: a 2-year EBRA-FCA-study. *Arch Orthop Trauma Surg*, 2014; **134**:1003–1008. doi: 10.1007/s00402-014-1984-x

31. Kaipel M, Grabowiecki P, Sinz K, et al. Migration characteristics and early clinical results of the NANOS® short-stem hip arthroplasty. *Wien Klin Wochenschr*, 2015; **127**:375–378. doi: 10.1007/s00508-015-0756-0

32. Schmidutz F, Graf T, Mazoochian F, et al. Migration analysis of a metaphyseal anchored short-stem hip prosthesis. *Acta Orthop*, 2012; **83**:360–365. doi: 10.3109/17453674.2012.712891

33. Schnurr C, Schellen B, Dargel J, et al. Low short-stem revision rates: 1-11 year results from 1888 total hip arthroplasties. *J Arthroplasty*, 2017; **32**:487–493. doi: 10.1016/j.arth.2016.08.009

34. Patel RM, Smith MC, Woodward CC, Stulberg SD. Stable fixation of short-stem femoral implants in patients 70 years and older. *Clin Orthop Relat Res*, 2012; **470**:442–449. doi: 10.1007/s11999-011-2063-z

35. Floerkemeier T, Budde S, Gronewold J, et al. Short-stem hip arthroplasty in osteonecrosis of the femoral head. *Arch Orthop Trauma Surg*, 2015; **135**:715–22. doi: 10.1007/s00402-015-2195-9

36. Suksathien Y, Sueajui J. The short stem THA provides promising results in patients with osteonecrosis of the femoral head. *J Med Assoc Thai*, 2015; **98**:768–774.

37. Malchau H, Herberts P, Eisler T, et al. The swedish total hip replacement register. *J Bone Joint Surg Am*, 2002; **84-A**(Suppl 2):2–20.
38. Hayashi S, Fujishiro T, Hashimoto S, et al. The contributing factors of tapered wedge stem alignment during mini-invasive total hip arthroplasty. *J Orthop Surg Res*, 2015; **10**:52. doi: 10.1186/s13018-015-0192-x

Chapter 15

Advances in Legg–Calvé–Perthes Disease

J. S. Bhamra, P. Singh, S. Madanipour and A. Malhi
Trauma & Orthopaedic Surgery Department,
University Hospital Lewisham High Street,
London SE13 6LH, United Kingdom
j_s_bhamra@hotmail.com

15.1 Background

Legg–Calvé–Perthes disease (LCPD) is characterised by avascular necrosis of the proximal femoral head. It has an insidious onset and commonly occurs in children aged between 4 and 10 years, with a 4:1 male-to-female ratio [1]. The incidence of this rare disease is approximated at 4 in 100,000 children.

Both hips can be affected in approximately 10%–12% of cases [2]. Many conditions are associated with LCPD, such as multiple epiphyseal dysplasia (MED) [1].

Even after a century worth of research, LCPD remains of unknown aetiology. Animal studies have, however, demonstrated the process occurs from the uncoupling of bone metabolism with increased resorption and delayed bone formation [3].

This book chapter is by no means an exhaustive literature review of the pathophysiology, surgical procedures and current management of LCPD. However, we present recent areas of research with current developments that may change the way we manage and treat LCPD in the foreseeable future.

15.2 Classifications Systems and Pathophysiology of Legg–Calvé–Perthes Disease

The classification systems here are only presented and not discussed in depth. These are well documented and presented elsewhere in the literature. Classification systems for LCPD are based on the pathology of the disease that defines each stage, radiographic changes [4] (ones that prognosticate the outcome [5–7]) and Herring's lateral pillar classification [8]. The Stulberg classification is commonly applied at skeletal maturity to prognosticate long-term outcome at maturity by categorising the severity of residual deformity and loss of hip joint congruency [9]. In a recent level II multicentre study, Larson et al., identified the extent of femoral head collapse and deformity with nonoperative treatment as the most important predictive factor for long-term outcome in their 20-year follow-up study [10].

15.3 Treatment Goals

The primary goal of treatment is to provide a pain-free hip whilst enabling the patient to maintain the full range of motion. Treatment options and subsequent management are individualised for each patient, with many factors governing the overall treatment. These include age at presentation (healing potential is closely linked to growth and remodelling), extent of epiphyseal involvement, presence of lateral epiphyseal extrusion, stage of the disease and the range of hip motion [11]. During the active stage of LCPD, strategies to minimise loads across the hip joint and allow the patient to maintain an active range of motion are utilised in an attempt to prevent progressive femoral head collapse and deformity [2, 12].

Evaluation of the surgical options in current literature suggests that there remains a lack of randomised control or multicentre trials for LCPD, and currently proven treatment for preventing femoral head collapse exists. Furthermore, the subsequent effect of surgical procedures on delaying development of subsequent osteoarthrosis and the need for reconstruction remains unclear [13]. The variability in disease sequelae highlights the difficulty in determining candidates for operative treatment. The lateral pillar and Catterall classifications can only be applied after significant deformity has already occurred.

Over the years, various treatment modalities have emerged in an attempt to deal with preventing femoral head collapse. Nonsurgical containment treatment with spica, casts and braces is reserved for the younger patients. Surgical containment does not appear to speed the healing process of the femoral head but results in a more spherical ossification of the head and seems to yield better overall results [1].

The healing potential under the age of five years is very good, with favourable outcomes. Modalities ranging from exercise to acupuncture and nonsurgical containment, such as adductor tenotomy, have been suggested. In contrast, Hardesty et al. suggest there is no benefit of bracing over treatment [14]. In children between five and seven years, if epiphyseal involvement is less than 50%, surgical containment procedures are recommended; pelvic osteotomies under six years and femoral osteotomies over six years are the treatments of choice. Wiig et al. confirmed this with their five-year follow-up study of 358 patients with proximal femoral varus osteotomies [15]. They also found no significant difference between physiotherapy and abduction orthosis in the over-six-year age group. Conversely, Kim et al. have suggested increased preservation of the femoral head with a greater varus angle following proximal femoral varus osteotomy [16]. Between the ages of 7 and 12 years, treatment remains controversial. For Catterall stages I and II, surgical containment is recommended and for the latter stages (III and IV), salvage procedures have been recommended to increase the load-bearing area. In the over-12-year age group, there is little or no remodelling potential. At this stage the only option left may be a salvage procedure such as shelf arthroplasty and Chiari osteotomy to improve hip congruity with a view to early total hip arthroplasty (THA) for poor functional outcome [2].

In a very recent study, Mosow et al. evaluated the outcome of combined pelvic and femoral varus osteotomies in children with skeletal maturity [17]. Their results did not demonstrate any significant functional or radiographic improvement compared to historic Salter osteotomy or proximal femoral osteotomy results. However, they reported better overall outcomes in children diagnosed and treated at a younger age.

Karimi et al. [18] performed a meta-analysis of outcomes of surgical and nonsurgical treatment of LCPD. They were unable to draw any significant conclusions for best treatment and concluded that there is lack of published evidence. They also included research that suggested no difference between interventions and no treatment [18]. Another study over a 20-year follow-up period by Larson et al. reported no difference between conservative treatment and no treatment with hip-related morbidity [10].

15.4 Current Trends in Total Hip Arthroplasty

THA for LCPD is technically challenging and difficult. Approximately 50% of untreated hips develop severe, debilitating arthritis by the sixth decade of life [2]. There have been no significant advances over the past decade despite ceramic, hybrid and modular prostheses all having undergone trials. THA for LCPD has less satisfactory outcomes than patients with primary osteoarthritis in general. Long-term follow-up is still lacking in these patients. There is a role for suitable preoperative planning with computed tomography (CT) or magnetic resonance imaging (MRI) scans. A few recently published studies provide satisfactory outcomes over reasonable time periods. Over a 10-year period, Al Khateeb et al. reported survivorship rates using cementless custom implants of 100% femoral and 79% acetabular components in a small study group of 15 patients with improved Harris Hip Scores [19]. Seufert et al. also report improved Harris Hip Scores using uncemented modular femoral components, with a reduced limp (94%) and improved leg length equality (100%), in 28 patients over a minimum two-year follow-up [20]. They suggest that using a femoral modular stem can accommodate metaphyseal/

diaphyseal size mismatch and allow for abnormal anteversion. Baghdadi et al. reviewed 99 primary total hip arthroplasties with a minimum two-year follow-up. Ten revisions were performed with an eight-year survival rate of 90% for cementless implants compared to 86% for hybrid implants [21]. They also reported an average improvement in Harris Hip Scores. A small number of studies have evaluated hip resurfacing as an option for treating LCPD rather than conventional THA. Improved range of motion and leg lengths with reduced impingement have been reported in favour of resurfacing [22, 23]. In contrast, patients with LCPD often have contributing disease factors for reduced resurfacing survivorship, such as abnormal anatomy, sizeable femoral and acetabular defects, femoral head deformity, small components and lower body mass index. Hence resurfacing may not be suitable [24].

15.5 Femoroacetabular Impingement and Legg–Calvé–Perthes Disease

The sequelae of LCPD can manifest as a spectrum of pathologies that contribute to femoroacetabular impingement and accelerated arthritic wear. There have been a few recent published level-IV reports. Anderson et al. reported their outcomes following surgical dislocation, osteochondroplasty, trochanteric advancement and treatment of intra-articular chondrolabral injury [25]. They reported some clinical and radiographic improvement in the osteochondroplasty group. However, there was no statistical significance. Clohisy et al. reported results of 16 patients who underwent hip preservation procedures [26]. They consequently combined surgical hip dislocation to treat intra-articular and extra-articular sources of femoroacetabular impingement and periacetabular osteotomy to address structural instability for residual Perthes-like hip deformities and associated acetabular dysplasia. All 16 hips had labral hypertrophy, with 13 of the 16 requiring treatment for articular cartilage lesions. Improved mean Harris Hip Scores with correlated radiographic outcomes in 14 patients at a median of 40-month follow-up were reported.

Further trials with longer follow-up periods are needed to draw any constructive conclusions. What appears to be apparent, however, is that some patients with residual hip deformities that have femoral head deformity and secondary remodelling of the acetabulum manifesting with mechanical dysfunction seem to show some improvement in the short term with safe surgical dislocation and addressing of femoroacetabular abnormalities.

15.6 Radiological Advances

Studies using perfusion MRI as a method for evaluating prognosis in LCPD have recently emerged.

Du et al. demonstrated that a magnetic resonance (MR) perfusion index from gadolinium-enhanced extraction MR images had correlation with the extent of deformity on anteroposterior (AP) radiographs [27]. There was substantial variability in the MR perfusion index amongst patients with early (Waldenstrom I or II) disease at the beginning of the study. However, at the two-year follow-up point, a high MR perfusion index was correlated with low radiographic deformity and vice versa. The authors suggest that the advantage of an MR perfusion index is the ability to prognosticate and plan before a deformity has occurred that could comprise future procedures.

This group has also studied revascularisation patterns of the femoral head as the natural history of LCPD progresses using serial perfusion MRI [28]. They demonstrate the femoral head revascularises initially in a horseshoe pattern from the periphery of posterior, lateral and medial aspects of the epiphysis, converging towards the anterocentral region of the epiphysis. The rate of revascularisation was found not to correlate with age or with lateral pillar involvement but was found to be significantly higher in Waldenstrom stage I disease as compared with stage II. Further results from this serial MRI study suggest that the revascularisation occurs from new vessels at the femoral neck that traverse the margin of the articular epiphyseal cartilage to reach the periphery, rather than penetrating from the metaphysis, as had previously been suggested. This may have implications for surgical planning with regard to safe dislocation and the benefits of physeal drilling [29].

The international Perthes study group has suggested a quantitative method for assessing hip synovitis in LCPD using gadolinium-enhanced MRI (Gd-MRI) [30]. They demonstrated that synovial volume enhancement was significantly increased during the initial and fragmentation stages of LCPD, with persistence of synovitis into the reossification stage. The group suggests that Gd-MRI could be useful in monitoring hip synovitis in patients undergoing treatment.

The use of serial perfusion MRI may indeed become the future 'gold standard' for evaluating LCPD. However, further studies are necessary to establish its role in assessing disease progression.

15.7 Use of Bisphosphonates in Legg–Calvé–Perthes Disease?

The use of bisphosphonates in LCPD in humans remains hypothetical. Clinical studies are yet to establish conclusive evidence for bisphosphonates. However, there have been a few successful trials in animal models that appear to be promising.

The rationale for bisphosphonate use is the potential for this class of drugs in preventing femoral head deformity secondary to collapse during the fragmentation phase of the disease sequelae. It is postulated that osteoclastic activity mediates the subchondral collapse by inhibiting osteoclastic bone resorption to slow resorption of necrotic bone and allow revascularisation with new bone formation before femoral head collapse occurs [31]. It is plausible that progression may be prevented in paediatric patients with LCPD [32, 33].

Ramachandran et al. evaluated prospective intravenous bisphosphonate therapy for nonidiopathic avascular femoral head necrosis in adolescents [31]. They identified 17 patients with osteonecrosis with a mean age of 12.7 years. The average duration of intravenous pamidronate or zoledronic acid therapy was 20.3 months, with a minimum follow-up of two years. Treatment was initiated within three months of initial traumatic event. Reported results are promising, with 14 of 17 (82%) being pain-free and 9 of 17 (53%) having minimal or no femoral head deformity (Stulberg classification). This study is limited by the lack of a control group but nevertheless has promising future implications.

Aya-ay et al. studied the effects of intraosseous administration of either radiolabelled (14C) or unlabelled ibandronate at one week, after surgical infarction of femoral heads of piglets. Femoral heads were assessed at 48 h, three weeks or seven weeks to determine the distribution and retention of the bisphosphonate using autoradiography and liquid scintillation analysis. The femoral heads injected with unlabelled ibandronate were also assessed for degree of deformity using radiography and histomorphometry [34]. Results showed significantly better preservation ($p < 0.001$) of the infarcted femoral heads treated with ibandronate. The study demonstrated that bisphosphonates may preserve femoral head structure after ischaemic necrosis [34].

Other piglet studies from the same group have demonstrated similar results, demonstrating the oral ibandronate preserves the trabecular structure of the osseous epiphysis and prevents femoral head deformity during the early phase of repair of ischaemic necrosis in the piglet model. This may translate to human trials and suggests that administration of bisphosphonates during the fragmentation stage of LCPD may prevent early flattening of the osseous epiphysis and this is of clinical importance [35].

Little et al. looked at the effects of zoledronic acid on femoral head sphericity in spontaneously hypertensive rat models [36]. They showed increased bone volume and trabecular mass compared to the control saline group. This suggests that bisphosphonates could be used to improve outcome and decrease future surgical procedures.

Young et al. reported their findings from a recent systematic review of bisphosphonate therapy for juvenile femoral head osteonecrosis [33]. They concluded that current published studies lacked consistency in patient data and drug protocol reporting, with many displaying a selection bias. Only three level-IV studies have evaluated the effect of bisphosphonates in clinical trials. One of these studies evaluated bisphosphonate use in the precollapse stage. Prevention of femoral head deformity was obtained in 9 of 17 patients [32, 37].

To date there are no randomised clinical trials of bisphosphonate therapy in LCPD. A role for bisphosphonates in the adult population also remains to be established.

15.8 Future Therapies: Preclinical Role of Morphogenic Protein-2 in Legg–Calvé–Perthes Disease?

Preclinical studies experimenting with bone morphogenic protein-2 (BMP-2) have implied a role in LCPD. BMP-2 is recognised as a potent osteoinductive agent. Experimental preclinical studies have evaluated its use in conjunction with bisphosphonates to accelerate apoptosis of osteoclasts to reduce bone resorption. Little et al. evaluated BMP-2 and ibandronate in non-weight-bearing piglet models with ischaemic osteonecrosis, suggesting that new bone formation is increased in this regimen [36]. However, this was associated with heterotopic ossification in the hip capsule. BMP-2 in LCPD appears promising but may only remain at the experimental phase due to its undesired side effects.

15.9 Future Therapies: Interleukin-6 and Synovitis

Hip synovitis is a common feature of LCPD. Kamiya et al. assessed the chronicity of the synovitis and presence of inflammatory cytokines in the synovial fluid during the active phase of the disease [38]. Serial MRIs (T2-weighted and gadolinium-enhanced images) showed an increase in synovial fluid, and multiple cytokine assays demonstrated significantly increased interleukin-6 (IL-6) protein levels in the affected hips (p = 0.0005). There was no rise in IL-1β and tumour necrosis factor–alpha (TNF-α) proinflammatory mediators in the assay. The role of IL-6 in the pathophysiology of synovitis in LCPD may be a promising field, and more clinical trials are needed.

15.10 Future Therapies: The Use of Mesenchymal Stem Cells?

Mesenchymal stem cells have become of great interest in the orthopaedic community recently due to the potential for regeneration and repair of damaged or abnormal tissue. This may become a key area for future developments in managing or limiting LCPD. However, there are no clinical or laboratory studies to date.

15.11 Conclusions

Treatment strategies for LCPD are still developing. As our understanding of the pathophysiology improves, so do the methods with which we evaluate and treat the disease. Throughout this chapter on recent advances in this field, a common stem is displayed suggesting that a magnitude of research, both clinical and laboratory based, is fundamentally necessary. Much of the literature consists mainly of level-IV case series or animal studies. Nevertheless, pioneering commendable work from units worldwide is furthering our knowledge of this challenging disease, and we look to the future to see how these treatment modalities impact our management.

References

1. Jaffe WL, Legg-Calve-Perthes disease. 2015. Available from: http://emedicine.medscape.com/article/1248267-overview

2. BMJ. Legg-Calvé-Perthes' disease. 2016. Available from: http://bestpractice.bmj.com/best-practice/monograph/751.html

3. Rowe SM, Lee JJ, Chung JY, Moon ES, Song EK, Seo HY. Deformity of the femoral head following vascular infarct in piglets. *Acta Orthop*, 2006; **77**(1):33–38.

4. The classic. The first stages of coxa plana by Henning Waldenström. 1938. *Clin Orthop Relat Res*, 1984; (191):4–7.

5. Catterall A. The natural history of Perthes' disease. *J Bone Joint Surg Br*, 1971; **53**(1):37–53.

6. Salter RB, Thompson GH. Legg-Calve-Perthes disease. The prognostic significance of the subchondral fracture and a two-group classification of the femoral head involvement. *J Bone Joint Surg Am*, 1984; **66**(4):479–489.

7. Herring JA, Kim HT, Browne R. Legg-Calve-Perthes disease. Part I: Classification of radiographs with use of the modified lateral pillar and Stulberg classifications. *J Bone Joint Surg Am*, 2004;86-A(10):2103-20.

8. Herring JA, Neustadt JB, Williams JJ, Early JS, Browne RH. The lateral pillar classification of Legg-Calve-Perthes disease. *J Pediatr Orthop*, 1992; 12(2):143–150.

9. Stulberg SD, Cooperman DR, Wallensten R. The natural history of Legg-Calve-Perthes disease. *J Bone Joint Surg Am*, 1981; 63(7):1095–1108.

10. Larson AN, Sucato DJ, Herring JA, Adolfsen SE, Kelly DM, Martus JE, et al. A prospective multicenter study of Legg-Calve-Perthes disease: functional and radiographic outcomes of nonoperative treatment at a mean follow-up of twenty years. *J Bone Joint Surg Am*, 2012; **94**(7):584–592.

11. Sinigaglia R, Bundy A, Okoro T, Gigante C, Turra S. Is conservative treatment really effective for Legg-Calve-Perthes disease? A critical review of the literature. *Chir Narzadow Ruchu Ortop Pol*, 2007; **72**(6):439–443.

12. Horn A, Eastwood D. Prevention of femoral head collapse in Legg-Calvé-Perthes disease: experimental strategies and recent advances. *Bone Joint 360*, 2017; **6**(1):3–6.

13. Brand RA. Legg-Calve-Perthes syndrome (LCPS): an up-to-date critical review Charles W. Goff, MD CORR 1962;22:93-107. *Clin Orthop Relat Res*, 2012; **470**(9):2628–2635.

14. Hardesty CK, Liu RW, Thompson GH. The role of bracing in Legg-Calve-Perthes disease. *J Pediatr Orthop*, 2011; **31**(2 Suppl):S178– S181.

15. Wiig O, Huhnstock S, Terjesen T, Pripp AH, Svenningsen S. The outcome and prognostic factors in children with bilateral Perthes' disease: a prospective study of 40 children with follow-up over five years. *Bone Joint J*, 2016; **98-B**(4):569–575.

16. Kim HK, da Cunha AM, Browne R, Kim HT, Herring JA. How much varus is optimal with proximal femoral osteotomy to preserve the femoral head in Legg-Calve-Perthes disease? *J Bone Joint Surg Am*, 2011; **93**(4):341–347.

17. Mosow N, Vettorazzi E, Breyer S, Ridderbusch K, Stücker R, Rupprecht M. Outcome after combined pelvic and femoral osteotomies in patients with Legg-Calvé-Perthes disease. *J Bone Joint Surg Am*, 2017; **99**(3):207–213.

18. Karimi MT, McGarry T. A comparison of the effectiveness of surgical and nonsurgical treatment of Legg-Calve-Perthes disease: a review of the literature. *Adv Orthop*, 2012; **2012**:7.

19. Al-Khateeb H, Kwok IH, Hanna SA, Sewell MD, Hashemi-Nejad A. Custom cementless THA in patients with Legg-Calve-Perthes disease. *J Arthroplasty*, 2014; **29**(4):792-796.

20. Seufert CR, McGrory BJ. Treatment of Arthritis associated with Legg-Calve-Perthes disease with modular total hip arthroplasty. *J Arthroplasty*, 2015; **30**(10):1743-1746.

21. Baghdadi YM, Larson AN, Stans AA, Mabry TM. Total hip arthroplasty for the sequelae of Legg-Calve-Perthes disease. *Clin Orthop Relat Res*, 2013; **471**(9):2980-2986.

22. Costa CR, Johnson AJ, Naziri Q, Mont MA. Review of total hip resurfacing and total hip arthroplasty in young patients who had Legg-Calve-Perthes disease. *Orthop Clin North Am*, 2011; **42**(3):419-422, viii.

23. Kim J, Cho YJ, Kim HJ. Role of total hip arthroplasty and resurfacing in Legg-Calve-Perthes disease. *J Pediatr Orthop*, 2011; **31**(2 Suppl):S241-S244.

24. Chaudhry S, Phillips D, Feldman D. Legg-Calve-Perthes disease: an overview with recent literature. *Bull Hosp Jt Dis*, 2014; **72**(1):18-27.

25. Anderson LA, Erickson JA, Severson EP, Peters CL. Sequelae of Perthes disease: treatment with surgical hip dislocation and relative femoral neck lengthening. *J Pediatr Orthop*, 2010; **30**(8):758-766.

26. Clohisy JC, Nepple JJ, Ross JR, Pashos G, Schoenecker PL. Does surgical hip dislocation and periacetabular osteotomy improve pain in patients with Perthes-like deformities and acetabular dysplasia? *Clin Orthop Relat Res*, 2015; **473**(4):1370-1377.

27. Du J, Lu A, Dempsey M, Herring JA, Kim HK. MR perfusion index as a quantitative method of evaluating epiphyseal perfusion in Legg-Calve-Perthes disease and correlation with short-term radiographic outcome: a preliminary study. *J Pediatr Orthop*, 2013; **33**(7):707-713.

28. Kim HK, Burgess J, Thoveson A, Gudmundsson P, Dempsey M, Jo CH. Assessment of femoral head revascularization in Legg-Calve-Perthes disease using serial perfusion MRI. *J Bone Joint Surg Am*, 2016; **98**(22):1897-1904.

29. Castaneda P. Can we solve Legg-Calve-Perthes disease with better imaging technology? Commentary on an article by Harry K.W. Kim, MD, MS, et al. "assessment of femoral head revascularization in Legg-Calve-Perthes disease using serial perfusion MRI". *J Bone Joint Surg Am*, 2016; **98**(22):e103.

30. Neal DC, O'Brien JC, Burgess J, Jo C, Kim HK, International Perthes Study G. Quantitative assessment of synovitis in Legg-Calve-Perthes

disease using gadolinium-enhanced MRI. *J Pediatr Orthop B*, 2015; **24**(2):89–94.

31. Ramachandran M, Ward K, Brown RR, Munns CF, Cowell CT, Little DG. Intravenous bisphosphonate therapy for traumatic osteonecrosis of the femoral head in adolescents. *J Bone Joint Surg Am*, 2007; **89**(8):1727–1734.

32. Lozano-Calderon SA, Colman MW, Raskin KA, Hornicek FJ, Gebhardt M. Use of bisphosphonates in orthopedic surgery: pearls and pitfalls. *Orthop Clin North Am*, 2014; **45**(3):403–416.

33. Young ML, Little DG, Kim HK. Evidence for using bisphosphonate to treat Legg-Calve-Perthes disease. *Clin Orthop Relat Res*, 2012; **470**(9):2462–2475.

34. Aya-ay J, Athavale S, Morgan-Bagley S, Bian H, Bauss F, Kim HK. Retention, distribution, and effects of intraosseously administered ibandronate in the infarcted femoral head. *J Bone Miner Res*, 2007; **22**(1):93–100.

35. Kim HKW, Randall TS, Bian H, Jenkins J, Garces A, Bauss F. Ibandronate for prevention of femoral head deformity after ischemic necrosis of the capital femoral epiphysis in immature pigs. *J Bone Joint Surg Am*, 2005; **87**(3):550–557.

36. Little DG, McDonald M, Sharpe IT, Peat R, Williams P, McEvoy T. Zoledronic acid improves femoral head sphericity in a rat model of perthes disease. *J Orthop Res*, 2005; **23**(4):862–868.

37. Little DG, Peat RA, McEvoy A, Williams PR, Smith EJ, Baldock PA. Zoledronic acid treatment results in retention of femoral head structure after traumatic osteonecrosis in young Wistar rats. *J Bone Miner Res*, 2003; **18**(11):2016–2022.

38. Kamiya N, Yamaguchi R, Adapala NS, Chen E, Neal D, Jack O, et al. Legg-Calve-Perthes disease produces chronic hip synovitis and elevation of interleukin-6 in the synovial fluid. *J Bone Miner Res*, 2015; **30**(6):1009–1013.

Chapter 16

Advances in Haemophilic Hip Joint Arthropathy

Muhammad Zahid Saeed, Amr Saad, Haroon A. Mann and Nicholas Goddard

Trauma and Orthopaedic Department, University College London Medical School, Royal Free Hospital, Pond Street, London NW3 2QG, United Kingdom
z.saeed@nhs.net, mzahidsaeed@hotmail.com

16.1 Recent Advances in Haemophilic Hip Joint Arthropathy

16.1.1 Introduction

Haemophilia is a hereditary X-linked recessive condition affecting males [1–3]. The deficiency or absence of coagulation factor VIII causes haemophilia A, and the deficiency or absence of coagulation factor IX causes haemophilia B. Haemophilia can lead to advanced arthropathy. Haemophilic arthropathy is permanent cartilage and bone destruction occurring in patients with haemophilia as a longstanding effect of repeated haemarthrosis. Haemarthrosis can

---

The Hip Joint in Adults: Advances and Developments
Edited by K. Mohan Iyer
Copyright © 2018 Pan Stanford Publishing Pte. Ltd.
ISBN 978-981-4774-72-7 (Hardcover), 978-1-351-26244-6 (eBook)
www.panstanford.com

be spontaneous or result from a minor injury. Approximately 50% of haemophilia sufferers will develop a severe arthropathy [1–5].

Haemophilic arthropathy can manifest as a monoarticular or oligoarticular condition affecting large joints. The most commonly involved joints are knees, elbows, ankles, hips and shoulders, in the order of frequency. The knee joint is the most commonly involved joint, and with the ankle joint and elbow joint, comprises almost 80% of the joints affected. The true incidence of haemophilic arthropathy of the hip is relatively rare, accounting for only approximately 4% of the cases. Nonetheless, degenerative disease of the hip makes a significant volume of the orthopaedic surgeon's work [1, 3, 4].

16.1.2 Epidemiology

Globally, it is estimated that 1 male in every 5,000 will have haemophilia A and 1 male in every 30,000 will suffer from haemophilia B. In the United States, 2.3% of children aged two to five years registered in the Universal Data Collection Project are known to have target joints. The target joint is defined as a joint in which four or more recurrent bleedings have occurred in the previous six months [7]. It has also been reported as a rare case report in haemophilia A' [8].

16.1.3 Pathophysiology

The pathophysiological mechanisms of haemophilic arthropathy affecting the hip joint have not been accurately identified. Our understanding of haemophilic hip arthropathy has been based mainly on the knowledge of the knee, ankle and elbow joint arthropathy. It is widely accepted that a relationship between recurrent intra-articular bleeding and the long-term development of joint damage exists [9]. However, our knowledge about the effects of blood on joint structures is limited, for example, the amount of blood or number of bleeding episodes that initiate joint damages [10, 12]. Reinecke and Wohlwill suggested a chemical effect in addition to the mechanical effect of haemarthrosis [10, 11].

In a haemophilic intra-articular and intraosseous haemorrhage occurs. The blood, an irritant, causes hyperplasia of the synovial membrane. Pigment and, subsequently, connective tissue are lined up and heal with fibrosis in the subsynovial tissues. In the meantime, the cartilage is worn at the margins of the joint and the central articular cartilage endures a variable amount of destruction. The distinctive change is the appearance of deep cavities, in the cancellous bone, and superficial depressions in the bone underlying the articular cartilage due to intraosseous haemorrhage [3, 6].

The characteristic features of haemophilic hip arthropathy before puberty are identical to those observed in Legg–Calvé–Perthes disease, and after puberty the features are similar to those of osteoarthritis [3, 6].

16.1.4 Clinical Features

Haemophilic arthropathy has bimodal presentation. The condition in children presents as a severe and very destructive hip disease, often associated with femoral head avascular necrosis. Most of the children suffering from haemophilia will have the first bleed into their joints before they celebrate their second birthday. In childhood, a single episode of haemarthrosis can cause increased intracapsular pressure, leading to avascular necrosis of the femoral head. The second peak presentation is in 20- to 50-year-old patients. In adulthood, haemarthroses are less common. The presentation in such cohort of patients is typically less flamboyant. Overall the joint destruction is less marked [3, 6].

The patients present with a broad range of clinical features depending upon the severity of the condition. Haemophilia is classified into three types according to the level of clotting factors in the blood: mild (>5%), moderate (1%–5%) and severe (<1%). Most of the patients present with severe haemophilia [4] (Table 16.1). The common symptoms are pain, stiffness, limping, clicking, locking and giving way. End-stage haemophilic arthropathy is associated with the loss of function and impaired quality of life.

Table 16.1 The classification of the severity of haemophilia [12]

Severity	Clotting factor level (% activity/[IU/ml])	Bleeding episodes
Mild	5–40 (0.05–0.40)	Severe, with major trauma or surgery
Moderate	1–5 (0.01–0.05)	Occasional/severe with trauma
Severe	1 (<0.01)	Spontaneous joints and muscle

16.1.5 Investigations

A plain radiograph reveals joint effusion in the presence of haemarthrosis, secondary degenerative changes, symmetrical loss of joint cartilage with periarticular erosions, subchondral cysts, osteophytes, sclerosis, periarticular osteoporosis and osteonecrosis [13]. A plain radiograph shows the distorted anatomy of the proximal femur (Figs. 16.1 and 16.2).

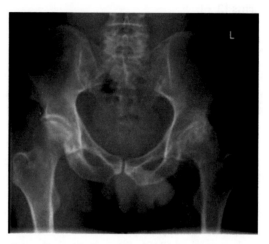

Figure 16.1 Preop X-ray: AP view of the pelvis.

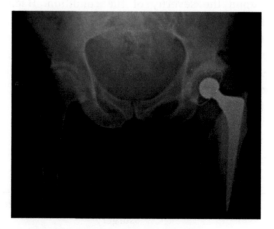

Figure 16.2 Postop THR X-ray: AP view of the pelvis.

As for radiological evaluation methods of haemophilic arthropathy, there are two major systems: Arnold–Hilgartner classification (Table 16.2) and the Pettersson score [14, 15].
The Arnold–Hilgartner classification is a plain radiograph grading system for haemophilic arthropathy.

Table 16.2 The Arnold–Hilgartner classification

Stage 0: Normal joint
Stage I: No skeletal abnormalities, soft-tissue swelling present
Stage II: Osteoporosis, no cysts, no narrowing of the cartilage space
Stage III: Early subchondral bone cysts, preservation of the cartilage space
Stage IV: Findings of stage III, but more advanced; narrowed cartilage space
Stage V: Fibrous joint contractures, loss of the joint cartilage space, extensive substantial disorganisation of the joint

The Pettersson score is the recommended additive scoring system by the World Federation of Hemophilia (WFH). The categories in this system have 0 (normal) to 2 (worse) points and are totally scored from 0 (normal) to 13 points (worst) [15].

The magnetic resonance imaging (MRI) scan is helpful in the detection of early disease. It shows synovitis, thickened synovium with low signal due to haemosiderin susceptibility effect, joint effusion, cartilage loss and erosions [14].

The bone scan is sensitive for detecting areas of disease over the entire skeleton, and follow-up scans can monitor the treatment response. Radioisotopes can be injected therapeutically into a joint to reduce bleeding and control synovitis. Rhenium186 is emerging as the preferred isotope over phosphorus32 and yttrium90 [14, 16].

16.1.6 Differential Diagnosis

- Juvenile rheumatoid arthritis
- Pigmented villonodular synovitis
- Synovial osteochondromatosis
- Amyloid arthropathy

16.1.7 Management

The major objectives of haemophilia treatment are reduction in spontaneous haemarthrosis, pain control and improvement in the quality of life. It is crucial for the haemophilia patients to adopt a healthy lifestyle. In all stages, conservative and prophylactic treatment of the affected joints should be attempted. Early Factor VIII or IX replacement can help prevent or delay joint destruction [1–4, 17, 18].

At an early stage, the treatment options are anti-inflammatory drugs, corticosteroids, joint infusion of hyaluronic acid or corticosteroids and rehabilitation. Research studies have revealed that radiosynoviorthesis can be effective in reducing bleeding and effusion in selected cases [14, 16]. Surgical options are synovectomy, arthrodesis, femoral osteotomy and total joint arthroplasty at present [1–4, 17, 18].

Total joint arthroplasty is the best choice and remains the gold standard surgical treatment [17, 18] (Figs. 16.2 and 16.3).

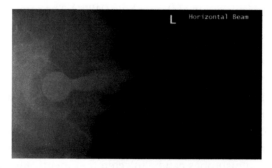

Figure 16.3 Postop THR X-ray: lateral view of the hip.

16.1.8 Multidisciplinary Team Approach

A surgical procedure in a patient with haemophilia can be a complex process. The young male patient's quality of life is adversely affected by the destruction of the hip joint. Following the established treatment protocols helps to achieve excellent functional outcomes [1, 3, 4].

The patient should be treated in a devoted haemophilia centre where a broad multidisciplinary team approach can be offered. Joint musculoskeletal clinics with haemophilia teams, allowing discussion of all aspects of patient management, can help manage this serious condition. Working in close collaboration with a haematology team to establish the factor replacement regime, tailored to the individual patient's needs, prior to orthopaedic surgery is recommended [1–4].

Specialist haemophilia nurses can play a vital role in managing patients in the orthopaedic ward during the pre- and postoperative periods. The addition of a dedicated musculoskeletal physiotherapist specialist can make a great contribution to the team [1–4].

16.1.9 Preoperative Screening Tests

Preoperatively, routine haematological and biochemical analysis should be performed. Preoperative management should consist of an assessment of the nature and severity of the bleeding disorder, the presence of inhibitors to coagulation factors, hepatitis and HIV status, HIV antibody, viral load, CD4 count, fibrinogen, prothrombin time/international normalised ratio (INR) and platelet count. Assessment of the cardiopulmonary status is essential [19].

The rate of postoperative infection is high in HIV-positive patients with haemophilia with CD4 counts <200 cells/mm. A preoperative CD4 count of >200 cells/mm [12] can be used to determine suitability for joint replacement surgery for HIV-positive patients. The risk-benefit ratio should be considered.

16.1.10 Surgical Techniques

Hip arthroplasty in this population is subject to the general operative problems associated with haemophilia, bleeding and concurrent viral disease being the two most notable [1–3]. In addition to thorough preoperative medical preparation of the patient, considerable attention should be given to surgical preparation. The anatomy of the proximal femur can be distorted and, in the most severe cases, structural changes are a challenge to the surgeon and include extremely small femoral medullary canal, valgus alignment and excessive anteversion of the head and neck, and protrusion of the acetabulum [3, 6].

Haemophilic patients are prone to developing contracture and extensive fibrosis, so a muscular release may be required on the operation table to achieve a satisfactory range of motion postoperatively. The quality of bone stock is poor. The strong bond of the implant to the host bone is aimed for long-term survival of the implant [3, 6].

The successful outcome depends on surgical factors, such as meticulous surgical technique, choice of implant, the choice between cemented and uncemented interfaces and correctly addressing the femoral and acetabular anatomical distortion. Polyarthropathy leads to uncharacteristic posture, tendering the hip replacement to higher stresses than typical of idiopathic osteoarthritis sufferers, leading to early failures [3, 6].

16.1.11 Perioperative Regime

Research studies have suggested that using the latest techniques of continuous infusion of the clotting factor has significantly helped to reduce the complication rates and has achieved excellent results that match those of the nonhaemophilic population undergoing total hip replacement [3].

Preoperatively, all patients should be given a bolus dose of clotting factor in order to increase levels to 100%. A continuous infusion of Factor VII, VIII or IX can then be started at the induction of anaesthesia at a rate of 4 U/kg/hour (Factor VIII) and 6 U/Kg/hour (Factor IX). Continuous infusion should be done for seven days postoperatively, aiming for factor levels of 100%. This should be followed by a daily bolus dose for a further seven days to maintain the levels at 100 U/dl and to allow postoperative rehabilitation without the risk of further bleeding from joints. After this prophylactic therapy should be continued for up to eight weeks during the rehabilitation phase [1, 3].

Postoperative pain management is important to facilitate the mobilisation and physical therapy of the replaced joint [3, 6, 19].

All the patients should receive mechanical thromboprophylaxis by wearing compression stockings unless it is contraindicated [3, 6, 19].

Haemophilia sufferers are likely to have significant loss of range of motion and strength and can benefit from a preoperative outpatient physiotherapy programme, which should be directed at maintaining joint mobility and improving muscle strength prior to the procedure. Range-of-motion exercises can be started immediately after surgery. All physiotherapy sessions should be pretreated with 20 IU/kg for up to six weeks postoperatively [1, 3, 6, 20].

16.1.12 Complications

Complications from joint replacement surgery can be classified as operative and postoperative. Following total hip replacement, haemophilia patients are prone to complications identical to those of nonhaemophilic patients. Bleeding leading to haematoma with swelling is the one complication for which people with haemophilia may have an increased risk. In severe cases the haematoma can be aspirated under ultrasound guidance with an aseptic technique. In rare cases, a return to the operating room for evacuation of a wound haematoma may be necessary.

It is speculated that haemophilia patients have a lower risk of deep vein thrombosis (DVT) because of their bleeding disorder. In fact haemophilia patients have the same risk of DVT as nonhaemophilic patients, because the coagulation factor level is normalised by administration of continuous concentrates at perioperative periods [3, 20–33].

The most common complication that can lead to a devastating failure is infection. This can occur any time from immediately after the operation through the rest of a person's life. There is less than 1% risk of infection. However, it is well established that people with haemophilia are at a higher risk of delayed secondary infection [3, 20–33].

A factual potential source of bacteraemia in a person with haemophilia is through contamination during intravenous factor replacement and dental procedure. For long-term prevention of infection, people with haemophilia should use international aseptic precautions with intravenous access and take prophylactic antibiotics at the time of dental procedures [35].

Long-term complications of prosthetic failure are low risk. They include loosening, prosthetic wear, breakage of the implants and periprosthetic fractures [3, 20–33].

16.1.13 Long-Term Prognosis

The outcome of patient series of hip arthroplasty in haemophilia patients is limited due to the fact that the number of patients is small and follow-up is relatively short. Traditionally, the results of hip arthroplasty in haemophilic arthropathy have been inferior to the excellent results achieved in patients with primary osteoarthritis of the hip joint [1–6, 11].

Failure rates of up to 20% have been reported in the literature. The main reasons for failure are deep sepsis, mechanical failure and technical difficulties. A few studies have shown promising results, but the number of cases has been relatively small with a short period of follow-up [3, 20–33]. As for durability, there are a few long clinical reports. Wang et al. reported a survival rate of 89% for 18 total hip replacements at 8.5 years when revision for aseptic mechanical failure was considered [34].

Goddard et al. [3] conducted the largest retrospective study till now about 34 total hip replacements. The follow-up ranges were from 1 to 17 years (mean 6.25 years). There were three cases of aseptic loosening at a mean of 7.67 years after surgery (range 3–13 years). The remaining patients had implants in situ with clinically and radiologically satisfactory outcomes [35]. The long-term results showed improved functional scores, good prosthetic survival and above all significantly low infection rates [36, 37].

16.2 Conclusions

Total joint arthroplasty for haemophilia is a challenging surgery. The cost of haemophilia treatment is a major economical concern. Nevertheless, total joint replacement has been successfully performed in people with haemophilia for years. The benefits of pain relief and improved function provided by total joint replacement make this procedure the most successful orthopaedic operation for managing chronic haemophilic arthropathy. To increase the chances of a successful outcome and decrease the risks, these procedures should be performed in hospitals where there are established haemophilia centres with all the specialists available [3].

References

1. Goddard NJ, Mann HA, Lee CA. Total knee replacement in patients with end-stage haemophilic arthropathy: 25-year results. *J Bone Joint Surg Br*, 2010; **89**:186–188.
2. Goddard N. Joint replacement. In *Textbook of Hemophilia*, 2nd Ed, Lee CA, Berntop EE, Hoots WK, eds. (Cornwall: Wiley-Blackwell), 2010; pp. 176–181.
3. Miles J, Rodriguez-Merchan EC, Goddard NJ. The impact of haemophilia on the success of total hip arthroplasty. *Haemophilia*, 2008; **14**(1):81–84.
4. Takedani H. Chapter 14: Total joint arthroplasty for hemophilia. In *Arthroplasty* (InTech), 2013; doi: 10.5772/53232
5. Weissleder R, Wittenberg J, Harisinghani MG. *Primer of Diagnostic Imaging* (Mosby Inc.), 2003.
6. Habermann B, Eberhardt C, Hovy L, Zichner L, Scharrer I, Kurth AA. Total hip replacement in patients with severe bleeding disorders. A 30 years single center experience. *Int Orthop*, 2007; **31**(1):17–21.
7. Valentino LA, Hakobyan N, Enockson C, Simpson ML, Kakodkar NC, Cong L, Song X. Exploring the biological basis of haemophilic joint disease: experimental studies. *Haemophilia*, 2012; **18**(3):310–318. doi: 10.1111/j.1365-2516.2011.02669.x
8. Iyer D, Brueton R. Slipped upper femoral epiphysis with haemophilia A. *Indian J Orthop*, 2007; **41**(3):250–251.
9. Rodriguez-Merchan EC, Goddard NJ, Lee CA. *Musculoskeletal Aspects of Haemophilia* (Blackwell Science), 2000; ISBN: 978-0-632-05671-1
10. Reinecke, Wohlwill F. Uber hämophile Gelenkerkrankung. *Archiv für Krlinische Chirurgie*, 1929; **154**:425.
11. Goddard NJ, Mann H. Diagnosis of haemophilic synovitis. *Haemophilia*, 2007; **13**(Suppl 3):14–19.
12. White GC 2nd, Rosendaal F, Aledort LM, et al. Definitions in hemophilia: recom- mendation of the scientific subcommittee on factor VIII and factor IX of the scientific and standardization committee of the International Society on Thrombosis and Haemostasis. *Thromb Haemost*, 2001; **85**:560.
13. Arnold WD, Hilgartner MW. Hemophilic arthropathy: current concepts of pathogenesis and management. *J Bone Joint Surg Am*, 1977; **59-A**:287–305.

14. https://radiopaedia.org/articles/haemophilic-arthropathy (last visited 29 March 2017).
15. Pettersson H, Ahlberg A, Nilsson IM. A radiologic classification of hemophilic arthropathy. *Clin Orthop Relat Res*, 1980; (149):153–159.
16. Kavakli K, Aydogdu S, Taner M, et al. Radioisotope synovectomy with rhenium186 in haemophilic synovitis for elbows, ankles and shoulders. *Haemophilia*, 2008; **14**(3):518–523.
17. Hilgartner MW. Current treatment of hemophilic arthropathy. *Curr Opin Pediatr*, 2002; **14**:46–49.
18. Rodriguez-Merchan EC. Management of the orthopaedic comlications of haemophlia. *J Bone Joint Surg Br*, 1998; **80**:191–196.
19. http://www1.wfh.org/publication/files/pdf-1210.pdf (last visited 29 March 2017).
20. Gilbert MS. Prophylaxis: musculoskeletal evaluation. *Sem Hematology*; 1993; **30**(Suppl 2):3–6.
21. Peacock K. Quality of life before and after surgery: mobility issues, the fear of surgery, inpatient recovery and outpatient rehabilitation. *Haemophilia*, 2005; **11**(Suppl 1):30–31.
22. Schulman S, Loogna J, Wallensten R. Minimizing factor requirements for surgery without increased risk. *Haemophilia*, 2004; **10**(Suppl 4):35–40.
23. Rooks DS, Huang J, Bierbaum BE, Bolus SA, Rubano J, Connolly CE, Alpert S, Iversen MD, Katz JN. Effect of pre-operative exercise on measures of functional status in men and women undergoing total hip and knee arthroplasty. *Arthritis Rheum*, 2006; **55**(5):700–708.
24. Lobet S, Pendeville E, Dalzell R, Defalque A, Lambert C, Pothen D, Hermans C. The role of physiotherapy after total knee arthroplasty in patients with haemophilia. *Haemophilia*, 2008; **14**(5):989–998.
25. DeKleijn P, Blamey G, Zourilzian N, Dalzell R, Cobet S. Physiotherapy following elective orthopaedic procedures. *Haemophilia*, 2006; **12**(Suppl 3):108–112.
26. Rodriguez-Merchan EC. Correction of fixed contractures during total knee arthroplasty in haemophiliacs. *Haemophilia*, 1999; **5**(1):33–38.
27. Sheth DS, Oldfield S, Ambrose C, Clyburn T. Total knee arthroplasty in hemophilic arthroplasty. *J Arthroplasty*, 2004; **19**(1):56–60.
28. Atilla B. Total knee arthroplasty in hemophilic arthropathy. *J Arthroplasty*, 2005; **20**(1):131.

29. Kim YH, Kim JS. Total knee replacement for patients with ankylosed knees. *J Bone Joint Surg Br*, 2008; **90**(10):1311–1316.
30. Vanarase MY, Pandit H, Kimstra YW, Dodd CA, Popat MT. Pain relief after knee replacement in patients with a bleed disorder. *Haemophilia*, 2007; **13**(4):395–397.
31. Bae DK, Yoon KH, Kim HS, Song SJ. Total knee arthroplasty in hemophilic arthropathy of the knee. *J Arthroplasty*, 2005; **20**(5):664–668.
32. Cohen I, Heim M, Martinowitz U, Chechick A. Orthopaedic outcome of total knee replacement in haemophilia A. *Haemophilia*, 2000; **6**(2):104–109.
33. Solimeno LP, Mancuso ME, Pasta G, Santagostino E, Perfetto S, Mannucci PM. Factors influencing the long-term outcome of primary total knee replacement in haemophiliacs: a review of 116 procedures at a single institution. *Br J Haematol*, 2009; **145**(2):227–234.
34. Wang K, Street A, Dowrick A, Liew S. Clinical outcomes and patient satisfaction following total joint replacement in haemophilia–23-year experience in knees, hips and elbows. *Haemophilia*, 2012; **18**(1):86–93.
35. Winston ME. Haemophiliac arthropathy of the hip. *J Bone Joint Surg Br*, 1952; **34-B**:412–420.
36. Habermann B, Eberhardt C, Hovy L, Zichner L, Scharrer I, Kurth AA. Total hip replacement in patients with severe bleeding disorders: a 30 year single center experience. *Int Orthop*, 2007; **31**(1):17–21.
37. Meehan J, Jamali AA, Nguyen H. Prophylactic antibiotics in hip and knee arthroplasty. *J Bone Joint Surg Am*, 2009; **91**(10):2480–2490.

Chapter 17

Direct Anterior Approach to the Hip Joint

John O'Donnell

Hip Arthroscopy Australia, 21 Erin St., Richmond, Victoria, 3121, Australia
john@johnodonnell.com.au

17.1 Introduction

The direct anterior approach to the hip joint was first described in *Der Grundriss der Chirurgie* (*The Compendium of Surgery*) in 1883 by German surgeon Dr Carl Hueter (Fig. 17.1) [1].

The approach is through the interval between the sartorius and rectus femoris anteriorly and the tensor fascia lata (TFL) posteriorly. It is the only commonly used approach that is both intermuscular and internervous (Fig. 17.2).

It was originally described for resection arthroplasty of the hip.

The approach was developed and enlarged by Marius Smith-Petersen, a Norwegian American, in 1917, and is often still referred to as the Smith-Petersen approach in the English-speaking world [2]. He used the approach to treat many hip conditions, including

---

The Hip Joint in Adults: Advances and Developments
Edited by K. Mohan Iyer
Copyright © 2018 Pan Stanford Publishing Pte. Ltd.
ISBN 978-981-4774-72-7 (Hardcover), 978-1-351-26244-6 (eBook)
www.panstanford.com

hip impingement. He also used the approach for the first hip arthroplasties – the vitallium-mould arthroplasty.

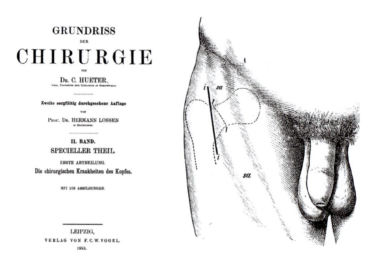

Figure 17.1 Hueter's original publication.

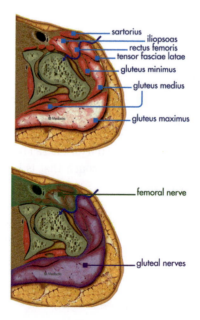

Figure 17.2 The approach is both intermuscular and interneural. The blue line is the line of the approach, passing between rectus femoris and sartorius, innervated by the femoral nerve, and the TFL and gluteal muscles, innervated by the gluteal nerves. Courtesy of Medacta.

The same approach was used by the Judet brothers in France in 1950 for hip hemiarthroplasty [3] and later total hip replacement and by Wagner in Germany for resurfacing.

However, its popularity went into temporary decline following the work of Charnley.

The technique of the Judet brothers utilised a modified traction table. Today many surgeons continue to use a leg positioner, but others have modified the technique so that the surgery is performed on a standard operating table.

17.2 Our Technique of Direct Anterior Approach THR Using a Leg Holder

This operation is much easier if performed with instruments that are made specifically for the direct anterior approach. They will include appropriate retractors and offset acetabular reamers and femoral broaches.

We use a combined spinal anaesthetic and general anaesthetic, which allows excellent operating conditions, minimises the risk of deep vein thrombosis (DVT) and aids post anaesthetic recovery.

We use preoperative skin washes and pre- and postoperative antibiotics (a total of three doses) to minimise infection risk.

DVT prophylaxis typically consists of the use of an intraoperative pneumatic foot pump on the nonoperated leg and postoperative bilateral foot pumps for the first night. We use aspirin but only use anticoagulants in higher-risk patients. Patients commence early mobilisation as soon as the spinal anaesthetic effect has ceased.

All patients receive nonsteroidal anti-inflammatory medication for prophylaxis against heterotopic bone formation unless there is a contraindication.

The patient is positioned supine as shown in Fig. 17.3. Care is taken to pad the foot on the operated leg. The nonoperated leg lies free. The leg holder is not critical, but it allows improved hip extension, rotation control and mild traction, which all make the operation easier.

We mark the operated hip to show the anterior superior iliac spine (ASIS), the interval between the sartorius and rectus femoris anteriorly and the TFL posteriorly, and the proximal femur. The incision starts 2 cm distal and posterior to the ASIS, and it continues obliquely distally and posteriorly over the TFL for 8–10 cm but can be shorter, if desired. The incision is extensile, if necessary (Fig. 17.4).

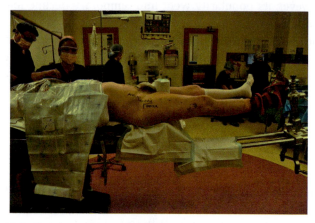

Figure 17.3 Patient positioned for right total hip replacement.

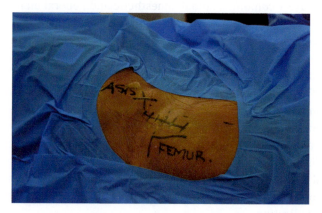

Figure 17.4 The line of the incision. ASIS X is the anterior superior iliac spine, the dotted line is the interval between the rectus femoris anteriorly and the tensor fascia lata posteriorly and the cross-hatched line is the incision line.

The subcutaneous tissue and deep fascia are incised in the same line, down to the fascia overlying the TFL. This is readily identified by the appearance of the fibres of the TFL running parallel to the incision and the bluish-red tinge of the muscle lying deep in the fascia.

The fascia is incised over a short distance and then cut with scissors, including a distal extension deep into the skin to allow greater mobilisation of the TFL.

The fascia of TFL anterior to the incision is separated from the muscle initially with a sharp dissection, and then blunt dissection is used to separate the TFL from the deeper, medial fascial layer covering the lateral side of the rectus. A large self-retaining retractor is used to retract the TFL posteriorly and display the rectus (Fig. 17.5).

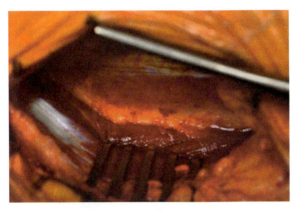

Figure 17.5 TFL retracted posteriorly to display the rectus femoris.

The rectus fascia is incised longitudinally and mobilised for a centimetre or two from the unnamed fascia lying deep in the muscle. Several small vessels enter the muscle here and can be diathermied. The self-retaining retractor is then moved to retract the rectus anteromedially from the deeper fascial layer. Beneath this fascia lie the large ascending branches of the medial circumflex femoral artery and vein. They are most often visible through the fascia. They must be identified and tied before they are divided, as they may bleed profusely (Fig. 17.6).

The fascia is next incised along the border of the rectus muscle. The femoral neck can be readily palpated through the remaining soft tissues at this stage. A blunt Hohman retractor is placed inferior to the femoral neck and a sharp retractor superiorly. These retractors allow excellent visualisation of the fat overlying the hip capsule.

We usually excise this fat to clearly identify the capsule, the gluteus medius above and the vastus lateralis below (Fig. 17.7).

Figure 17.6 The branches of the medial circumflex vessels, which must be ligated.

Figure 17.7 The hip capsule, with gluteus medius to the left and vastus lateralis to the right.

The capsule is incised in a V shape, based superiorly. The first incision is immediately adjacent to the edge of the rectus medially from above the neck to below it. The reflected head of the rectus may be divided to improve visualisation, but this is not generally necessary. It is important to take this capsule incision distally beyond the neck as this will allow easier removal of the femoral head later. The second limb of the V incision is along the intertrochanteric line of the femur. This is done more easily after placing a suture at the apex of the V for retraction, which can also be used later to close the capsule. I find it easiest to make this part of the capsulotomy with a diathermy, passing from inside the capsule to outside. The capsule incision is continued proximally to the junction of the neck and shaft. The two retractors are then moved so that they each lie within the capsule, to allow excellent visualisation of the neck (Fig. 17.8). The lesser trochanter (LT) is still not visible, but the intertrochanteric line of the femur is visible and is known to run to the superior edge of the LT.

Figure 17.8 The capsule is retracted superiorly (with a stitch in the apex of the capsulotomy), and retractors are placed inside the capsule to display the femoral neck.

Gentle traction is then applied to the leg until the femoral head is seen to start to subluxate out of the acetabulum just a little. A femoral neck osteotomy is cut from the junction of the neck and shaft superiorly along the intertrochanteric line or a little above, according to the preoperative plan.

Care must be taken that the inferior retractor lies between the femoral neck and the capsule and that the osteotomy is made towards this retractor, so the osteotomy will then be safely above the LT. Also, the saw cut should stop just as the saw passes through the bone, as there are several vessels just a little deeper, which can bleed quite a lot. The osteotomy is completed with an osteotome, which is also used to lever open the osteotomy.

The leg is then externally rotated about 30° and a corkscrew instrument is inserted up the centre of the femoral neck. After firm insertion, the instrument can be angled anteriorly and proximally to display the remaining intact posterior capsule. Incision of this capsule allows free mobility of the head. Several rotations of the head will tear off from the ligamentum teres, and the head can then be dislocated.

All retractors are then removed and replaced with a modified Charnley retractor, which is placed to retract on the capsule anteriorly at the edge of the acetabulum and posteriorly on the V-shaped piece of the capsule attached to the femur superiorly (Fig. 17.9).

Figure 17.9 View of the acetabulum.

It is generally not necessary to place any additional retractors around the acetabulum, but they can be added if desired by the surgeon to more fully see the edge of the acetabular bone.

Standard reaming of the acetabulum is then performed. Before inserting the acetabular component we inject a mixture of a local anaesthetic – adrenaline, ketorolac and tranexamic acid – into the soft tissues around the acetabulum and femur. This greatly aids postoperative pain relief, as well as minimises blood loss. Care is taken when inserting the acetabular component not to leave any protruding metal edge anteriorly, as this may cause psoas impingement and pain later.

The proximal femur must next be mobilised so that it can be delivered into the wound, to allow broaching. This will most often require some further capsule release.

The leg is externally rotated, and the remaining inferior part of the capsule (pubofemoral ligament) is incised fully. The LT will now be visible, and the height of the osteotomy can be checked. Further the neck bone can be removed at this stage if necessary.

If the proximal femur can now be lifted up with a bone hook, no further capsule release is necessary. If not, a second additional release can be performed immediately adjacent to the greater trochanter (GT) over a length of 1–2 cm, and to a depth where fat can be seen.

A bone hook is used to pull the GT laterally away from the acetabulum, the femur is externally rotated and the 'hip' extended. It is very important that this step be performed with no traction on the leg, and the extension should be performed slowly, as it is possible to fracture the GT, particularly if the posterior capsule is tight or the bone is osteoporotic.

The proximal femur should now be easily accessible (Fig. 17.10) to allow standard broaching and insertion of the femoral component. We use a Charnley curette to first pass along the calcar and enter the femoral canal. The curved, blunt end minimises any risk of femoral perforation. Standard femoral broaching is then undertaken.

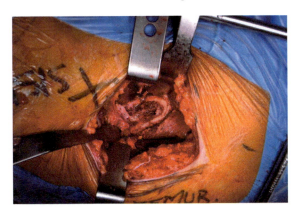

Figure 17.10 The proximal femur is delivered up into the wound.

Additional soft-tissue local anaesthetic injections can be performed at this stage.

After insertion of either trial components or the actual prosthesis, standard methods are used to check leg length and hip stability (Fig. 17.11).

Figure 17.11 All components in place.

The wound is closed very simply. The capsulotomy is closed using the previously placed stitch and 1 or 2 more stitches, taking care not to catch the psoas tendon.

The fascia over the TFL is closed with a dissolving suture, and then a standard soft-tissue and skin closure is performed.

17.3 Postoperative Care

The hip will be very stable. As soon as the spinal anaesthetic has sworn off, and full power returned to the legs, the patient begins to mobilise, initially with the aid of a frame for support. This can rapidly be changed to crutches, and the patient walks with full weight bearing as tolerated.

No particular movement restrictions are required other than to avoid forced extension, particularly with external rotation, for four weeks. We encourage patients to discard the crutches as soon as they are walking well and comfortably but to avoid high-torque activities for six weeks.

We ask the patient to avoid resisted hip flexion and straight leg raising for four to six weeks to minimise the risk of rectus or psoas tendinitis.

Hydrotherapy can be particularly helpful and generally commences when the wound is fully healed at around two weeks.

17.4 Indications for the Direct Anterior Approach

There are no specific indications, or contraindications for this approach. It is more a matter of surgeon skill and preference.

During the first cases, it is easier for the surgeon if the patient is not a heavily muscled male, and does not have short, varus femoral necks, as these cases are more difficult due to tighter access.

17.5 Specific Complications

The same general complications apply with the anterior approach as all others. The major complications that seem to be more prevalent, at least initially, are femoral perforation with the broach or prosthesis, GT fracture and calcar fracture. These are all primarily due to inadequate mobilisation of the proximal femur and can be generally avoided when care is taken to have excellent mobility and visibility before commencing broaching of the femur.

If there is any small split created in the calcar during broaching, this can be fixed with a minifragment screw or a cerclage wire.

Generally, a fracture of the GT requires no treatment as the soft-tissue attachments are intact and will hold the fragments in apposition. If the GT fragments separate, then the trochanter will need to be internally fixed.

17.6 Conclusion

We have found the anterior approach to the hip provides excellent access to the hip and allows early patient mobilisation with great safety, and it is extremely well accepted by patients. It can be readily extended and is suitable for both primary and revision hip replacements.

References

1. Hueter C. Funfte abtheilung: die verletzung und krankheiten des huftgelenkes neunundzwnzigtes capitel. In *Grundriss der Chirurgie*, 2nd Ed, Hueter C, ed. (Leipzig: FCW Vogel), 1883; pp. 129–200.

2. Smith-Petersen MN. A new supra-articular subperiosteal approach to the hip joint. *J Bone Joint Surg Am*, 1917; **15**:592–595.
3. Judet J, Judet R. The use of an artificial femoral head for arthroplasty of the hip joint. *J Bone Joint Surg Br*, 1950; **32-B**:166–173.

Chapter 18

Total Hip in a Day: Setup and Early Experiences in Outpatient Hip Surgery

Manfred Krieger and Ilan Elias
Orthopedic Clinic, August-Bebel-Straße 59,
GPR Hospital Rüsselsheim, Germany
krieger@krieger-rkf.de, ilanelias@hotmail.com

18.1 New Hip in a Day: Setup and Initial Clinical Experiences in Germany

18.1.1 Introduction

Total joint replacement of the hip is one of the most frequently performed surgeries in the orthopaedics, with approximately 230,000 primary operations in 2010 in Germany [1] and more than 310,000 in the United States [1, 2]. In our clinic, the authors operate on approximately 300 patients with elective total hip arthroplasty (THA) per year. In the recent years, we have adapted the rapid recovery track for our total joint replacement patients [3, 4]. This

---

The Hip Joint in Adults: Advances and Developments
Edited by K. Mohan Iyer
Copyright © 2018 Pan Stanford Publishing Pte. Ltd.
ISBN 978-981-4774-72-7 (Hardcover), 978-1-351-26244-6 (eBook)
www.panstanford.com

means that patients who receive a knee or hip replacement will be discharged after a maximum of three to four days of the surgery.

In this chapter, we will present our initial experiences with the first 10 cases (five females and five males; average age 60 years) of patients that underwent ambulatory elective hip replacement surgery, so-called hip in a day. Inspired by our Dutch colleagues around Dr Stephan B. Vehmeijer at Reinier de Graaf Hospital, Delft, which was the first group in Europe to perform THA in an outpatient setting [5, 6], we have initiated an ultrafast recovery track, which is in fact an ambulatory setting of patients undergoing primary cementless THA under general anaesthesia. This selected patient group was discharged the same day as the surgery.

An ambulatory setting of patients undergoing THA is not a brand-new concept and was to the best of our knowledge first published by Dr Richard Berger and his colleagues out of Chicago, Illinois. This study group operated on 100 patients as early as in the year 2000. Of these patients 97 were discharged the same day as the surgery [7–9]. In the meantime a few other study groups have published their results and experiences with hip replacement surgery and patient discharge on the same day [10–13]. However, this is still an uncommon pathway in orthopaedic clinics and remains a challenging endeavour.

Achieving all prerequisites for a patient to be discharged on the day of surgery requires intense coordination within the entire medical team in the hospital with respect to content and communication. Suitable for this programme are active patients without any major concomitant diseases and with a stable and reliable home environment, where adequate care for these patients once they are home is ensured. All patients complete a preoperative patient training programme and receive emotional and structural preparation, both for the surgery and postoperative care. Further, a special cleansing and body care regimen is also explained and prepared, and assistive devices are provided (underarm crutches, grips and rails, raised toilet seat, etc.) if necessary. Once patients have completed the specific preoperative preparation and the patient training successfully, they go back home and return to the hospital early on the day of surgery.

18.1.2 The Course of the Day of Surgery

The schedule on the day of the surgery is as follows: The patient is admitted to the ward at approximately 6:30 a.m. The patient is accompanied by a coach (Fig. 18.1), who has been designated by the patient long before the surgery date and who has attended the patient training together with the patient. The task of the coach is to provide support as well as emotional and practical care for the patient.

Figure 18.1 A female patient is entering the hospital early in the morning.

The patient receives a bed in the day ward (Fig. 18.2).

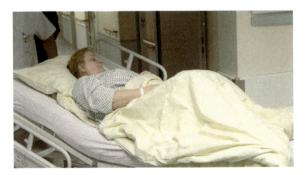

Figure 18.2 The patient receives a bed in the day ward.

Routine preop prep is performed as for all patients, and the patient is brought into the operating room, where he or she is placed into the necessary position and anaesthesia is administered. Either spinal anaesthesia or general anaesthesia may be appropriate. In our hospital, general anaesthesia is preferred and was done in the 10 cases of our hip-in-a-day patients. Ideally, the operation is scheduled as the first case of the day (incision at 8:00 a.m.) so that rehabilitation can start as soon as the patient is awake and stable, usually as early as 9:30 a.m. Standard general anaesthesia is followed by a laryngeal mask airway. No endotracheal tube or inhalation anaesthetic was used in these cases. Furthermore, we use short-acting anaesthetics with total intravenous anaesthesia (TIVA) based on propofol-remifentanil (Ultiva) for fast-recovery eligibility. The primary objective of the general anaesthesia for our purposes is to achieve full muscular relaxation with Esmeron (rocuronium bromide) during the operation without any postop myalgia, adequate postoperative nausea and vomiting (PONV) prophylaxis with ondansetron and dexamethasone, and postoperative bleeding and pain management.

For bleeding prevention intravenous (IV) administration of 1 g tranexamic acid is administered. Additionally for pain, the patient receives 2 g of metamizole, 50 mg of dexketoprofen and 3 or 4 mg piritramide all intravenously. We also routinely administer a single shot of antibiotics (cefuroxime 1.5 g) intravenously (Table 18.1).

Our technique for THA is a minimally invasive single-incision anterior approach (Fig. 18.3).

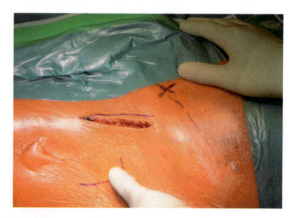

Figure 18.3 Intraop single-incision anterior approach combined with intraoperative local infiltration anaesthesia – the so-called LIA (200 ml 0.2% Naropin [ropivacaine] solution: 150 ml with 1 mg epinephrine and 50 ml without any epinephrine for subcutaneous infiltration).

Table 18.1 Overview of standard medication administered to the hip-in-a-day patient

Dipidolor 2 mg IV Tranexamic acid 1 g IV Cefuroxime 1.5 g IV	Novalgin (metamizole) 2 g IV Sympal (dexketoprofen) 50 mg IV Dipidolor (piritramide) 4 mg IV	Novalgin (metamizole) 500 mg Dipidolor (piritramide) 2 mg IV, as needed Arcoxia (etoricoxib) 120 mg 0-0-1-0	Novalgin (metamizole) 500 mg 2-2-2 Omeprazole 20 mg 1-0-0 Arcoxia (etoricoxib) 120 mg 1-0-0-0
PONV prophylaxis: Dexamethasone 4 mg IV Ondansetron 4 mg IV Saline solution 1500 ml IV	LIA 150 ml Naropin (ropivacaine) + 1 mg epinephrine 50 ml Naropin without epinephrine s.c. Esmeron (rocuronium)	Targin (oxycodon) 10 mg at home	Targin (oxycodon) 10 mg only in the evening and as needed

The use of local infiltration anaesthesia (LIA) is well documented in the literature [19–21] and results in early full-weight-bearing functional and pain-free mobilisation. The patients are allowed their first liquids and energy supply in the form of a sweet ice cream as early as 30 minutes postop. If this is tolerated well, which usually depends exclusively on the medication administered, the first mobilisation can follow immediately. Approximately 45 minutes after extubating, the patient can stand up and walk two to three steps with help from another person, such as a nurse or a physician (Fig. 18.4).

In the recovery unit we frequently do IV administration of 2 mg of piritramide up to three times that day, as well as 500 mg metamizole oral in the late afternoon. Immediately following the patient's return out of the operating room, the second mobilisation is conducted about 90 minutes postop by a physical therapist, followed by two further mobilisations with an increasing radius of motion, walking down the room or hallway (Fig. 18.5).

Figure 18.4 Patient stands up first time after surgery with the help of the surgeon.

Figure 18.5 Patient walks down the hallway of the ward with the help of a physical therapist.

A standard lunch is served about three hours postop. In general, energy supply plays a major role overall in the outpatient postop rehabilitation. Also very important are periods of rest, during which the patient relaxes or naps. Approximately five hours postop, stair climbing with an underarm crutch is practiced – the first time with help, the second time independently. The underlying principle for the outpatient setup is,

'Eat, sleep and train'. Supper is served in the early evening, and the wound is checked through the transparent sterile bandaging to make sure it is dry. A Redon drainage is routinely omitted. If everything is tolerated well, the patient can be discharged by the surgeon and the physical therapist jointly. Obviously, it is important to prescribe adequate postoperative pain medication that prevents dizziness, faintness or nausea and vomiting. In addition to a small dose, for example, 3 mg Dipidolor (piritramide), at about 9:00 a.m., a nonsteroidal anti-inflammatory drug (NSAID) is given in the hospital and Targin (oxycodon) 10/5 is prescribed for use at home at 9:00 p.m. The outpatient follow-up check-ups as well as follow-up physical therapy are arranged prior to the surgery date.

18.1.3 Discussion

An endoprosthetic hip replacement represents one of the most beneficial, gratifying and successful surgical procedures in the field of orthopaedics. In recent years, there has also been a significant reduction in overall surgical trauma as well as a considerable reduction of the average time spent in the hospital, due to improvements in surgical techniques, advanced implants and improved management of the patient, including anaesthesiological management [14]. The average time spent in the hospital after THA can be quite varied within the overall European culture, with averages in Germany being relatively high when compared to other Western countries [15].

While years of advocacy for fast-track surgery have resulted in reducing the average hospital stay of hip replacement patients to approximately 2–3 postop days in the Scandinavian countries [3], in the United States on average 3 nights [1, 2] and within the territory of the Federal Republic of Germany, hospitalisation of almost 8–12 days are very common [15]. In our orthopaedic surgery department at the GPR Rüsselsheim Hospital, Germany, we have been able to reduce the length of stay for THA to approximately three to five postop days for several years now, including more than 100 individual cases of only one to two days, by relying on the clinical studies of our Scandinavian and Dutch colleagues [3–6]. By changing and optimising the anaesthesiological setup, using minimally invasive surgical techniques, improving postoperative pain therapy and optimising physical therapy, we have now been able to perform single-day outpatient hip replacement procedures in selected patients for about one year. This means that these patients stay in the hospital only for the day of surgery and leave on the same evening. So far, there have been no readmissions, complications due to too early discharge or any other adverse events related to shortened length of stay. All patients that were initially identified for the ambulatory pathway did in fact leave the same day. There were no clinical reasons for prolonged length of stay. During the follow-up course all 10 patients stated that they were very satisfied with the ambulatory concept and that they would recommend it to a friend or family member.

Only healthy patients who have a good family caregiver environment at home are suitable for same-day hip replacements. Unsuitable for these same-day procedures are patients with concomitant conditions such as cardiovascular diseases, diabetes mellitus and neurologic impairments. Patients with a health status of ASA-III (an American Society of Anesthesiologists classification) or greater or patients with inadequate care available at home, for example, patients without a partner or a caregiver, are also not suitable.

Performing a hip replacement in a single-day requires vigorous efforts, which includes the commitment of the patient, the patient's social circle and the entire clinical organisation, which must be available to the patient during that entire day and even through the night via telephone. To optimise the preoperative preparations, we implemented the 'Rapid-Recovery-School' for our patients (Table 18.2).

Table 18.2 Numerus professionals teaching in a predetermined sequence at the Rapid-Recovery-School in our hospital. Each course is 10 minutes

Sequence	Professional
1st	Orthopaedic surgeon
2nd	Anaesthesiologist
3rd	Physical therapist
4th	Station desk personnel
5th	Physician assistant or nurse
6th	Medical supply store representative
7th	Social service worker

This includes a patient-teaching-and-training course one week prior to surgery. In this patient training, each of the professional groups the patient will encounter during the single-day surgery holds a 10-minute presentation (Fig. 18.6).

Figure 18.6 Dr Thilo Hartmann, anaesthesiologist, is giving a 10-minute lecture in the Hip-in-a-Day school.

This helps, the patient to prepare for the operation physically, for example, hygiene; practically as well as emotionally. Everyone, from the surgeon to the anaesthesiologist to the hospital's social services representative, presents the important topics that will prepare the patient and make a worry-free recovery of the patient at home possible.

Patient training includes a multidisciplinary approach consisting of the following professionals in a predetermined sequence: (i) the orthopaedic surgeon, (ii) the anaesthesiologist, (iii) physical therapist, (iv) station desk personnel, (v) physician assistant or nurse, (vi) medical supplies store representative and (vii) the social service worker (Table 18.2).

Otherwise, preoperative preparation is the same as for patients who will remain in the hospital for a few days. This includes routine anaesthesiological evaluation and preoperative diagnostic laboratory work and X-ray tests, including the digital planning of the endoprosthetic hip implant (Fig. 18.7).

In addition to the standard medications, we routinely administer 120 mg Arcoxia (etoricoxib) at 7:00 p.m. and Targin 10/5 at home at about 9:00 p.m. once the patient gets home. Management of antithrombotic therapy consists of Clexane (enoxaparin) injections or Xarelto (rivaroxaban) for 30 days after surgery.

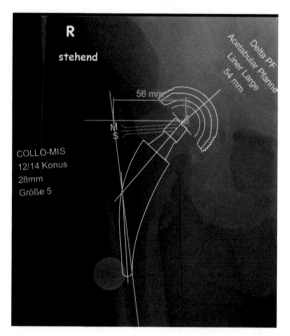

Figure 18.7 Preop X-ray digital planning of a neck-preserving short stem for the right hip.

The surgical procedure for hip replacement is performed via an anterior single-incision approach as the current standard procedure in our hospital [16] (Fig. 18.3).

This means that the hip joint is approached from the intermuscular spaces between the M. tensor fasciae latae muscle and the sartorius or rectus muscle. No muscles, tendons or ligaments are dissected! After clamping or ligating the femoral circumflex artery, the hip joint can be exposed from the front with very little bleeding. Following preparation of the acetabulum, the hip joint replacement is performed. On the basis of the corresponding preoperative planning and trial positioning, the final hip endoprosthesis is implanted and its position checked (Fig. 18.8).

Both short-stem implants as well as traditional cemented or uncemented shafts can be implanted via the anterior approach. During the surgery, we administer systemic IV infusion of 1 g tranexamic acid, which reduces blood loss. We administer exclusively via IV infusion, not via topical injections of tranexamic acid since it was shown to be more effective than topical injections [17, 18]. We also routinely infiltrate a total of 200 ml 0.2% Naropin solution into the treatment area as intramuscular infiltration as a standard procedure, of which 50 ml, injected subcutaneously, does not contain epinephrine and another 150 ml contains 1 mg epinephrine for topical pain therapy and to reduce bleeding. There are several papers showing the positive results of this pain management [19–21]. We also experienced that our patients benefit from the LIA and tranexamic acid administration, as we can relinquish the use of a Redon drainage. This allows the patient to be mobilised immediately after surgery, which is one of the key factors for early discharge.

Figure 18.8 X-ray showing a neck-preserving short stem in an optimal position in the femur and cup in the acetabulum.

After this, the operation is completed by closing up the layers of the wound – tensor fascia latae suture, subcutaneous suture and closure of the skin via intracutaneous suture or staples – with additional use of skin adhesive (Dermabond). This serves to further seal the wound and reduces postoperative secretion significantly. Finally, the surgical area is covered by sterile transparent dressing, which shall remain on the skin ideally for 12 days.

In a recent presentation at the American Academy of Orthopaedic Surgeons (AAOS) meeting in San Diego Dr Taunton from the Mayo

Clinic in Rochester, Minnesota, reported about the short-term benefit of anterior hip replacement. With the anterior procedure, less morphine was needed, ratings of pain were lower and the distance walked during the first walking session with a physical therapist was farther than with the posterior procedure. Radiographic parameters and complications were comparable between the two groups [22]. The authors have previously evaluated the very good clinical outcome of minimally invasive THA using a neck-preserving short stem in 270 patients with either an anterolateral or with an anterior tissue-sparing approach to the hip [16, 23].

Certainly, the anterior approach is challenging and needs a learning curve but was proven to have clinical advantages [22, 24]. Minimally invasive surgery (MIS) techniques for THA have the potential for reduced tissue trauma, leading to a more rapid recovery and return to function, compared with traditional approaches to THA. However, to achieve these potential benefits, all other aspects of patient care need to be modernised. Development and implementation of these newer anaesthetic and rehabilitation protocols allow MIS-THA to be done safely. In their study, Dr Sher et al. identified characteristics associated with same-day discharge after THA as well as assessed risk factors for complications in this select patient population. They came to the conclusion that patient characteristics, comorbidities and severe adverse events before discharge can be used to assess the potential for discharge within 24 hours [10].

Klein et al. evaluated the safety of 549 outpatient THAs that were operated in an ambulatory surgical centre. The average age of the patients was 54.4 years (range 27–73). The average ASA score was 1.6 (range 1–3). Of the 549 patients, 3 (0.5%) were admitted from the surgery centre to the local hospital due to adverse events. The authors came to the conclusion that same-day-discharge THA at an ambulatory surgical centre is safe, reproducible and effective when performed on the appropriately indicated patient [13].

What makes it in fact possible to discharge the patient the same day as the surgery? It is a combination of anaesthesia and immediate postop rehab protocols combined with optimal pain management advancements in multimodal opioid-sparing analgesia [25, 26], minimal invasive surgical techniques (e.g., approach to the joint), no dissection of muscles and tendons and certainly the selection of the implant, which has to be suitable for implantation via the intermuscular gaps. Furthermore, no standard urine catheters are used and no opioid medications are administered while the patient is still in the hospital. Subsequently, it is critical to carefully select the patient, who should be an active and motivated individual and also has a stable and caring family and friend environment. The final discharge criteria are that patient has no orthostatic hypotension, is able to walk down the hallway at least 150 m with or without an assistive device and is able to walk up and down the stairs without any major problems.

When all these points are coming together and no immediate postoperative adverse events occur, we are very confident to let the patient go home the same day as the surgery (Figs. 18.9–18.11).

Figure 18.9 The patient is leaving the ward in the early hours of the evening.

Figure 18.10 The patient is leaving the hospital around 7 p.m.

Figure 18.11 The patient has arrived home the same day as the hip replacement surgery.

Due to the increased pressure of work and inadequacy of time, this ambulatory setting for hip replacement is used and its initial report is herewith presented.

This concept will certainly undergo many modifications and changes over time. However, in our preliminary experience and on the basis of the work of some other orthopaedic institutions around the world [13–22], outpatient total hip replacement is feasible and well tolerated by a selected group of patients. Nevertheless, further scientific work is required to optimise standard protocols for rapid recovery and outpatient settings in THA. There is a need for high-quality prospective cohort and randomised trials to definitively assess the safety and effectiveness of outpatient THA [12]. Our research group will conduct a large German-government-funded multicentre research study 'PROMIS', in order to give valid answers to at least some of these unknowns. We are very confident that in future years fast-track and outpatient THA will be a standard and routine surgery for a selected group of patients.

Appendix: Our Standardised Protocol for Pain Management

Preoperative:

- Dipidolor 2 mg IV and more as needed
- Tranexamic acid 1 g IV

- Cefuroxime antibiotic 1.5 g IV
- 1500 ml isotonic saline solution

Perioperative

- LIA 200 ml (150 ml with epinephrine, 50 ml without epinephrine)
- Novalgin 2 g IV
- Sympal 50 mg IV
- Dipidolor 3–4 mg IV

Postop day of surgery:

- Novalgin 500 mg and Arcoxia 120 mg at around 7 p.m. before discharge
- Targin (oxycodon) 10 mg at around 9 p.m. at home; if necessary the same for the second night

Postop the following days:

- Arcoxia 120 mg 1-0-0
- Novalgin 500 mg 2-2-2
- Omeprazol 20 mg 1-0-0
- Oxycodon in the evening only if needed

References

1. Wengler A, Nimptsch U, Mansky T. Hip and knee replacement in Germany and the USA: analysis of individual inpatient data from German and US hospitals for the years 2005 to 2011. *Dtsch Arztebl Int*, 2014; **111**(2324):407–416.
2. Wolford ML, Palso K, Bercovitz A. Hospitalization for total hip replacement among inpatients aged 45 and over: United States, 2000–2010. *NCHS Data Brief*, 2015; (186):1–8.
3. Husted H, Solgaard S, Hansen TB, Søballe K, Kehlet H. Care principles at four fast-track arthroplasty departments in Denmark. *Dan Med Bull*, 2010; **57**(7):A4166.
4. Kehlet H, Søballe K. Fast-track hip and knee replacement — what are the issues? *Acta Orthop*, 2010; **81**(3):271–272.
5. Hartog YM, Mathijssen NM, Vehmeijer SB. Total hip arthroplasty in an outpatient setting in 27 selected patients. *Acta Orthop*, 2015; **86**(6):667–670.

6. Den Hartog YM, Mathijssen NM, Vehmeijer SB. Reduced length of hospital stay after the introduction of a rapid recovery protocol for primary THA procedures. *Acta Orthop*, 2013; **84**(5):444–447.

7. Berger RA, Jacobs JJ, Meneghini RM, Della Valle C, Paprosky W, Rosenberg AG. Rapid rehabilitation and recovery with minimally invasive total hip arthroplasty. *Clin Orthop Relat Res*, 2004; (429):239–247.

8. Berger RA. A comprehensive approach to outpatient total hip arthroplasty. *Am J Orthop (Belle Mead NJ)*, 2007; **36**(9 Suppl):4–5.

9. Berger RA, Cross MB, Sanders S. Outpatient hip and knee replacement: the experience from the first 15 years. *Instr Course Lect*, 2016; **65**:547–551.

10. Sher A, Keswani A, Yao DH, Anderson M, Koenig K, Moucha CS. Predictors of same-day discharge in primary total joint arthroplasty patients and risk factors for post-discharge complications. *J Arthroplasty*, 2016; pii:S0883-5403(16)30902-0.

11. Nelson SJ, Webb ML, Lukasiewicz AM, Varthi AG, Samuel AM, Grauer JN. Is outpatient total hip arthroplasty safe? *J Arthroplasty*, 2016; pii:S0883-5403(16)30875-0.

12. Pollock M, Somerville L, Firth A, Lanting B. Outpatient total hip arthroplasty, total knee arthroplasty, and unicompartmental knee arthroplasty: a systematic review of the literature. *JBJS Rev*, 2016; **4**(12).

13. Klein GR, Posner JM, Levine HB, Hartzband MA. Same day total hip arthroplasty performed at an ambulatory surgical center: 90-day complication rate on 549 patients. *J Arthroplasty*, 2017; **32**(4):1103–1106.

14. Husted H, Lunn TH, Troelsen A, Gaarn-Larsen L, Kristensen BB, Kehlet H. Why still in hospital after fast-track hip and knee arthroplasty? *Acta Orthop*, 2011; **82**(6):679–684.

15. Bleß HH, Kip M. *Weißbuch Gelenkersatz: Versorgungssituation bei endoprothetischen Hüft- und Knieoperationen in Deutschland* (Berlin: Springer), 2017; ISBN: 978-3-662-52904-1

16. Elias I, Krieger M. Collo-MIS. In *Kurzschaftendoprothesen an der Hüfte*, Jerosch J, ed. (Springer-Verlag), 2016; pp. 286–301.

17. Zhang P, Liang Y, Chen P, Fang Y, He J, Wang J. Intravenous versus topical tranexamic acid in primary total hip replacement: a meta-analysis. *Medicine (Baltimore)*, 2016; **95**(50):e5573.

18. Xie J, Hu Q, Huang Q, Ma J, Lei Y, Pei F. Comparison of intravenous versus topical tranexamic acid in primary total hip and knee arthroplasty: an updated meta-analysis. *Thromb Res*, 2017; **153**:28–36.
19. Morin AM, Wulf H. Lokale Infiltrationsanästhesie für Hüft- und Kniegelenksendoprothesen – Eine kurze Übersicht über den aktuellen Stand. *Anästhesiol Intensivmed Notfallmed Schmerzther*, 2011; **46**:84–86.
20. Kuchálik J, Granath B, Ljunggren A, Magnuson A, Lundin A, Gupta A. Postoperative pain relief after total hip arthroplasty: a randomized, double-blind comparison between intrathecal morphine and local infiltration analgesia. *Br J Aenesth*, 2013; **111**(5):793–799. doi: 10.1093/bja/aet248
21. Villatte G, Engels E, Erivan R, Mulliez A, Caumon N, Boisgard S, Descamps S. Effect of local anaesthetic wound infiltration on acute pain and bleeding after primary total hip arthroplasty: the EDIPO randomised controlled study. *Int Orthop*, 2016; **40**(11):2255–2260.
22. Taunton M, Short-term benefit found for anterior hip replacement American Academy of Orthopaedic Surgeons (AAOS) 2017 annual meeting: abstract 011. Presented March 14, 2017.
23. Elias I, Krieger M, Laufer A, Elli A, Rinaldi G. 5 years outcomes of total hip arthroplasty using a femoral neck preserving short stem. The 36th annual meeting of the Israeli Orthopedic Association, Tel-Aviv, Israel Nov 29-Dec 2, 2016.
24. den Hartog YM, Mathijssen NM, Vehmeijer SB. The less invasive anterior approach for total hip arthroplasty: a comparison to other approaches and an evaluation of the learning curve: a systematic review. *Hip Int*, 2016; **26**(2):105–120.
25. Berger RA, Sanders SA, Thill ES, Sporer SM, Della Valle C. Newer anesthesia and rehabilitation protocols enable outpatient hip replacement in selected patients. *Clin Orthop Relat Res*, 2009; **467**(6):1424–1430. doi: 10.1007/s11999-009-0741-x
26. Kehlet H, Aasvang EK. Regional or general anesthesia for fast-track hip and knee replacement: what is the evidence? *F1000Res*, 2015; **4**. pii: F1000 Faculty Rev-1449.

Preoperative	Perioperative	Postop day of surgery	Postop days after surgery
Dipidolor 2 mg IV			
Tranexamic acid 1 g IV			
Cefuroxime 1.5 g IV			
Isotonic saline solution 1500 ml IV			

Appendix

MIS techniques for total hip arthroplasty THA have the potential for reduced tissue trauma, leading to more rapid recovery and return to function than with traditional approaches to THA. However, to achieve these potential benefits, all other aspects of patient care need to be modernised. Development and implementation of these newer anaesthetic and rehabilitation protocols allow MIS-THA to be done safely.

Minimally invasive surgical techniques have become an important component of modern hip replacements. These techniques require minimisation of damage to periarticular soft tissues and conservation of bone substance to the extent possible. The prerequisite for these requirements is the development of an endoprosthesis whose stem is designed both to conserve bone mass and to largely avoid damage to soft tissues, which permits faster restoration of hip function.

Quick functional rehabilitation is supported by the selection of innovative articular interfaces (materials) and especially by the selection of implant designs that support muscle- and bone-saving surgical techniques. These tissue-sparing, minimally invasive techniques facilitate the patient's functional rehabilitation immediately postop and potentially a higher long-term survival rate of the prosthesis.

Another benefit of this type of hip arthroplasty is that fully weight bearing follow-up therapy adjusted to pain is possible and desirable on the day the surgery is performed. Due to the limited soft-tissue trauma, which results in a significantly shorter recovery phase, outpatient physical therapy can begin after an average hospitalisation of three days. Overall, almost all study participants exhibited early recovery with good results.

Chapter 19

Advances in Osteoarthritis of the Hip

Pratham Surya, Sriram Srinivasan and Dipen K. Menon
Trauma & Orthopaedics Department, Kettering General Hospital NHS Foundation Trust, Rothwell Road, Kettering NN16 8UZ, United Kingdom
pratham.surya@gmail.com, sriramharish@yahoo.com, dipenmenon@gmail.com

19.1 Introduction

Osteoarthritis (OA), or degenerative joint disease (DJD), is a chronic systemic disorder of the bone and joint characterised by the progressive wear of articular cartilage. This can be classified into primary and secondary OA depending on its causal factors. OA is considered the most common form of joint disorder, affecting synovial joints. There is a strong correlation between increasing age and the prevalence of OA [1]. There is evidence of age-related changes in articular cartilage and chondrocyte function that can contribute to the development and progression of OA [2]. More than 40% of the population gets affected by OA. The joints involved in OA are the hip, knee, ankle, shoulder and small joints in the fingers and toes. The most common joint affected by OA is the knee joint.

The Hip Joint in Adults: Advances and Developments
Edited by K. Mohan Iyer
Copyright © 2018 Pan Stanford Publishing Pte. Ltd.
ISBN 978-981-4774-72-7 (Hardcover), 978-1-351-26244-6 (eBook)
www.panstanford.com

Hyaline cartilage is a highly specialised connective tissue that is smooth, elastic and firm, covering the articulating ends of the component bones in diarthrodial joints. Articular cartilage has viscoelastic properties that allow deformation under load-bearing conditions primarily due to alterations in fluid flow through a solid matrix [3]. Articular cartilage has excellent shock absorptive properties and helps in load transfer across a joint. The articular cartilage layer is smooth, allowing almost frictionless motion between the joint surfaces [3].

Conditions that produce increased load transfer or altered patterns of load distribution accelerate the development of OA [4]. Current knowledge segregates the risk factors into two fundamental mechanisms: abnormal loading of normal articular cartilage and normal loading of abnormal cartilage [4]. The findings of premature chondrocyte senescence and apoptotic acceleration in OA substantiate that the disease is age dependent, mechanically driven and chemically mediated (particularly by reactive oxygen species [ROS]) [5]. OA is associated with synovial inflammation and hypertrophy. The activated synovial tissue and chondrocytes can release cytokines (interleukin-1 [IL-1]) beta, tumour necrosis factor–alpha and other inflammatory mediators, such as matrix metalloproteinases, IL-6, IL-8, prostaglandin E2 and nitric oxide. These cytokines activate degradative (catabolic) biochemical pathways [5]. In addition to these degradative biochemical pathways there can be biomechanical derangements (e.g., obesity and joint malalignment) that predispose one to and perpetuate OA [5].

OA of the hip can adversely affect activities of daily living, the ability to work and the quality of life (QoL) in individuals. There is no known cure as yet for OA, but its symptoms can be managed by activity modification, physiotherapeutic, pharmacological and surgical treatment [6, 7]. OA can be classified into nongenetic (high body weight, age, gender, joint trauma, career related to heavy weight lifting, functional stress, occupation related) and genetic, which results from minor contributions from several genetic loci [8] as a result of genetic mutations [9].

In Fig. 19.1 is an image of the hip joint showing the smooth articular cartilage, bone, synovial membrane and joint space.

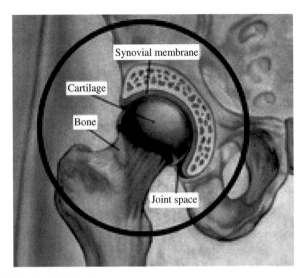

Figure 19.1 The hip joint.

19.2 Epidemiology [10]

The most common form of arthritis is OA. OA has a high prevalence and often results in disability. OA is responsible for significant impairment in the QoL in affected individuals. It is the main indication for the need for a joint replacement. More transparent prevalence and incidence data can be found in epidemiological studies.

- OA occurs in people of all ages but most commonly affects people over the age of 65 years.
- It affects approximately 27 million Americans. Knee OA affects 70 million Europeans.
- One in every four adults will be affected by OA of the hip.
- OA is the most common of the more than 100 types of arthritis, and the hip joint is the second-most affected joint in the body.

A study reported on the prevalence of OA in the Indian population. It concluded that out of 4909 participants, about 28.7% were suffering from OA [11]. The incidence of OA has been found to be higher in big cities and villages than in the smaller towns and small cities. The percentage of people suffering with the disease in big cities and villages is approximately 32%, whereas the percentage decreases by almost half, to 17%, in small-town populations [12]. The incidence of OA differs as the definition of OA changes according to the specific joints under investigation and the features of the study participants. Radiographic knee OA in adults of over 45 years was found to be 19.2% in the Framingham Study of age-standardised incidence of the disease. It was found to be 27.8% in the Johnston County OA Project. In the third survey of National Health and Nutrition Examination Survey (NHANES III) almost 37% of the population aged more than or equal to 60 years was found to be suffering from knee OA [6].

19.3 Aetiology of Primary Hip OA

Various factors are involved in the aetiology of primary hip OA, such as genetic factors (polygenic disease), body weight, occupational and anatomical factors [11]. Primary OA of the hip is a disease that primarily affects Caucasians. A flow chart describing these factors is shown in Fig. 19.2.

OA is a disease with varied pathophysiology. A workshop on aetiopathogenesis of OA concluded that OA can be idiopathic (aetiology unknown). It was previously believed to be caused by wear and tear of the joint. Currently scientists and researchers believe it to be a disease of the joint. Some factors that have a role in development of hip OA are as follows:

- **Genetic:** OA has a significant heritable component. Genes associated with OA tend to be associated with the process of synovial joint development. Mutations in these genes might directly cause OA [9]. Early-age-onset OA is caused by mutations in matrix molecules often associated with chondrodysplasias. Middle-age onset of OA is caused by mutations that predispose the joint to injury or malalignment and late-age-onset OA to mutations that regulate subtle aspects of joint development and structure [9]. Some inherited traits may result in a rare defect where the body does not produce collagen, which is a component protein in cartilage. Recently, researchers found that a gene called *FAAH* (fatty acid amide hydrolase), which mediates pain, is found to be higher in patients suffering from OA than in a normal person [13]. Other genes that are linked to OA are vitamin D receptor, oestrogen receptor-1 and inflammatory cytokines such as IL-1, IL-4 and matrilin-3.

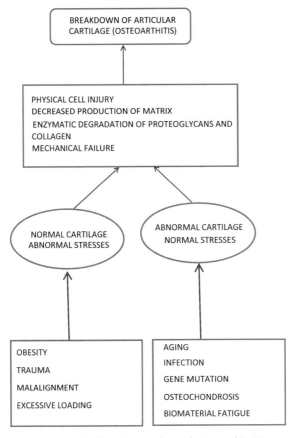

Figure 19.2 Factors involved in the aetiology of primary hip OA.

- **Obesity:** A high body mass index (BMI) increases the pressure on lower limb weight-bearing joints such as the hip and knee. It is also proven that an excess of fat tissue induces the production of inflammatory cytokines, which can cause further damage to the joints.
- **Overuse and damage:** Occupational hazards such prolonged weight bearing and heavy lifting can lead to damage of the cartilage. Intra-articular fractures lead to OA.

- **Developmental or acquired deformities:** Some examples are hip dysplasia, Perthes disease of the hip (Fig. 19.3) and slipped upper femoral epiphysis (SUFE).

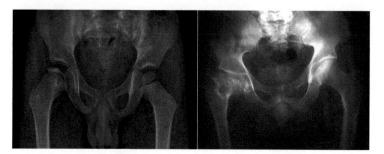

Figure 19.3 Perthes disease in a skeletally immature individual leading to OA in adulthood.

- **Joint disorders in athletes:** Injuries involving a joint are very common in athletes. Joint injuries occur in professional players, especially soccer players. In every thousand hours on the sports field 10 to 35.5 injuries are reported [13]. Articular cartilage has very poor regenerative properties, leading to OA. A study conducted on soccer players indicates that 32%–49% of the players have OA. In a study it was determined that there is a threefold increased risk of hip OA in elite players than comparatively less elite ones [13].
- **Other or miscellaneous factors:** Hormonal disturbances, like an excess of growth hormone, lead to OA. Other joint and metabolic disorders may lead to an increased risk of OA.

19.4 Basic Science of Cartilage and Changes in Hip Arthritis

It is important to understand the normal structure of articular cartilage before describing the pathology of OA. The cartilage can be divided into cellular and extracellular components. The water content of the cartilage is between 70% and 80%.

- **Cellular components**
    - **Chondrocytes:** They constitute up to 2% of the cartilage. They are highly specialised cells that respond to growth

factors, mechanical loads, piezoelectric forces and hydrostatic pressures and are metabolically highly active. Chondrocytes play an important role in synthesising the extracellular matrix. However, they have poor regenerative potential.

- **Extracellular components**
    - **Collagen:** Collagen constitutes the majority of the dry weight of articular cartilage, varying from 50% to 65%. The important collagen in the articular cartilage is Type 2, which constitutes up to 90%–95% of the collagen. The main function of collagen is to provide mechanical integrity to the cartilage to resist tensile and shearing forces.
    - **Proteoglycan:** These are made up of units of core protein linked to glycosaminoglycan (GAG). There are two main types of GAGs in the articular cartilage, which are chondroitin sulphate and keratan sulphate (Fig. 19.4). GAG has carboxylate and sulphate ions, which are negatively charged. These negatively charged ions attract water and bind to it. This is important in building up hydrostatic pressure to resist the compressive forces on the articular cartilage.

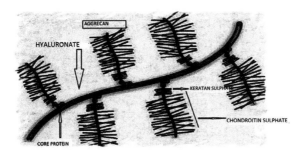

Figure 19.4 Diagrammatic representation of chondroitin sulphate and keratan sulphate in articular cartilage.

The structural changes of the cartilage in arthritis are given the Table 19.1. The main factor attributed to the development of OA is the increase in water content in the cartilage, which softens it and causes its damage.

Table 19.1 Structural changes in the articular cartilage in OA

Factors	Changes
Water	Increased
Chondrocytes	Increased and are in clusters
Young's modulus of elasticity	Decreased
Chondroitin to keratan sulphate ratio	Increased
Collagen	Disorganised
Metalloproteinase	Increased

Loss of articular cartilage is the hallmark of OA, but researchers are of the opinion that the subchondral bone (Figs. 19.5 and 19.6) also has an important role in the occurrence and prevalence of the disease. Cross et al. in their study suggest that hip and knee OA is the eleventh-leading cause of disability [15]. Orthopaedic surgeons, cellular and molecular biologists, biochemists, biomaterial scientists and engineers are working on ways that could possibly help in defining pathways of OA pathogenesis and aetiology and subsequently to identify novel, potential targets and therapeutic interventions for OA therapy [16]. The main aim behind the present contribution is to bring new, critical aspects of OA research like clinically relevant, scientific and translational breakthroughs.

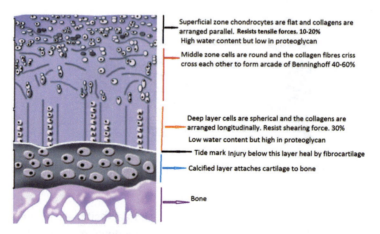

Figure 19.5 Diagrammatic representation of the microstructure of articular (hyaline) cartilage.

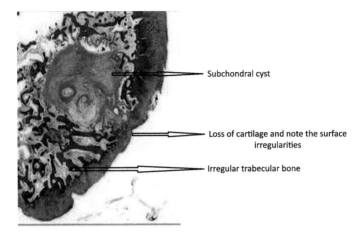

Figure 19.6 Histopathology of articular cartilage in OA.

The role of the subchondral bone in pathophysiology is emphasised in recent developments. The subchondral bone is located immediately under the cartilage. The subchondral bone and articular cartilage form an osteochondral unit.

Radin and his colleagues first proved that the subchondral bone has a major role in the development of OA [17]. They proposed that recurrent stresses acting on the joint cartilage lead to trabecular microfractures in the subchondral bone. Healing of the microfractures leads to stiffness in the subchondral bone. This increase in stiffness causes tension and shear stress on the cartilage. This ultimately leads to joint degeneration. However, there are differences in opinion, as a few other studies suggest that there is a decrease in stiffness in the subchondral bone in OA.

Recent developments throw light on the two different stages of OA (early- and late-stage OA).

19.5 Early-Stage OA

In early OA, there is a combination of joint pain and the presence of lesions of both cartilage and the subchondral bone [18], as seen in arthroscopy and magnetic resonance imaging (MRI). Scintigraphic analyses of bone reveal that following a rupture of the anterior cruciate ligament, there is a subchondral bone deformity, which may lead to cartilage damage, ultimately leading to OA [19].

19.6 Late-Stage OA

Most of the joint structure is affected in late-stage OA and is associated with joint effusion and stiffness. Patients with late-stage OA are candidates for surgical intervention.

19.7 Symptoms of OA

19.7.1 Initial Symptoms

The most common symptoms of hip OA are joint pain and stiffness. In the advanced stages, there may be hip deformity or leg length discrepancy. The pain is usually localised in the groin, thigh or buttock. The pain is aggravated by activities that require lifting, walking and twisting of the body. Patients can present with rest pain, start-up pain as well as nocturnal pain. The pain tends to worsen with activity, especially following a period of rest. The hip can stiffen after a period of rest, and this has been called the gel phenomenon. OA can cause morning stiffness, but it usually lasts for less than 30 minutes, unlike rheumatoid arthritis, which can cause stiffness for 45 minutes or more. Patients may report joint locking or instability. These symptoms impair the QoL in affected individuals [19, 20].

19.7.2 Progression of Disease

Although the disease progression is slow, it ultimately leads to a malfunctioning joint. Pain occurring in early OA can be reduced by life style modifications, such as rest, analgesia and the avoidance of factors that aggravate the symptoms. As the disease progresses it fails to respond to conservative management. In more advanced disease, surgical treatment will need to be considered. A total hip replacement can consistently relieve pain in patients with advanced OA.

19.8 Diagnosis [7, 11]

OA can often involve joints in an asymmetric pattern. A patient may have severe, debilitating OA of one hip, with almost normal function and radiological appearance of the opposite hip. Physical examination is important in making the diagnosis. Pain on hip movement (particularly internal rotation with the hip flexed) and limitation of active and passive movement range are common in OA of the hip. Patients with more advanced disease present with hip deformities and alterations in their leg lengths.

The severity of radiological OA can be classified by the Kellgren and Lawrence system (shown in Table 19.2).

Computed tomography (CT) scans may be used in cases of severe bone deformity, malalignment and dysplasia to plan surgery. There is no established role of the MRI scan in the diagnosis of OA. Studies are yet to prove its role in diagnosing early stages of OA. But it can be useful in detecting conditions that predispose a person to OA of the hip, like avascular necrosis, labral pathology and femoroacetabular impingement.

Table 19.2 Kellgren and Lawrence system to classify severity of radiological OA

Grade 0	No features of OA on X-ray
Grade 1	Joint space narrowing and possible osteophytic lipping
Grade 2	Joint space narrowing, subchondral sclerosis and osteophytes
Grade 3	Severe narrowing of the joint space, subchondral sclerosis, cysts, multiple osteophytes and possible bone deformation
Grade 4	All of the Grade 3 features but with defined deformity and possible bony ankyloses

19.9 Management of OA

OA can manifest in a variety of ways in different patients. For some, the progression of OA may be slow, while other patients may have rapidly progressive disease.

19.10 Animal Models of OA

Animal models are essential tools that help investigate the progression of OA. Invasive models that trigger the development of OA by inducing an intra-articular injury in the animal may confound the results because of the effect of the injury on the joint. Noninvasive models, on the other hand, avoid this problem. The latter procedures use mechanical induction without causing any cut in the skin or disruption of the joint. A few noninvasive mouse models have been used for research purposes, and they represent the most unique models for studying different aspects of the disease process.

19.10.1 Invasive Model

Advantages are that rapid induction is possible with a high level of reproducibility. The technique is readily available, and there are widely described systems in this area.

Limitations are the risk of infection, the need for expertise and that early changes are often undetectable because of rapid induction.

19.10.2 Noninvasive Model

Advantages include rapid induction, high reproducibility, low infection risk and a long time frame of OA changes, allowing for early evaluation and intervention

Limiting factors are that the technique is not readily available, expertise is needed and there is limited literature available in this area.

19.11 Nonpharmacological Management of Hip OA

Nonpharmacological or nondrug treatment aims to achieve relief from the symptoms of OA without the use of any chemical substance. Activities that involve lifting heavy weight should be avoided to prevent further damage to the hip joint. Obese people can work towards weight reduction as this can slow the progression of arthritis. Regular exercise for 20–30 minutes a day to maintain musculoskeletal fitness is recommended. Exercises like stretching, Pilates, yoga and swimming increase core stability and endurance of the musculoskeletal system. Patient education is highly recommended. Using a walking stick; wearing shoes with thick, soft, shock-absorbing soles and using other related mechanical aids can help to manage the symptoms of OA. The local application of warmth or cold (thermotherapy) and transcutaneous nerve stimulation have also been shown to have salutary effects on the management of pain and joint stiffness in OA. Physiotherapy involves the use of acupuncture, massage and other therapies, which can be beneficial in relieving the symptoms of OA, facilitating exercise treatment.

There is limited evidence that aquatic exercises can benefit patients with hip and knee OA. Although there are not many high-quality studies conducted in this field, a few studies support the benefit of aquatic exercises. Moderate differences have been observed in function (SMD 0.26, 95% confidence interval [CI] 0.11 to 0.42) and similar effects have been noticed in QoL (SMD 0.32, 95% CI 0.03 to 0.61). Pain was found to be reduced by 3% (absolute reduction) and 6.6% (relative reduction). However, this treatment was not very effective in improving joint stiffness and walking ability. One may consider aquatic exercises as a treatment modality, although there are very limited data for further recommendations [19].

19.12 Pharmacological or Drug Treatment [13, 14]

Drugs that reduce pain and provide symptomatic relief are nonsteroidal anti-inflammatory drugs (NSAIDs), paracetamol and capsaicin cream.

- **Hyaluronic acid**: This provides lubrication to joints and helps in easing joint movement.
- **Long-acting (depot) steroid injection**: This reduces inflammation within the hip joint.
- **Platelet-rich plasma therapy**: A sample of the patient's own blood is centrifuged to harvest concentrated blood platelets, and this is injected into the affected joint.

Clinical recommendations that can be practiced are:

- Physical therapy using land- or water-based exercises can help reduce pain and improve function and movement in patients suffering with OA of the hip.
- The first-line drug for early-stage or mild OA is paracetamol.

- NSAIDs are superior to paracetamol. They are used for the treatment of the symptoms arising from moderate OA.
- Intra-articular corticosteroid injections are helpful in reducing hip joint pain in the short term (few weeks to months). They can be used for diagnostic purposes to localise pain, in addition.
- Viscosupplementation: Intra-articular hyaluronic acid injections of the hip can potentially delay the need for surgical intervention.
- Combined treatment with glucosamine and chondroitin can be efficacious in diminishing pain in OA patients, although this may not provide consistent results.
- Total hip replacement is the preferred treatment option in patients with chronic, symptomatic hip OA, who fail conservative management.

The recommended dosage of Paracetamol is 650 mg to 1000 mg. The dose of celecoxib is 200 mg once a day. Diclofenac and diclofenac misoprostol combination can be taken two to three times in a day – a dose of 50 mg and 200 mcg, respectively. Ibuprofen can be taken thrice daily – a dose strength of up to 600 mg.

19.13 Surgical Treatment

Total hip arthroplasty (THA) is an operation that was developed by the late Sir John Charnley in the late 1950s. THA is a commonly performed orthopaedic procedure that can greatly improve the QoL in patients suffering from OA of the hip. The procedure and implants have evolved over the years, with both cemented and uncemented hip implants currently in use. Recent trends in the development of this operation include the use of minimally invasive exposures, such as the anterior and SuperPath approaches, and enhanced recovery using a combination of appropriate analgesia, perioperative medications, early mobilisation and physiotherapy. Patients undergoing minimally invasive procedures can be expected to start walking as early as two hours following the operation and can be discharged home as early as two to three days after the procedure.

THA is not recommended in the early stages of hip OA. It is reserved for patients with advanced symptomatic disease when conservative options have failed. Balancing the risk of adverse outcomes following the procedure with the benefits derived from the procedure is important in the decision-making process when THA is considered in a patient.

19.14 Drugs Assessed for Disease-Modifying Potential in OA [20]

Agents	Proposed mechanisms of action
Glucosamine sulphate	May stimulate proteoglycan synthesis by chondrocytes; mild anti-inflammatory properties
Chondroitin sulphate	Stimulates RNA synthesis by chondrocytes, partially inhibits leukocyte elastase, overcomes dietary deficiency of sulphur-containing amino acids
Doxycycline	Inhibits matrix metalloproteases
Diacerein	Inhibits IL-1 synthesis and activity
Risedronate	Reduces bone marrow lesions in the subchondral bone
Hyaluronic acid (HA)	Promotes endogenous HA production, antioxidant function in joints, may provide joint lubrication (which is a hypothesis not proved)
Avocado/soybean unsaponifiables	Repress chondrocyte catabolic activities, inhibit inflammatory mediators

19.15 New Targets for Treating OA

There are a number of potential targets that can be manipulated to alter the course of OA. Research is ongoing to develop effective drugs in this area. Calcitonin is in trial clinically as it has shown remarkable effects on bone remodelling. Vitamin D gets depleted in OA, and vitamin D replacement can benefit patients with OA. Some protease enzymes deplete certain enzymes and can have an adverse effect on articular cartilage. Inhibition of such proteases can be the potential target for treatment of OA. One such enzyme inhibitor, ADAMTS5, is being investigated [21].

Pharmacological manipulation of nerve growth factor (NGF) is another potential target area that is under trial. NGF is overexpressed in joint tissues and believed to have a role in the progression of OA. Tanezumab is a molecule that can inhibit NGF and its receptors. Tanezumab treatment has produced remarkable results in early clinical trials, providing hope for OA patients. Nitric oxide is a highly reactive cytotoxic free radical that is implicated in tissue injury, including cartilage. Inhibition of nitric oxide synthesis (nitric oxide synthase) is a therapeutic strategy in altering the course of OA. Synovitis is a common feature in OA. A mediator called bradykinin is released in synovitis. An antagonist of bradykinin is being developed. A recent phase 2 study showed its effectiveness in OA patients when compared with a placebo [22].

Surgery: Joint replacement surgery results in effective reduction in pain and improved function. For advanced symptomatic hip arthritis, total arthroplasty is the definitive procedure [23]. Current research is being directed towards alternative bearing surfaces, minimally invasive techniques and enhanced recovery from the operation.

19.16 Conclusion

A deeper understanding of the pathology of OA is needed to develop effective therapeutic strategies that influence the course of the disease. Currently available treatments, surgical or conservative, are inadequate to prevent the development of OA. There is a strong need for the development of molecular, biological and mechanical interventions to delay or prevent the onset of OA [24].

References

1. Lawrence JS, Bremner JM, Bier F. Osteo-arthrosis. Prevalence in the population and relationship between symptoms and x-ray changes. *Ann Rheum Dis*, 1966; **25**(1):1–24.
2. Buckwalter JA, Mankin HJ. Articular cartilage: tissue design and chondrocyte-matrix interactions. *Instr Course Lect*, 1998; **47**:477–486.
3. Cohen NP, Foster RJ, Mow VC. Composition and dynamics of articular cartilage: structure, function, and maintaining healthy state. *J Orthop Sports Phys Ther*, 1998; **28**(4):203–215.

4. Goldring MB, Goldring SR. Articular cartilage and subchondral bone in the pathogenesis of osteoarthritis. *Ann N Y Acad Sci*, 2010; **1192**(1):230–237.

5. Krasnokutsky S, Attur M, Palmer G, Samuels J, Abramson SB. Current concepts in the pathogenesis of osteoarthritis. *Osteoarthritis Cartilage*, 2008; **16**(3):S1–S3.

6. http://www.arthritis.org/about-arthritis/types/osteoarthritis/causes.php

7. Ling SM, Bathon JM. Osteoarthritis in older adults. *J Am Geriatr Soc*, 1998; **46**(2):216–225.

8. Dieppe PA, Lohmander LS. Pathogenesis and management of pain in osteoarthritis. *Lancet*, 2005; **365**(9463):965–973.

9. Sandell LJ. Etiology of osteoarthritis: genetics and synovial joint development. *Nat Rev Rheumatol*, 2012; **8**(2):77–89.

10. Cooper C, Javaid MK, Arden N. Epidemiology of osteoarthritis. In *Atlas of Osteoarthritis* (Springer Healthcare Ltd.), 2014; pp. 21–36.

11. Brandt KD. Non-surgical treatment of osteoarthritis: a half century of "advances". *Ann Rheum Dis*, 2004; **63**(2):117–122.

12. Pal CP, Singh P, Chaturvedi S, Pruthi KK, Vij A. Epidemiology of knee osteoarthritis in India and related factors. *Indian J Orthop*, 2016; **50**(5):518.

13. Lee HH, Chu CR. Clinical and basic science of cartilage injury and arthritis in the football (soccer) athlete. *Cartilage*, 2012; **3**(1 Suppl):63S–68S.

14. Felson DT, Lawrence RC, Dieppe PA, Hirsch R, Helmick CG, Jordan JM, et al. Osteoarthritis: new insights. Part 1: The disease and its risk factors. *Ann Inter Med*, 2000; **133**(8):635–646.

15. Cross M, Smith E, Hoy D, Nolte S, Ackerman I, Fransen M, et al. The global burden of hip and knee osteoarthritis: estimates from the global burden of disease 2010 study. *Ann Rheum Dis*, 2014; **73**(7):1323–1330.

16. Cucchiarini M, Girolamo LD, Filardo G, Oliveira JM, Orth P, Pape D, et al. Basic science of osteoarthritis. *J Exp Orthop*, 2016; **3**(22):1–18.

17. Radin E, Paul I, Rose R. Role of mechanical factors in pathogenesis of primary osteoarthritis. *Lancet*, 1972; **299**(7749):519–522.

18. Zhang Y, Jordan JM. Epidemiology of osteoarthritis. *Clin Geriatr Med*, 2010; **26**(3):355–369.

19. Bartels EM, Lund H, Hagen KB, Dagfinrud H, Christensen R, Danneskiold-Samsøe B. Aquatic exercise for the treatment of knee and hip osteoarthritis. *Cochrane Database Syst Rev*, 2007; (4):CD005523.

20. Patient education: osteoarthritis symptoms and diagnosis (beyond the basics) http://www.uptodate.com/contents/osteoarthritis-symptoms-and-diagnosis-beyond-the-basics

21. Anandacoomarasamy A, March L. Current evidence for osteoarthritis treatments. *Ther Adv Musculoskelet Dis*, 2010; **2**(1):17–28.

22. Lee HH, Chu CR. Clinical and basic science of cartilage injury and arthritis in the football (soccer) athlete. *Cartilage*, 2011; **3**(1 Suppl):63S–68S.

23. Sinusas K. Osteoarthritis: diagnosis and treatment. *Am Fam Phys*, 2012; **85**(1):49–56.

24. Dieppe P. Osteoarthritis some recent trends. *J Indian Rheumatol Assoc*, 2005; **13**:107–112.

Chapter 20

Advances in Primary and Revision Hip Arthroplasty

Shibu P. Krishnan and G. Gopinath
Trauma & Orthopaedics, Buckinghamshire NHS Trust, United Kingdom
shibupkrishnan@hotmail.com, shibupkrishnan@yahoo.co.in

Total hip arthroplasty (THA) remains one of the most successful and cost-effective forms of treatment ever introduced. Advances in hip arthroplasty surgery focus on improving accuracy of reconstruction, survival of the prostheses and recovery rates after surgery; minimising complications and making subsequent revision surgery easier. Advances in THA aimed at improving prosthetic survival include advances in bearing surfaces. Attempts to improve accuracy of reconstruction include the use of patient-specific instrumentation, navigation and computer guidance. Attempts to accelerate recovery rates include mainly mini-invasive techniques and enhanced recovery programmes. This chapter intends to provide an update on the relevant advances in hip replacement surgery.

The Hip Joint in Adults: Advances and Developments
Edited by K. Mohan Iyer
Copyright © 2018 Pan Stanford Publishing Pte. Ltd.
ISBN 978-981-4774-72-7 (Hardcover), 978-1-351-26244-6 (eBook)
www.panstanford.com

20.1 Update on Polyethylene

The polyethylene acetabular component forms part of the traditional bearing couple, that is, hard on soft, which articulates with a metal or ceramic head. A major factor influencing prosthetic survival is wear of the polyethylene with its debris, causing osteolysis and loosening of hip components. Advances in reducing the wear characteristics of polyethylene focused on three stages: manufacturing, sterilisation and shelf life. The types of polyethylene include high-density polyethylene (HDP), ultra-high-molecular-weight polyethylene (UHMWPE), highly cross-linked polyethylene (HCLPE) and vitamin E–doped polyethylene (E-Poly). The direct compression moulding technique has been identified as the preferred manufacturing process to achieve consistently lower wear rates. Inferior wear characteristics are achieved with other manufacturing techniques, such as ram bar extrusion with secondary machining, hot isostatic pressing into bars with secondary machining and compression moulding into bars with secondary machining. The calcium stearate component of the lubricant used to protect the processing equipment was recognised to produce unfused polyethylene particles, thus diminishing the mechanical properties of the final product. It was therefore avoided in the manufacturing process.

Cross linking of UHMWPE improves resistance to bearing wear, adhesive wear as well as abrasive wear. Cross linking is achieved by using gamma or electron beam radiation at approximately 10 mrad. However, increased cross linking reduces mechanical properties and leads to reductions in tensile strength (force required to pull to a point where it breaks), fatigue strength (highest stress it can withstand for a given number of cycles without breaking), ductility (ability to deform under tensile stress without fracture) and fracture toughness (force required to propagate a crack). Reduction in such mechanical properties by increased cross linking is deleterious in clinical practice since the HCLPE acetabular component can fracture when subjected to excess mechanical loading; for example, edge loading of the acetabulum by the prosthetic femoral neck leads to fracture of the acetabular liner. HCLPE is also disadvantageous in the hip since the average size of particles released during wear is smaller compared to that from conventional PE.

A secondary process used to improve wear and oxidative degradation of polyethylene is heat annealing. Heat annealing involves heating polyethylene close to its melting point. This is done to cross link the remaining free radicals after initially cross linking using ionic radiation. Any remaining free radicals can be oxidised during the sterilisation process and during the shelf life. Oxidation of polyethylene molecules results in chain scission (fragmentation and shortening of large polymer chains), which leads to lowering of the molecular weight of the polymer, reduced yield strength, reduced elongation to break (brittle), reduced ultimate tensile strength and reduced toughness. Controversy remains regarding heat annealing since an annealing cycle above the melting point effectively removes the free radicals but weakens the polyethylene microstructure [1, 2]. On the other hand, heating below the melting temperature preserves the microstructure but allows an unknown quantity of free radicals to remain trapped in the final product.

Polyethylene can get oxidised during its shelf life. This can be minimised by reducing the free radical content and exposure to oxygen. Packaging the products in an oxygen-free environment and a reduced shelf life minimise this risk.

20.1.1 E-Poly

Infusion of vitamin E into irradiated polyethylene is aimed at reducing the oxidation potential of the material. The vitamin E molecule is made up of two ring structures and a carbon chain. The carbon chain makes the vitamin E molecule hydrophobic, which allows it to be readily infused into the polyethylene. When a molecule of vitamin E encounters a free electron in the polyethylene, it donates a hydrogen atom from the –OH group on the ring structure, which, in effect, transfers the free radical from the polyethylene chain to the vitamin E molecule. Unlike the remelted material, E-Poly HCLPE still has detectable levels of free radicals, but the key to this technology is the location of those free radicals. After the infusion process, the free radicals detected in the polyethylene are likely associated with the ring structures on the vitamin E molecule, not the polyethylene molecule. Therefore, if oxygen was introduced into the system, the oxygen molecules would only react with the vitamin E molecules, leaving the polyethylene molecules untouched. In addition, the free radicals associated with the vitamin E molecules are part of the electron field of the ring structures, making it more difficult for oxygen to react with the free radicals.

Manufacturers use different techniques to produce the polyethylene components. Most evaluation procedures available in literature assess the microstructural properties of the liners, which are hard to visualise or compare in their 3D molecular assembly. Currently there is not enough data to make quantitative structural comparisons among the liners available from different manufacturers.

20.2 Ceramics in Total Hip Arthroplasty

Advances in ceramics include improved bearing surfaces, improved prosthetic design for implant fixation and the development of specific taper designs. Ceramic bearing surfaces offer low wear rate and high biocompatibility. Concerns regarding the use of ceramics include their brittle nature, their mechanical properties in high load applications and 'squeaking'.

20.2.1 Alumina Matrix Composite (Biolox Delta Ceramic)

Biolox delta is a zirconia-toughened alumina ceramic. An earlier design – Biolox forte – was made of alumina with a small proportion of magnesium oxide. The addition of zirconium to create Biolox delta increased the fracture toughness of alumina. Biolox delta contains 82% alumina, 17% zirconium oxide, 0.3% chromium oxide and 0.6% strontium oxide [3]. It was initially introduced for use as femoral heads only and more recently expanded to use as Biolox-delta-on-Biolox-delta bearings in total hip arthroplasty (THA). Because of its increased fracture toughness and burst strength, it is expected to reduce the estimated fracture risk of alumina ceramic (Biolox forte) heads (0.02% is the fracture risk for Biolox forte heads). A wider range of head sizes is available with Biolox delta compared to Biolox forte.

20.2.2 Oxidised Zirconium-Bearing Surface (Oxinium)

A low-friction zirconium oxide surface can be produced over metallic zirconium alloy by a thermally driven oxygen diffusion process. This is not a ceramic coating but the transformation of the surface (5–10 μm thick). This layer is much harder and more scratch resistant than the underlying metallic zirconium alloy as well as cobalt-chrome femoral heads. However, compared to alumina ceramics, the oxinium surfaces are less hard and less scratch resistant.

The disadvantages of oxinium include inadequate in vivo data on their durability and concerns regarding failure of the oxide surface, thus exposing the inferior zirconium bearing surface. The oxinium surface has been shown to have damage in cases with recurrent hip dislocations and therefore both the femoral head and acetabular liner components should be replaced during revision surgery.

20.2.3 Hydroxyapatite Coating

Coating of uncemented femoral stems and acetabular shells with hydroxyapatite (HA) improves osseointegration of the component [4]. HA has inferior fatigue properties, and therefore techniques have been adopted to develop thinner coatings of HA (30–100 μm) on a titanium implant using techniques such as plasma spray, sputtering and pulse layer deposition. The shear strength of HA-plasma-spray-coated titanium implants is comparable to the shear strength of the cortical bone. Adverse reactions due to third body wear on polyethylene components have been reported with first-generation thicker HA coatings and are expected to be reduced with the current thinner and more uniform HA-coated designs.

20.3 Current Status of Metal-on-Metal Bearing Surfaces

The risks of metal-on-metal (MOM) bearings outweigh the benefits achieved. Metal/ceramic on highly cross linked UHPWPE and ceramic-on-ceramic articulations show excellent reductions in wear rates and are safer alternatives to MOM bearings. Soft-tissue and bony destruction caused by MOM hips produce challenging revision scenarios with inferior outcomes. Surgeons expose themselves to litigations with adverse clinical outcomes caused by MOM bearings.

20.4 Computer-Assisted THA

Computer-assisted orthopaedic surgery (CAOS) in THA is aimed at improving the accuracy of component positioning along with restoration of offset and limb lengths, thereby improving outcomes and minimising complications. The different techniques of CAOS THA include passive, active and semiactive systems [5]. Passive systems include the use of navigation aids that assist the surgeon in preoperative planning and informs on implant positioning during surgery. Active systems include the use of surgical robots that autonomously perform the surgery as planned by the surgeon. A semiactive system provides guidance to the operator with a haptic robot arm for positioning of the implant that was planned preoperatively. A detailed description of CAOS THA has been dealt with in a separate chapter.

Computed tomography (CT)-based navigation may be superior than an imageless system in patients with abnormal anatomies, such as hip dysplasia and posttraumatic deformities, or in revision surgery [6]. Despite the best computer guidance, deviations in the positioning of components can occur during impaction of press-fit components due to sizing errors.

The positioning of implants, both acetabular and femoral components, and restoration of limb lengths and offset are improved consistently with the use of CAOS THA when compared to freehand techniques, even when intraoperative methods of landmarking are used [7, 8]. Lewinnek et al. [9] looked at factors influencing hip dislocation and found that anterior dislocations were associated with increased acetabular component anteversion. They found no significant correlation between the cup-orientation angle and posterior dislocation. The dislocation rate for cup orientation with an anteversion of 15° ± 10° and a lateral opening of 40° ± 10° degrees was 1.5%, while outside this 'safe' range the dislocation rate was 6.1%. CAOS THA improved the surgeon's ability to place the acetabular component within the desired alignment as defined by Lewinnek et al.

Some of the disadvantages of CAOS THA are a steep learning curve, increased surgical time and cost. Short- to medium-term outcome data are comparable for freehand and computer-assisted techniques. Controversy remains as to whether the small gains in the accuracy of reconstruction achieved by CAOS THA translate to significantly improved outcomes in the long term.

20.5 Noncemented (Biologic) Fixation of Hip Components

The techniques used include the use of porous-coating, grit-blasting, HA-coating and trabecular metal technology.

20.5.1 Porous Coating

An optimal number of pores are fabricated on the surface of the metallic components so that the host bone can grow into these pores and secure the prosthesis. An ideal porous-coated implant with the maximum biological fixation is thought to have an optimum pore size of 50–350 μm (ideally 50–150 μm), and 40%–50% of its porous-coated surface is porous. A greater than 40%–50% of porosity of the porous-coated surface reduces the shear strength of the material and should be avoided. A higher pore depth increases the strength of the bone interlock. The strength of fixation is also dependent on a gap less than 50 μm between the prosthesis and the bone.

20.5.2 Grit Blasting

The surface of the metallic components is roughened by an abrasive spray of particles, which create pits with peaks and valleys. The surface roughness created is defined as the average distance from the peak to the valley of the pits on the surface. The rough surface stimulates on-growth of host bone on to the prosthesis. Higher surface roughness produces higher interface strength. With grit blasting, bone fixation occurs only on the surface. Therefore, larger areas (e.g., the entire prosthesis) need to be grit-blasted to produce sufficient fixation.

20.5.3 Hydroxyapatite Coating

Additional coating of the grit-blasted or porous-coated implants with HA has been shown to enhance prosthesis-bone interface stability [10]. HA is an osteoconductive material and readily receives osteoblasts, providing a bidirectional (bone to prosthesis and prosthesis to bone) closure of the interface between prosthesis and bone. A thicker coating of HA should be avoided since it can delaminate and weakens the bone. For the same reasons, HA may be better suited for use over porous-coated implants since the bone ingrowth is deeper than the on-growth that occurs with grit-blasted surfaces. A high crystallinity of HA and an optimum thickness of 50 µm are thought to be ideal.

20.5.4 Trabecular Metal Technology

Conventional porous-coated materials have their porosity restricted to a maximum of 40%–50% due to reductions in shear strength. This limits their maximum interfacial strength that can develop by bone ingrowth. Trabecular metal is a new porous biomaterial made of tantalum [11], which has a high and interconnecting porosity with a very regular pore size and shape. Tantalum metal (atomic number 73) is highly biocompatible and corrosion resistant and has a high fatigue strength and compression modulus, which allows it to bend before breaking. It has a low modulus of elasticity, which is similar to cancellous bone, that allows normal physiological loading and therefore reduced stress shielding. Its properties allow it to be moulded into complex shapes and used either as bulk implants or for surface coating.

Studies have demonstrated an increased rate of bone ingrowth and a higher interfacial shear strength with trabecular metal components [11]. This is attributed to a higher proportion of permissible porosity as well as a higher osteogenic response with trabecular metal components compared to the conventional porous-coated prosthesis. A higher porosity in tantalum can potentially weaken its mechanical properties. However, the mechanical properties can be controlled during the manufacturing process by adding tantalum to any weaker struts.

20.6 Enhanced Recovery Programmes for Primary THA

Enhanced recovery programmes (ERPs) aim to improve patient experience and outcomes, reduce morbidity and complications as well as hasten recovery following THA. Multimodal intervention is thought to reduce stress-induced organ dysfunction and associated morbidity that leads to subsequent hospital admissions [12]. It has been shown to reduce the mortality rate and blood transfusion requirements [13].

ERP is a multidisciplinary, multimodal approach that focuses on all aspects of patient pathway and includes behavioural, pharmacological and procedural modifications. It relies on multidisciplinary cooperation between patients, surgeons, anaesthetists, nursing team, physiotherapists, occupational therapists, clerical staff and pharmacists.

What does ERP involve?

20.6.1 Preoperative Measures

This includes the use of preoperative physiotherapy, optimisation of health and comorbidities, nutritional optimisation (including correction of anaemia), management of expectations by patient education and counselling as well as the organisation of discharge plan prior to hospital admission. Pre-emptive analgesia before surgery has been shown to reduce the overall usage of perioperative pain medications and reduce morbidity and the length of hospital stay [14].

20.6.2 Intraoperative Measures

This includes the use of standardised surgical techniques, reduced surgical times, optimised anaesthesia, local infiltration anaesthesia, etc. Good postoperative pain control and less postoperative narcotic side effects allowing early mobilisation within hours of surgery have been achieved with the advent of local infiltrative anaesthesia [15].

20.6.3 Postoperative Measures

These include early physiotherapy and mobilisation, effective pain management (including reduced opioid use to minimise nausea, dizziness, etc., which hamper mobilisation), restoration of hydration and feeding, promotion of 'wellness' model of care and robust discharge planning (including clear instructions to facilitate an independent rehabilitation plan).

20.7 Modular Femoral Stem Designs in THA

Modern femoral stem designs are modular compared to old monoblock stem designs (e.g., the Charnley monobloc stem) [16]. This allows independent intraoperative adjustments in vertical and horizontal offsets, the leg lengths and version of the neck. Such adjustments permit accurate reconstruction of the hip biomechanics and limb lengths, especially in patients with complex anatomies. However, it affords additional risks, such as an increased risk of mechanical failure (e.g., fatigue fracture) and release of metal debris at the junctions leading to adverse soft-tissue reactions. Modularity can be in the head-neck junction, ini the neck-stem junction or within the stem itself. The modular head-neck junction is referred to as a 'Morse taper'. Various stem designs use different types of taper junctions based on the Feldmühle specifications, ranging from 9/10 to 14/16 taper. These numbers refer to the larger and smaller diameters in the coupling mechanism of the taper junction. Dimensional mismatch and material combinations determine the relative movement and corrosion at the interface. Larger taper designs cause impingement, and smaller ones risk fatigue fractures and fretting corrosion.

Advantages of modular stem designs:

- Improves the biomechanics of the reconstructed hip
- Allows for mixing of materials
- Facilitates revision surgery
- Offers potential for reduced inventory in the theatre

Disadvantages of modular stem designs:

- Increased risk of mechanical failure

- Cold welding and taper damage at junctions
- Dissociation of components
- Higher rate of corrosion
- Metal debris–related issues

While modularity affords clear advantages, it also brings in additional risks. It is prudent to advance the stem designs by introducing incremental changes over a period with stringent monitoring using mechanisms such as beyond compliance.

20.8 Cementless Acetabular Cup Designs

The goal of cementless acetabular fixation is to achieve initial mechanical stability between component and bone by press-fit fixation with optional supplemental stabilisation using screws. In the long-term it relies on biological fixation by bone growth. Micromotion at the interface of >40 μm generates fibrous tissue and leads to loosening [17].

Modern cementless acetabular cups can be hemispherically expanded (e.g., Trident, Stryker, Mahwah, New Jersey) hemispherical shell) or equatorially expanded (peripheral self-locking) shells (e.g., Trident, Stryker, Mahwah, New Jersey). The hemispherical shells are inserted into an acetabulum that is typically under-reamed by 1–2 mm, while the peripheral self-locking designs are oversized at the periphery (equator) and therefore commonly inserted after using a same-sized reamer to prepare the acetabulum.

The advantages of cementless designs include the ability to fine-tune the cup's position during surgery to the desired position after the final trial of the components, use various liner options (neutral or elevated) – due to its modularity – to improve stability, make subsequent revisions easily by exchanging the liners, easily remove the cup using special instruments (e.g., Zimmer xplant) without significant bone loss, etc.

The recognised disadvantages include added risk for acetabular fracture, loosening of component due to inadequate initial fixation or failure of biological integration, uncoupling of the liner due to an inadequate locking mechanism or inaccurate seating and risks associated with supplemental screw fixation – catastrophic haemorrhage, backside wear of the liner, etc.

20.9 Bearing Surfaces in Modern THA

Current bearing surfaces aim to reduce wear- and debris-related issues, such as osteolysis and adverse tissue reaction. The modern bearing couplings include metal or ceramic-on-conventional-polyethylene or ceramic-on-HCLPE, ceramic-on-ceramic and MOM bearing surfaces. These newer bearings have been shown to reduce both wear and osteolysis. The efforts to further increase the longevity of THA should therefore focus on other unresolved areas, such as mechanical failure, stress shielding and its consequences, impingement and joint laxity, and bearing noise. Current data suggest that MOM and metal-on-conventional-polyethylene inserts have inferior survivorship, at 15 years [18]. Ceramic-on-ceramic, ceramic- or metal-on-HCLPE and ceramic-on-conventional-polyethylene inserts have comparable survivorship at 15 years. Longer-term randomised control trials (RCTs) are required to see whether there is a survivorship advantage for any of these bearing surfaces.

20.10 Dual-Mobility Cup

The dual-mobility concept was first introduced by Professor Bousquet in France (1976). The idea is to marry two established concepts. Firstly, Charnley's low-friction arthroplasty principle reduces the wear between the small femoral head and the polyethylene hemisphere. Secondly, the large-head concept of Mckee Farrar is used in incorporating a larger polyethylene hemisphere to reduce the risk of dislocation.

The design consists of two bearings: a small head, which articulates with the trunnion of the femoral stem, and a larger polyethylene hemisphere, which articulates with a smooth metal liner on the acetabular cup. It provides two points of articulation: one between the small femoral head and the large polyethylene head and the other between the polyethylene head and the metal liner of the acetabulum. The main advantage of a dual-mobility cup is reduced dislocation risk; however, other perceived advantages include a reduction in impingement risk, lower friction and less wear. The potential disadvantages of dual mobility include intraprosthetic dislocation (separation of the bearing surfaces), which makes it extremely difficult to reduce with closed reduction techniques; accelerated wear imparted by two articulating surfaces and the lack of long-term in vivo data.

References

1. Gomoll A, et al. Quantitative measurement of the morphology and fracture toughness of radiation crosslinked UHMWPE. *Orthopaedic Research Society*, February 2001.
2. Gillis A, et al. An independent evaluation of the mechanical, chemical, and fracture properties of UHMWPE crosslinked by 34 different conditions. *Orthopaedic Research Society*, February 1999.
3. Kuntz M. Validation of a new high performance alumina matrix composite for use in total joint replacement. *Semin Arthroplasty*, 2006; **17**:141–145.
4. Geesink RG, de Groot K, Klein CP. Bonding of bone to apatite-coated implants. *J Bone Joint Surg Br*, 1988; **70**:17–22.
5. Chang JD, Kim IS, Bhardwaj AM, Badami R. The evolution of computer-assisted total hip arthroplasty and relevant applications. *Hip Pelvis*, 2017; **29**(1):1–14.
6. Kalteis T, Handel M, Bäthis H, Perlick L, Tingart M, Grifka J. Imageless navigation for insertion of the acetabular component in total hip arthroplasty: is it as accurate as CT-based navigation? *J Bone Joint Surg Br*, 2006; **88**(2):163–167.
7. Gandhi R, Marchie A, Farrokhyar F, Mahomed N. Computer navigation in total hip replacement: a meta-analysis. *Int Orthop*, 2009; **33**(3):593–597.
8. Manzotti A, Cerveri P, De Momi E, Pullen C, Confalonieri N. Does computer-assisted surgery benefit leg length restoration in total hip replacement? Navigation versus conventional freehand. *Int Orthop*, 2011; **35**(1):19–24.
9. Lewinnek GE, Lewis JL, Tarr R, Compere CL, Zimmerman JR. Dislocations after total hip-replacement arthroplasties. *J Bone Joint Surg Am*, 1978; **60**(2):217–220.
10. Hamadouche M, Witvoet J, Porcher R, Meunier A, Sedel L, Nizard R. Hydroxyapatite-coated versus grit-blasted femoral stems. A prospective, randomised study using EBRA-FCA. *J Bone Joint Surg Br*, 2001; **83**(7):979–987.

11. Bobyn JD, Stackpool GJ, Hacking SA, Tanzer M, Krygier JJ. Characteristics of bone ingrowth and interface mechanics of a new porous tantalum biomaterial. *J Bone Joint Surg Br*, 1999a; **81**:907-914.
12. Kehlet H, Wilmore DW. Multimodal strategies to improve surgical outcome. *Am J Surg*, 2002; **183**(6):630-641.
13. Malviya A, Martin K, Harper I, et al. Enhanced recovery program for hip and knee replacement reduces death rate: a study of 4,500 consecutive primary hip and knee replacements. *Acta Orthop*, 2011; **82**(5):577-581.
14. Dalury DF, Lieberman JR, Macdonald SJ. Current and innovative pain management techniques in total knee arthroplasty. *Instr Course Lect*, 2012; **61**:383-388.
15. Kerr DR, Kohan L. Local infiltration analgesia: a technique for the control of acute postoperative pain following knee and hip surgery: a case study of 325 patients. *Acta Orthop*, 2008; **79**:174-183.
16. Krishnan H, Krishnan SP, Blunn G, Skinner JA, Hart AJ. Modular neck femoral stems. *Bone Joint J*, 2013; **95-B**(8):1011-1021.
17. Kim JT, Yoo JJ. Implant design in cementless hip arthroplasty. *Hip Pelvis*, 2016; **28**(2):65-75.
18. Yin S, Zhang D, Du H, Du H, Yin Z, Qiu Y. Is there any difference in survivorship of total hip arthroplasty with different bearing surfaces? A systematic review and network meta-analysis. *Int J Clin Exp Med*, 2015; **8**(11):21871-21885.

Chapter 21

Advances in Adult Dysplasia

Kaveh Gharanizadeh
Shafa Yahyaeian Hospital, Baharestan Square, Tehran, Iran
Orthopedic Bone and Joint Research Center,
Iran University of Medical Sciences (IUMS), Iran
kaveh.gharani@gmail.com

21.1 Adult Hip Dysplasia

The term 'hip dysplasia' includes a wide variety of pathomorphologic disorders that usually occur due to a delay in the growth and development of the hip, and mainly the acetabulum. The pathologic disorder is a dynamic and mechanical process that would result in insufficient coverage of the head by the acetabulum and joint instability [1].

In its severe form, it can result in true dislocation of the head during the prenatal period; but in its mild form, subluxation or dysplasia will develop. Dislocation is defined as the absence of contact between head and acetabulum, which can happen before or after birth and can be high or low (Figs. 21.1 and 21.2).

The Hip Joint in Adults: Advances and Developments
Edited by K. Mohan Iyer
Copyright © 2018 Pan Stanford Publishing Pte. Ltd.
ISBN 978-981-4774-72-7 (Hardcover), 978-1-351-26244-6 (eBook)
www.panstanford.com

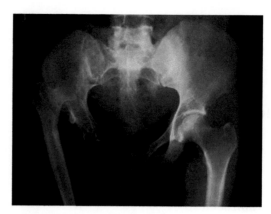

Figure 21.1 Unilateral high-riding dislocation of the hip. Obvious proximal femur dysplasia with noncircular head and small acetabular socket. Right hip in an adduction deformity.

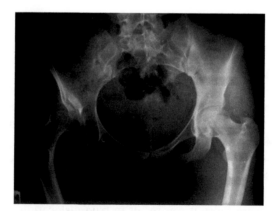

Figure 21.2 Unilateral low dislocation. Femoral head articulates with a pseudo acetabulum, which partially covers the true acetabulum. Note the nonspherical head and obscured lesser trochanter due to high femoral anteversion.

Subluxation means that the femoral head is out of the centre of rotation of the acetabulum but there is still some contact between the head and the acetabulum (Fig. 21.3).

In milder forms, it presents as instability, where the hip joint can be reduced congruously in abduction, but in adduction or external rotation the head becomes incongruous due to inadequate coverage of the anterolateral part of the acetabulum (both panels of Fig. 21.4).

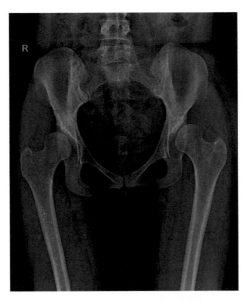

Figure 21.3 Bilateral hip subluxation more severe on the left side, with a broken Shenton line. Rim overstress depicted by reactional sclerosis.

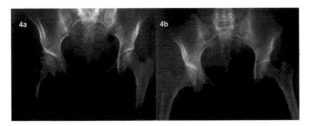

Figure 21.4 (4a) Right dysplastic hip without evidence of subluxation. (4b) Improved congruity and containment in an abduction view.

Structural instability has recently been suggested [1, 2] to describe the potential instability of the hip joint because of the small osseous socket of acetabular dysplasia and to differentiate it from soft-tissue instability and severe laxity (an ill-defined entity of the hip), which is mostly diagnosed in sport medicine [2].

On the other hand in the extreme form of subluxation, the femoral head cannot be reduced by abduction because of overriding of the head on the acetabular rim. In this situation it is difficult to differentiate severe subluxation from true dislocation (Fig. 21.5).

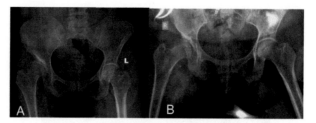

Figure 21.5 (A) Severe subluxation of the right hip that makes it really difficult to differentiate it from low dislocation. (B) In the abduction view, there is good congruity and containment. The Shenton line is restored, indicating that the original pathology is subluxation, not dislocation.

And at the end the term 'dysplasia' means there is inadequate coverage of the head by the acetabulum but there are no signs of subluxation (Fig. 21.4a).

21.2 Hip Dysplasia

Hip dysplasia (HD) can be primary or secondary.

Primary HD: It can be due to residual developmental dysplasia of the hip (DDH), or it can be due to adolescent onset dysplasia [1].

Secondary HD: It can be secondary to femoral head damage during growth period, with open physis, such as Legg–Calvé–Perthes disease (Fig. 21.6).

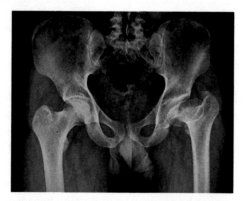

Figure 21.6 Secondary hip dysplasia due to Legg–Calvé–Perthes disease. Note the acetabular retroversion (cross-over sign) and severe head deformity (coxa magna, plana and breva).

Septic joint and posttraumatic injury or neuromuscular disorders like cerebral palsy (CP), poliomyelitis and myelomeningocele result in muscular imbalance (Figs. 21.7 and 21.8).

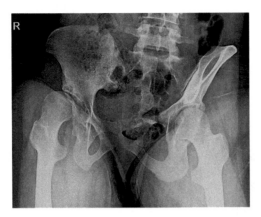

Figure 21.7 Neurologic dysplasia in spastic cerebral palsy. A 22-year-old girl suffering from progressive hip subluxation with coxa valga and an anteverted femoral neck.

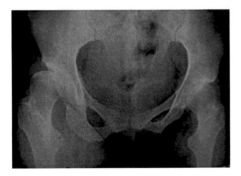

Figure 21.8 Right hip secondary neurologic dysplasia due to poliomyelitis. Femoral side dysplasia including excessive anteversion and coxa valga contributing in more hip instability.

21.3 Epidemiology and Natural History

The incidence of DDH is different according to various races. For instance, it is relatively common among Caucasians but it is almost absent in native Africans. It is really rare in China but relatively common in Japan [3, 4].

Overall, a prevalence of about 0.1% has been reported [5, 6]. Female sex [7, 8], breach presentation [9], a positive family history [9] and primigravida are among the risk factors [10–12].

Thanks to widespread diagnostic screening and the use of ultrasound [13], the number of true dislocations have decreased dramatically, but still a significant number of cases of adult acetabular dysplasia is being reported. DDH is still considered the most common reason for secondary hip osteoarthritis (OA) [14] and involves about 20%–40% of the cases [15–17] (Fig. 21.9).

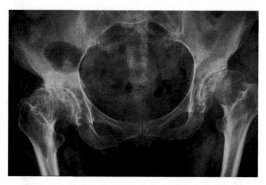

Figure 21.9 Bilateral severe osteoarthritis (Tonnis stage III) secondary to a bilateral dysplastic hip in a 38-year-old lady.

According to extensive studies, in asymptomatic patients the prevalence of DDH is between 3.6% and 5.4% [18, 19].

21.4 Pathophysiology and Aetiology

For a better understanding of the pathophysiology of HD, it is better to have a look at the formation and development of the hip. After the formation of the hip anlage in the sixth week of embryologic development, a cellular structure named scleroblastema appears in the proximal of the lower limb bud. It contains a central spherical segment composed of two parts. The internal part forms the femoral head, and the external part forms the acetabulum, which is composed of three disk-like masses, of the ilium, the pubis and the ischium [20–22].

The triradiate cartilage, that appears as a T-shaped structure at the sixteenth week of gestational age, is responsible for the development, growth and depth of the acetabulum [23, 24]. The triradiate cartilage usually closes during the ages between 14 and 16 years [25]. However, secondary growth centres appear in the rim of the acetabulum, which would shape the final acetabulum. For example, os acetabular is an epiphyseal centre adjacent to the pubis, which appears at seven years of age and is closed at nine years of age [24]. It is responsible for the formation of the entire anterior wall of the acetabulum. Other secondary epiphyseal growth plates of the acetabulum close up to 18 years of age [25].

Complete development of the acetabulum is dependent on the continuous mechanical pressure of the well-seated femoral head inside the acetabulum. The triradiate cartilage forms the shape and depth of the acetabulum in response to the moulding effect of the femoral head. On the other hand the secondary growth centres are responsible for fine-tuning of the details. These centres activate after the closure of the triradiate cartilage [7, 24].

Both neglected DDH and residual DDH are related to genetic factors or intrauterine problems like breach position. But the main reason of adolescent onset dysplasia is delays in the triradiate cartilage growth and also defects in the development of secondary growth plates that would appear during 12–18 years of age [25, 26].

Furthermore, the prevalence of this type of disorder is relatively higher in women (around 88%) compared to DDH. In patients with DDH the probability of DDH in the first-degree relatives of the patients is much higher than the probability of early OA that would require total hip replacement (THR) (59% versus 23%). In contrast in adolescent onset dysplasia there is a higher probability of the occurrence of early THR due to OA compared in HD in the first-order family members (50% to 16%) [27]. In addition in adolescent onset dysplasia there is a higher probability of bilateral dysplasia (61%–84%) than the left side, while in DDH the prevalence of left side involvement is much more common [28, 29].

In the course of DDH treatment, residual dysplasia can happen due to various reasons and it has been reported in up to 2.7% of the cases [30]. For example, during the treatment of DDH, femoral head deformity secondary to avascular necrosis (AVN) could happen and insufficient remodelling of the acetabulum next to a deformed femoral head can cause residual dysplasia.

HD secondary to DDH usually becomes symptomatic earlier, as in 36% of symptomatic HD cases there is a history of treatment of DDH in childhood [31].

For these reasons it is necessary to follow up with the DDH patients until skeletal maturity to make sure there is no residual dysplasia.

21.5 Pathomorphology

Regardless of what has caused HD, it would result in the same biomechanical and pathomorphological changes according to the severity of the HD.

In patients suffering from dysplasia or subluxation, the acetabulum is shallow and it would result in eccentric loading due to insufficient coverage on the femoral head [32]. A decrease in the contact surface will increase contact stress and rim stress, and by distortion of the biomechanics it would lead to progressive degeneration of the joint [1, 33]. Besides it has been proven that the labrum can become hypertrophic in order to compensate for the lack of coverage that would play a major role in the load bearing of the hip [34–36]. Magnetic resonance imaging (MRI) studies have also proven that degeneration, tear or avulsion of the hypertrophic labrum is quite common [37] (Fig. 21.10).

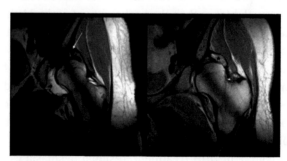

Figure 21.10 Radial cut MR arthrography. Hypertrophied labrum in DDH. The arrow depicts a labral tear. The * shows labral degeneration.

A tear of the labrum has several biomechanical insults. In addition to a decrease in joint stability, it reduces the sealing effect of the labrum, which compromises the amount of lubrication and distribution of the joint force [38, 39]. The importance of the labrum is such that Klaue et al. [40] consider labral ruptures as the precursor of OA in HD. Both reasons justify the cause of premature failure of arthroscopic labral debridement [39]. A degenerated labrum can get ganglion cysts or suffer from stress rim fracture, which is called os acetabular [40].

On the other hand the acetabulum does not developed normally in complete hip dislocation and therefore remains small and triangular in shape [41] (Figs. 21.1 and 21.11).

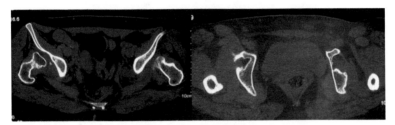

Figure 21.11 Axial CT scan of a bilateral high-riding dislocation. Excessive femoral anteversion is obvious. The acetabulum is narrow and nondeveloped and is triangular in shape.

The femoral head will remain small and nonspherical, and the limb will remain functionally shorter and is associated with weak abductors. In a unilateral dislocation, knee alignment tends to valgus deformity due to adductor contracture of the hip (Fig. 21.12).

And lumbar hyperlordosis is seen in both unilateral and bilateral hip dislocation due to hip flexors contractures [1–18].

It has been traditionally accepted that acetabular defects are on the anterolateral side because of increased anteversion and lack of lateral coverage of the acetabulum [42]. However, it has been proven that acetabular dysplasia is a global and 3D defect both in terms of shape and volume [43].

Steppacher et al. [44] used radial cut MRI and showed that in HD acetabular cartilage, the surface area is on average 16% less [45] than that of a normal acetabulum (Fig. 21.13).

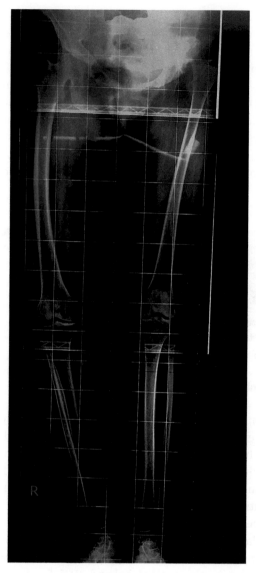

Figure 21.12 Left lower-limb valgus alignment due to a unilateral hip dislocation.

Mavcic et al. [46] showed that the cumulative contact stress over time is the best predictor of secondary OA. Linear extrapolation data demonstrated that the normal hips would only achieve dysplastic-level cumulative contact stresses at the age of 90. This increased pressure can result in premature aging of the acetabular cartilage.

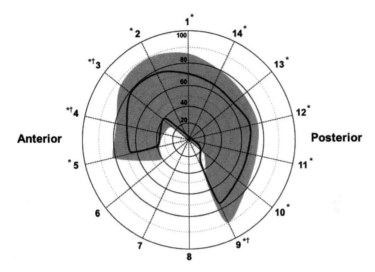

Figure 21.13 Diagram simulating the lunate surface area. The dark background shows normal acetabular articular surface, and the black shape inside outlines the smaller surface area in dysplastic hips.

It has been shown recently that between 17% and 34% of dysplastic hips suffer from acetabular retroversion. Ten per cent of them had only anterior defects, and less than 5% of defects were only lateral [47, 48]. Neppel et al. [49] in a recent study using 3D CT scans on 103 dysplastic hips have shown that lack of lateral coverage is seen in all patients; however, 30% of them were suffering from anterior defects, 34% from posterior defects and 36% from global defects (Fig. 21.14).

The existence of the cross-over sign was associated with a poor prognosis.

In secondary dysplasia due to Perthes disease 42% retroversion [50, 51] (Fig. 21.15) was reported, and 100% of proximal femoral focal deficiency (PFFD) cases or dysplasia secondary to trauma suffer from retroversion [52, 53]. Although there is a lower incidence of HD in men, acetabular retroversion and head deformity are more common [54].

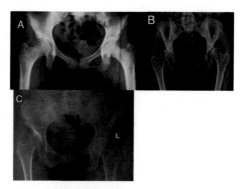

Figure 21.14 Acetabular deficiency in hip dysplasia is different. (A) Retroverted acetabulum with a positive cross-over sign, which shows posterior insufficiency. (B) classic extreme anteverted acetabulum with no crossing of the acetabular walls even in the lateral edge of the acetabulum. This means anterior insufficiency. (C) Severe acetabular dysplasia with subluxation. A very oblong and steep acetabulum with global acetabular deficiency.

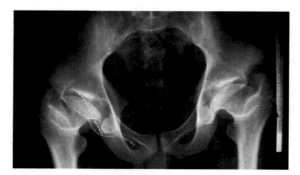

Figure 21.15 Sequel of bilateral severe Perthes disease. Extreme bilateral acetabular retroversion. The dashed lines show the posterior wall totally medial to the anterior wall.

21.6 Femur Morphology

Secondary deformity can appear on the femoral side, too. More subluxation results in more femoral deformity [55, 56]. The head is usually elliptical. Increased anteversion is a fixed finding in HD, and on average it is about 22.2° compared to the normal anteversion of 14.3° [57–59] Figs. 21.1 and 21.16.

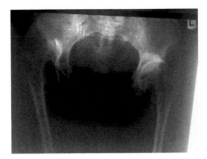

Figure 21.16 Right high dislocation and left dysplastic hip. Both developed severe OA at age 40.

The neck–shaft angle (NSA) can be normal, valgus or varus. If severe coxa valga exists, there is a possibility of fovea alta. In the normal position, the fovea capitis is located out of the weight-bearing (WB) area of the femoral head, which is determined by the fovea angle delta. The fovea is approximately located along the axial axis of the femoral neck in the normal population. If a line is drawn from the centre of the femoral head to the medial part of the sourcil of WB and another line is drawn from the centre of the head to the cranial part of the fovea, the resulting delta angle should be around 26 ± 10 (Fig. 21.17) [60]. This angle in DDH tends to zero or minus. Practically in the fovea alta, part of the WB surface of the femoral head will be filled with this fossa that eventually decreases the contact area between head and the acetabulum. MRI is the best option to view all the pathologies and measure the delta angle (Fig. 21.17).

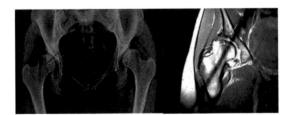

Figure 21.17 Fovea alta. On the radiography, the white line is from the centre of the head to the medial border of the weight-bearing sourcil. The black line is from the centre of the head to the cranial part of the fovea. The resulting delta angle is negative in this case. On the radial MR arthrogram arrows show high fovea location and countercoup lesion on the acetabular side, which is due to ligamentum teres impingement.

The neck is usually shorter and wider, in which the combination of short neck and coxa valga results in a shorter abductor lever arm [61]. The femoral canal is narrow and hypoplastic, particularly in medial to lateral diameter [55].

Clohisy [62] has explained femoral problems in 92% of patients. Forty-four percent of patients had valgus, and 4% had varus NSA. A decrease in the head-neck offset was also seen in 75% of patients; cam lesions were reported in 42% of them [63]. More subluxation is associated with a much more decrease in the femoral head offset [64].

Although it has previously been reported that in high-riding unilateral DDH, the femoral length is 5–10 mm longer than the normal side [65], according to a study done by the author on 60 patients suffering from unilateral DDH, which is not published yet, 10% of the patients had similar femoral lengths and 58.3% of the patients had a more than 5 mm difference in which the normal side was longer in 26.6% and the dislocated side was longer in 31.6%. About 32% of the patients had a less than 5 mm difference. On average in about three-fifths of the cases, a more than 5 mm difference was seen, in which approximately a little more than half of the cases had longer femoral lengths on the dislocated side and at the same rate but slightly fewer patients had longer femurs on the normal side.

21.7 Natural History of Hip Dysplasia

It can vary according to the severity of dysplasia and the type of dislocation.

Primary hip OA is very rare, and 90% of the patients with OA under 50 years of age suffer from underlying conditions that HD comprise half of them [66, 67]. Nevertheless, THR secondary to HD is not relatively common. In the European joint registry 1.8%–8.6% of the cases of THR reported are due to HD [68], and in the Mayo Clinic joint registry the incidence of HD was reported to be 7% in 31,000 cases of THR during the years 1969–2010 [69].

In an extensive study on the Greek population, Hartofilakidis et al. [70] showed that in an adult population suffering from HD, 47.7% were dysplastic, 23.9% were low dislocation and 8.8% were high dislocation. The average age for the appearance of symptoms was 34.5 years in the first group, 32.5 years in the second group and 31.2 years in the third group in the case of having a false acetabulum, and without a false acetabulum it was at 44 years of age. In the case of unilateral high dislocation, secondary compensatory problems would appear as a knee valgus of the involved side and functional limb shortening. The contralateral knee will also become symptomatic due to long-lasting overload [71]. In high dislocation, low back pain occurs due to compensatory lumbar hyperlordosis [72]. However, patients with bilateral high dislocation can tolerate the situation until the fifth or sixth decades of their lives without any surgery [69].

There have been some studies done on the natural history of dysplastic patients. The study of 286 patients showed that in patients whose lateral centre edge angle is more than 16 (LCEA > 16), the femoral head migration index is <30% and the Tonnis roof angle is <15°, they will not suffer from advanced OA until 65 years of age [33].

A prospective study on 20 dysplastic patients (32 hips) with a 22-year follow-up showed that 21 hips developed advanced OA.

Terjesen et al. studied 60 hips with a history of treatment for DDH in childhood, with an average of 45 years of follow-up, and showed that OA was developed in 80% of subluxated hips, 22% of those who have LCE = 10–19 and in 5% of those who have LCE > 20 [73].

Albinana et al. did a follow-up of 40 years in 72 patients with DDH and showed that the probability of the need for THR would be 29% if LEC = 15°–20° and 49% if LEC < 15° [74].

Cooperman followed 32 dysplastic hips without subluxation for 20 years and showed that eventually all of them developed OA [75].

In general, a combination of dysplasia and subluxation is expected to develop advanced early OA and probably will require THR by up to 45 years of age (Fig. 21.18) [71].

On the other hand mild and borderline dysplasia is usually benign and can remain asymptomatic and without OA for many years [32, 33].

In a recent extensive study on patients who have undergone unilateral THR while the other side was normal, Wyles et al. [76], with an average follow-up of 20 years (10–35 years), showed that

DDH can accelerate the incidence of OA on the nonaffected side. On the contrary, femoroacetabular impingement (FAI) had no relation with progressive OA (the same as normal hips) but the combination of DDH with cam lesion has the most progression to OA. The likelihood of radiographic degeneration was increased in patients with the following findings: femoral head lateralisation > 8 mm, femoral head extrusion index > 20%, acetabular depth-to-width index < 0.30, LCEA < 25 and Tonnis angle > 8.

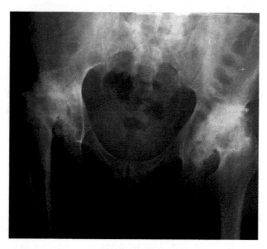

Figure 21.18 Bilateral advanced secondary OA due to subluxation.

In addition to the severity of dysplasia, in a recent study done on patients undergoing periacetabular osteotomy (PAO), the role of the level of activity of patients was studied and it was shown that both the severity of dysplasia and the level of activities are independently effective in the time of appearance of symptoms but BMI was not an independent factor. Patients with severe dysplasia (LCEA < 5) or with extreme activity become symptomatic four years earlier and patients with a combination of both experienced pain seven years earlier and seek treatment around the ages of 20–25 [77].

It is worthy of attention that OA criteria were based on radiographic findings using the Tonnis classification [78], but according to recent studies and with the use of MRI it has been shown that even at stage 0 of radiology, where everything is reported to be normal, there is a noticeable likelihood of chondral damage.

21.8 Clinical Examination and Diagnosis

Groin pain is the most common presenting symptom of HD in adult patients, which usually starts insidiously [40]. The pain is usually mechanical and increases with activity. If the labrum is torn symptoms like clicking or locking can appear. Up to 77% of patients may present with a Trendelenburg limp, wherein the stance phase pelvis drops to the opposite side. Some patients have an abductor lurch, in which the trunk bends towards the affected side during WB (stance phase).

The range of motion (ROM) is usually close to normal in patients without dislocation or subluxation. But patients suffering from relatively severe subluxation will experience some limitations and stiffness in abduction and extension of hip due to flexion and adduction contractures. On the other hand patients with an anterolateral acetabular deficiency usually have a higher flexion and internal rotation and quite reverse, patients with a posterosuperior defect have a higher flexion and external rotation in physical examination [79]. The anterior impingement sign (flexion, adduction and internal rotation [FADDIR] test) may be positive due to labral damage, a decrease in the head-neck offset or acetabulum retroversion. The posterior impingement sign will be positive in dysplastic patients who have extra-articular impingement between the greater trochanter (GT) and ischium due to a higher femoral anteversion combined with the coxa valga, which that would also result in a decrease in the offset [80, 81]. In the case of simultaneous anteversion of head and acetabulum, there will be anterior instability, particularly if the combination of acetabular and femoral anteversion reaches up to 60° (McKibben Index) [57, 58]. In this case there will be pain and a feeling of instability during extension and external rotation of the hip (positive apprehension test).

Patients with true dislocation have different signs and symptoms. In the unilateral form, functional shortening of the limb is a big problem and an obvious limp (due to a combination of limb discrepancy and Trendelenburg limp) is a major complaint. They usually develop genu valgum, which may not be clear at the first look because of the adducted position of the hip, but by keeping the hip in a neutral position the real amount of the genu valgum will appear (Fig. 21.12).

Back pain and contralateral knee pain also appear gradually because of coronal malalignment and overloading, respectively.

Patients with high dislocation and without a false acetabulum have a much higher rotational ROM than normal. A fixed finding in all patients with dislocation is decreased abduction (adduction contracture).

Bilateral dislocations can usually be well tolerated until the fourth to sixth decades of life, especially if the patients have no false acetabulum. They have a waddling gait from childhood, with compensatory lumbar hyperlordosis. They may present with low back pain or with hip problems.

21.9 Imaging

The first step in the diagnosis of patients with HD is standard radiography that includes an anteroposterior (AP) view of the pelvis and a lateral hip view. In addition, a false profile view [82] and dynamic (functional) views such as a pelvic AP view in abduction and extension and also in abduction and 20° flexion have their own indications [83].

Various radiographic criteria exist in the definition of HD and its severity. The first group of criteria represents pathomorphology of the acetabulum or the femoral head alone, and the second group is composed of those that represent the relationship between these two. It is not necessary to emphasise that all criteria are only reliable in standard radiologic views. In summary, for a standard AP view the tube must be 120 cm away from cassette and should be centred on the intersection of the lines from the anterosuperior iliac spines and the upper edge of the pubic symphysis [84]. In the radiography the obturator foramina must be symmetric. Various criteria have been studied for dysplasia; the most practical ones will be described here.

21.9.1 Acetabular WB Index or Tonnis Angle

It shows the slope of the sourcil (WB surface) of the acetabulum [85]. A line is drawn from the medial to the lateral edges of the sourcil, and another line is the interteardrop or the interischial line (horizontal reference line). The normal range of it is about 0°–5°, and more than 10° is considered dysplastic (Fig. 21.19).

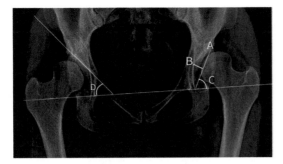

Figure 21.19 Acetabular depth–width ratio (ADR): A is from the inferior margin to the lateral border of the acetabulum, and B is from the centre of line A to the depth of the acetabulum. B/A should be more than 30%. Angle C depicts the acetabular sharp angle; the angle between the horizontal interteardrop line and the line drawn from the lower edge of the acetabulum near the teardrop to the outer edge of the sourcil. Normally it should be less than 43°–45°. Angle D is the acetabular WB index (Tonnis angle). A line is drawn from the medial to the lateral edges of the sourcil, and another line is the interteardrop (horizontal reference line). The normal range of it is about 0°–5°.

21.9.2 Acetabular Sharp Angle [86]

It is obtained from the angle between the horizontal line of the interteardrop, with the line drawn from the lower edge of the acetabulum near the teardrop to the outer edge of the sourcil. Although Sharp has determined the normal ranges as 33°–38°, in different studies 43°–45° is described as the cut-off point (Fig. 21.19) [87, 88].

21.9.3 Acetabular Depth–Width Ratio

The acetabular depth–width ratio (ADR) is the proportion of depth to width of the acetabulum. A line from the lower part of the acetabulum near the teardrop is drawn to the outer edge of the sourcil, and from the centre of that line a vertical line will be drawn to the depth of the acetabulum and the ratio of the second line to the first line will be calculated. The cut-off point for HD is <30% (Fig. 21.19) [75, 89].

21.9.4 Femoral Side Criteria

On the femoral side there is no distinguished separate criterion.

21.9.5 Neck–Shaft Angle

In a normal hip the NSA or centrum collum diaphyseal angle (CCD angle) is reported to be 125°–135° [90–92]. In HD it can be normal, varus or valgus. The head is usually not spherical and becomes more elliptical with increase in the severity of dysplasia [55, 56].

21.9.6 Fovea Alta

In severe coxa valga, the fovea can be in cranial position and can even go under the WB area. As a result, the delta fovea angle tends to zero or even minus and the fovea may articulate with the lunate surface. This situation leads to a decreased WB surface area and the acceleration rate of OA. Also potentially, round ligament impingement may occur (Fig. 21.17) [60].

The neck is also shorter and wider, and a cam lesion may be seen. In the lateral view a decreased femoral head neck offset ratio may be obvious (in which less than 17% is pathologic and an offset of less than 8 mm is considered abnormal) (please refer to Chapter 15 of *The Hip Joint*[1]).

Criteria of the relation between the femoral head and the acetabulum:

- **LCEA:** One of the most well-known criteria is LCEA of Wiberg, in which a line from the centre of the head is vertically drawn on the horizontal reference line and then another line from the centre of the head is drawn to the outer edge of the WB area (sourcil) (Fig. 21.20).

  A centre edge angle (CEA) ≥ 25 is normal, CEA < 20 is dysplastic and LCEA 20°–25° is considered as borderline [32].

- **Femoral head extrusion index:** First of all the width of the femoral head is measured and then the width of the part of the head that is outside of the acetabulum is measured [33]. Then the second one is divided to the head width. In normal cases only less than 25% of the head must be uncovered (Fig. 21.20) [93, 94].

---

[1]Biring GS. Arthroscopy of the hip joint. In *The Hip Joint*, Iyer KM, ed. (Pan Stanford), 2016.

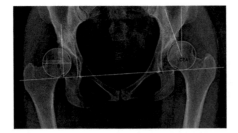

Figure 21.20 Lateral centre edge angle (LCEA), on the left side. A line from the centre of the head is vertically drawn to the horizontal reference line; then another line from the centre of the head is drawn to the outer edge of the WB area (sourcil). The normal LCEA is 25–40. In this case the LCEA is negative. Femoral head extrusion index, on the right side: B outlines the width of the femoral head, and A shows the width of the part of the head that is outside of the acetabulum. A/B should be less than 25%. In this case it's about 60%.

- **Anterior centre edge angle of Lequesne and de Seze (ACEA)**: It shows the amount of anterior coverage of the head by the acetabulum and is measured in a false profile view. In summary a false profile view is done in a standing position, the cassette is located vertically and the patient stands sideways. The desired hip should be placed next to the cassette. The pelvis is then rotated 25° backwards in a way that the back of the patient is at an angle of 65° to the cassette and then the X-ray beam is centred on the medial part of the contralateral groin (Fig. 21.21) [82].

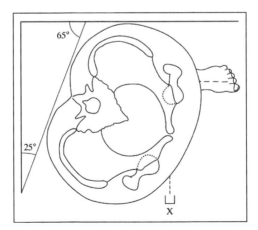

Figure 21.21 False profile view technique. See the text for details.

- **Correct positioning is checked using the following criteria:** The anterosuperior part of the pubis projects near the anterior edge of the GT; the distance between the two femoral heads is between two- and three-thirds of the diameter of the targeted femoral head; the centre of the femoral head, the axis of the femoral neck and the axis of the femoral shaft are approximately on the same vertical line; the lesser trochanter is slightly prominent posteriorly, which indicates neutral rotation of the lower limb; the posterior margin of the GT projects just behind the posterior edge of the femoral neck.

  In this view a vertical line is drawn from the centre of the femoral head and another line from the centre of the femoral head to the anterior edge of the sourcil. Angles less than 20° show a decrease in the anterior coverage of the head by the acetabulum (Fig. 21.22) [82].

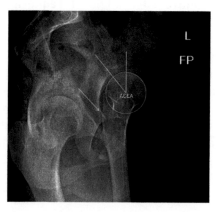

Figure 21.22 Anterior centre edge angle of Lequesne and de Seze (ACEA): a vertical line is drawn from the centre of the femoral head and another line from the centre of the femoral head to the anterior edge of the sourcil. The normal value is 20°. In this case the femoral head is subluxated with ACEA = −30. The arrow demonstrates an empty posteroinferior facet of the acetabulum.

- **Joint congruency:** Joint congruency means that the centre of rotation of the femoral head and acetabulum are the same. Yasunaga et al. [95] have explained some criteria and determined four grades (Fig. 21.23, Table 21.1):
  - Excellent: Femoral and acetabular subchondral plates are parallel, and the joint space is spared.

o Good: Subchondral plates of the head and the acetabulum are not parallel, but the joint space is maintained.
   o Fair: There is relative narrowing of the joint space.
   o Poor: There is relative loss of joint space.
- **Medial clear space:** It is the distance between the ilioischial line and femoral head. It shows that the head is located in the centre of the acetabular cavity. Numerical values of more than 10 mm are a sign of head lateralisation (Fig. 21.23) [96].
- **Ménard–Shenton line:** The reliable criteria for subluxation is an intact Shenton line. It is a line that is drawn from the medial border of the femoral neck to the upper margin of the obturator foramina (inferior border of superior Ramus) (Fig. 21.23) [95].

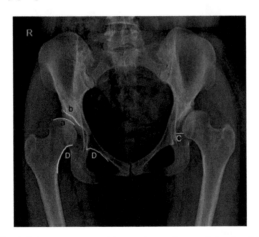

Figure 21.23 Joint congruency: the arches of a and b should be parallel normally. This case shows incongruity. C, a medial clear space, is the distance between the ilioischial line and the femoral head that normally should be less than 10 mm. D is Ménard–Shenton line. It is a line that is drawn from the medial border of the femoral neck to the upper margin of the obturator foramina. In this case it is broken, which shows subluxation.

21.9.7 CT Scan

The possibility of exact assessment of dysplasia was recently achieved through 3D CT scan reconstructions. In addition to the version and coverage it can demonstrate the volume of the acetabulum and virtual congruency of the joint in different femoral head positions [97–101]. Combined with arthrography, CT scan can demonstrate cartilage damages. The drawback is radiation exposure to patients and false negative results in the initial stages of cartilage damage [102].

21.9.8 MRI

MRI and especially MR arthrography can clearly show decreased femoral neck offset, deformity of the head and fovea alta (decreased delta fovea angle), and hypertrophic labrum and its pathologies, which are important in appropriate management [103, 104] (Fig. 21.24).

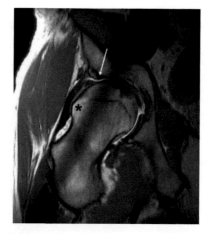

Figure 21.24 Radial-cut MR arthrogram in dysplasia. The * illustrates an obvious cam lesion and reduced femoral head offset. A tear of the hypertrophied labrum is depicted by the arrow.

With new advances in biochemical MRI techniques, such as delayed gadolinium-enhanced MRI of cartilage (dGEMRIC), T2 mapping and T1 rho, the application of MRI is evolving. These techniques are able to show the biochemical characteristic of the cartilage, especially the quantity of the proteoglycan and collagen damage. dGEMRIC is receiving the most attention among them, and quantitative changes of the proteoglycan content is in correlation with the severity of the dysplasia (according to LCEA in radiography) and also with the patient's symptoms. It can predict the results of osteotomy surgery [105, 106].

In the dGEMRIC studies done before arthroscopy, 323 milliseconds was the cut-off point that shows chondropathy and early OA initiation.

21.10 Classification

Table 21.1 shows the Tonnis classification for OA, Hartofilakidis classification for the grade of the dislocation and Yasunaga grading for congruency.

Table 21.1 Classifications of adult hip dysplasia

Hartofilakidis classification: Level of dislocation **Dysplasia (Type A):** Femoral head within the acetabulum despite some subluxation; segmental deficiency of the superior wall and inadequate depth of the true acetabulum
Low dislocation (Type B): Creation of a false acetabulum superior to the true acetabulum by the femoral head; complete absence of the superior wall and inadequate depth of the true acetabulum
High dislocation (Type C): Femoral head completely uncovered by the true acetabulum and migrated superoposteriorly; complete deficiency of the acetabulum and excessive anteversion of the true acetabulum
The classification of Yasunaga: Joint congruency 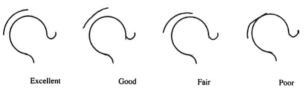
Excellent: Femoral and acetabular subchondral plates parallel and the joint space spared
Good: Subchondral plates of the head and the acetabulum not parallel but the joint space maintained
Fair: Relative narrowing of the joint space
Poor: Relative loss of joint space – modified Severin classification 10

Tonnis classification for osteoarthritis
0: No signs of OA
1 (Mild): Increased sclerosis, slight narrowing of the joint space and no or slight loss of head sphericity
2 (Moderate): Small cysts, moderate narrowing of the joint space and moderate loss of head sphericity
3 (Severe): Large cysts, severe narrowing or obliteration of the joint space and severe deformity of the head
Modified Severin classification: Morphology of the hip after conservative treatment
Excellent IA CEA > 19°, age 6–13 years; CEA > 25°, age > 14 years
Good IB CEA 15°–19°, age 6–13 years; CEA 20°–25°, age > 14 years
II Moderate deformity of the femoral head, femoral neck or acetabulum but otherwise the same as grade II
Fair III Dysplastic hip, no subluxation; CEA < 20°, age > 14 years
Poor IV Subluxation
V Femoral head in a false acetabulum
VI Redislocation

[a]CE: Centre edge

21.11 Management

Management of adult HD can vary according to its type (dysplasia, subluxation or dislocation), grade of OA, patient's age and other demographic characteristics of the patients.

Bilateral high dislocations, which usually become symptomatic in the fifth or sixth decade of a patient's life, do not have any treatment other than THR, and there is no indication for hip preserving surgery (Fig. 21.25).

In unilateral dislocation, symptoms appear much earlier. In a patient with OA and aged above 30 years, thanks to the new bearing surfaces in arthroplasty and technical advancements, the best management is THR. Meanwhile Ganz et al. [107] suggested a modified version of the old method of capsular arthroplasty known as Codivilla–Hey Groves–Colonna surgery. By using surgical hip dislocation in patients below the ages of 20–25 who suffer from severe subluxation or high dislocation and have damaged acetabulum cartilage, they reported good clinical outcomes of nine patients with a mean follow-up of 1 to 27 years. This method is suggested for very young patients in order to delay joint replacement. Contraindications are bilateral DDH, severe damage of femoral head cartilage and head deformity (nonspherical head) that results in stiffness. (For technical details please refer to Ref. [19].) The author has experience of one case, a 19-year-old girl (Fig. 21.26).

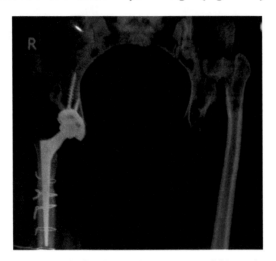

Figure 21.25 Bilateral high-riding dislocations. Total hip arthroplasty with subtrochanteric shortening osteotomy was done on the right side. On the left side the femoral head is malformed and the subtrochanteric area has a varus deformity. The only option in a bilateral dislocation is a hip replacement.

In patients suffering from subluxation or dysplasia, the treatment strategy is made according to the ability of joint reduction, its congruity and the severity of the OA.

Since the main problem in HD is on the acetabular side, correction of the acetabulum plays the main role in the restoration of the normal biomechanics of the hip joint. In general there are two major types of surgery to improve femoral head coverage and joint stability: reorientation osteotomies such as triple osteotomy [85,

114], acetabular rotational osteotomies [115, 116] and PAO [117] and salvage or bony augmentation osteotomies like shelf [118] and Chiari [119] (Fig. 21.27). In reorientation surgery correction is done by total rotation of the acetabulum in which hyaline cartilage is in contact with femoral head cartilage and has the best biomechanical results.

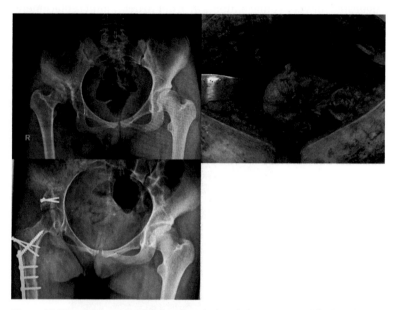

Figure 21.26 A 19-year-old girl with right hip dislocation. Modified Codivilla–Hey Groves–Colonna chosen because of the age of the patient. Congruous reduction was impossible. During surgery in addition to subtrochanteric shortening, the capsular flap was wrapped around the femoral head totally and shelf augmentation was done because of the small true acetabulum even after reaming. A two-year postoperative radiograph shows a stable joint and a preserved joint space. The functional HHS is 90, with a follow-up of two years and excellent results. Because the results of total hip arthroplasty (THA) in the population younger than 50 years represent a decline after 10 years [108], this hip-preserving surgery with good medium- to long-term results is advisable. Performing subtrochanteric osteotomy concomitant with hip surgical dislocation has significantly decreased complications like osteonecrosis of the femoral head, stiffness and redislocation, which have been the main reasons of failure of the original Codivilla–Hey Groves–Colonna procedure [109–113]. This surgery can be useful for neurologic and arthrogryposis patients, just like idiopathic DDH. The author believes since arthroplasty in neurologic patients has higher complications, attempts for hip preservation are more reasonable [107].

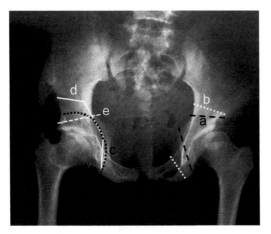

Figure 21.27 Different types of pelvic osteotomies: a: Tonnis triple osteotomy; b: steel triple osteotomy; c: rotational osteotomy; d: PAO; e: Chiari osteotomy.

But in salvage surgeries, part of the joint capsule turns into fibrocartilage using a new bony roof that will be in contact with the head that can lead to OA through the years. Meanwhile the labrum and cartilage of the acetabular rim, which are probably already damaged, will remain in the new WB area.

Triple osteotomy is performed in various methods, including steel and Tonnis operations. In general pubis, ischium and ilium bones are osteotomied in this method. In the steel method the cuts are away from the acetabulum and strong ligaments of the pelvis (sacrospinous and sacrotuberous) remain attached to the osteotomy fragment, preventing the need for significant correction; therefore it is mostly useful before skeletal maturity.

On the contrary, in the Tonnis method osteotomies are close to the joint and as a result there is a better possibility of correction due to the lack of a pelvic ligament connection [120]. In a long-term study for 25 years, 18 out of 33 cases (55%) who had undergone steel osteotomy developed OA. Long-term results in the Tonnis method were reported with 89%–93% success rate and in Germany and France is still being performed [121, 122]. The main problem in this method is fracture of the posterior column of the acetabulum and difficulties in mobilising the patient within the first six weeks after the surgery.

21.12   Rotational Osteotomy

This method has a long history in Japan and is still the most common type of reorientation surgery in that country. It was first introduced by Ninomiya and Tagawa [116] as a spherical osteotomy that has an extensive contact surface for quick healing and leaves the pelvic ring untouched [123]. This osteotomy passes the growth plates of the triradiate; therefore it is only performed after skeletal maturity [124, 125]. Although it requires a bone graft to preserve the correction, it makes a considerable amount of correction possible. There is also a risk of osteonecrosis of the acetabular fragment [126, 127] and intra-articular fracture [128], and the osteotomied fragment cannot be medialised [129]. There is also no possibility of joint arthrotomy.

The author has no experience in this operation, and readers can refer to original articles [115, 116]. Good results are reported in patients without OA or with early stages of OA (see Chapter 10 of *The Hip Joint*[2]). However, if preoperative OA exists, poor results are reported [130]. According to a long-term study, the success rate was 80% for patients without OA and only 27% for patients with OA [131].

21.13   Periacetabular Osteotomy

Bernease PAO was first performed by professor Ganz in 1984 [117]. Since then, PAO has been the most common type of osteotomy performed for the treatment of adult HD in English-speaking countries, North America and Switzerland [2]. Major benefits have been explained for this type of surgery [132], including the use of a single incision (anterior Smith–Peterson approach), allowing simultaneous arthrotomy and addressing labral pathologies or offset correction in the head-neck junction. The posterior column remains intact, which makes the osteotomy stable enough to be fixed with only a few screws. There is no need for immobilisation after surgery, and quick rehabilitation of the patient is possible. The shape of the true pelvis and the pelvic outlet doesn't change, which would allow vaginal delivery in women. Abductors are minimally manipulated. A significant amount of reorientation of the acetabulum is possible

---
[2]Iyer KM, ed. Osteotomies around the hip joint. In *The Hip Joint* (Pan Stanford), 2016.

to correct the version and lateral coverage and also to medialise the head in order to decrease joint reaction force [133].

Details of the surgical technique are completely explained in the literature [117, 134]. In brief, exposure is obtained by Smith–Peterson anterior incision and then four orthogonal osteotomy cuts are made and a controlled fracture of the acetabular fragment is performed while protecting the posterior column (Fig. 21.28).

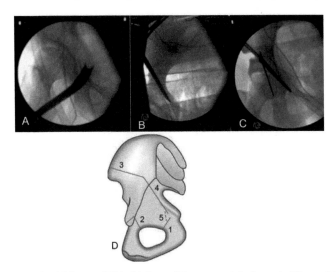

Figure 21.28 PAO cuts. (A) ischial cut, (B) retroacetabular cut, (C) provisional fixation and (D) an overview of all the cuts.

Because the osteotomy fragment is free from any ligament attachments, three rotational degrees of freedom and a huge amount of reorientation is possible, including medial or lateral displacement of the head (centre of rotation of the joint), adjusting the version and also correction of lateral coverage. Technically, the main disadvantage is its long learning curve due to the complexity and difficulty of the surgery compared to other types of osteotomy [2, 135–138].

21.14 Indications of Reorientation Osteotomy

The best candidates for PAO are dysplastic patients under the age of 40 with no OA (Tonnis grade of OA < 1) with possible congruous reduction of the hip joint and a BMI less than 30 [139]. Functional or dynamic views are the best method to check the possibility of joint congruity before surgery. A comparison of an AP pelvis with an AP pelvis in abduction can show the improvement of congruity, containment and the available joint space (Fig. 21.29).

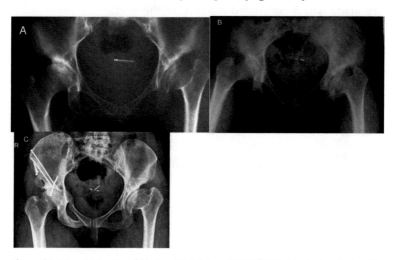

Figure 21.29 A 26-year-old lady with bilateral DDH. (A) Right hip dysplasia with overstress of the rim. Left hip with a history of inadequate Chiari osteotomy and a double-contoured acetabulum. (B) Improvement of joint congruency in an abduction view. On the right side good congruency confirms that isolated PAO is enough. On the left side although the congruency is improved the cranial part of the acetabulum is the remodelled fibrocartilage portion of the previous Chiari osteotomy. (C) Four years after right PAO. Perfect correction and normal version achieved with no pain and Merle d'Aubigne score = 18.

In severe anterolateral deficiency with subluxation, the joint space narrowing is obvious in an AP view of the pelvis. If clear improvement of joint space is visible in an abduction view, it indicates that anterolateral subluxation is the cause of the incongruency (in these cases the joint space is spared also in the false profile view) and that PAO alone will be sufficient [83] (Fig. 21.5).

But persisting joint space narrowing may reveal permanent cartilage loss and advanced OA.

In the case of excessive anteversion of the femur, abduction alone cannot result in complete congruity of the joint and the joint space may still be narrow. Combining internal rotation or flexion of

10°–15° to the abduction view can display better congruency and joint space improvement. Then it confirms the necessity of simultaneous proximal femur osteotomy (either derotation and/or varus) [83] (Fig. 21.30).

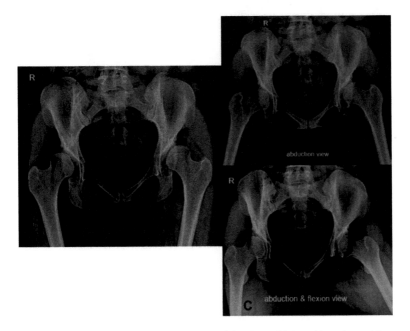

Figure 21.30 Functional dynamic views. (A) Bilateral hip subluxation with a broken Shenton line. (B) Although in the abduction view congruity is improved it is still not completely contained, especially on the cranial part. (C) Adding flexion to the abduction on the left side shows perfect congruity and containment, indicating that apart from PAO, proximal femur osteotomy is necessary.

The same scenario is in severe coxa valga and fovea alta. They may need femoral-side correction, too.

If obvious OA (Tonnis grade >= 2) exists then hip-preserving surgeries are not indicated. If after dynamic radiography there is still doubt about the degree of OA, MR arthrography can determine the amount of cartilage damage and reveal intra-articular problems like FAI. If articular cartilage is normal but there is no possibility for containment and congruity, correct decision making depends on the severity of pathologies of the femoral head and extra-articular and intra-articular problems. New surgical methods with the use of surgical hip dislocation and extended femoral head retinacular flap dissection are available [140]. They include femoral head–reducing osteotomy (Fig. 21.31).

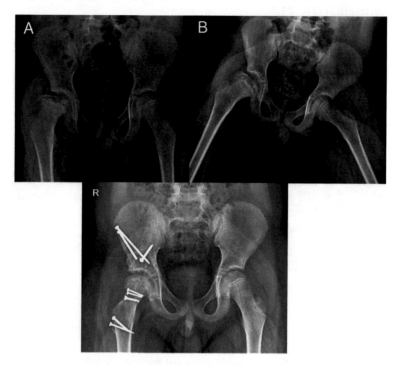

Figure 21.31 An eight-year-old boy with right hip Perthes disease. (A) Coxa magna with central necrosis and extruded head. (B) In an abduction view pelvis tilted and no true hip abduction possible because of the head deformity that doesn't go inside the socket, indicating hinge abduction. (C) Containment and correction achieved by head-reducing osteotomy and relative neck lengthening first and triple osteotomy in the same session. Two years after surgery radiography shows perfect congruency and excellent function. Merle d'Aubigne score = 17.

Femoral neck osteotomy, relative neck lengthening and trimming of the head make it possible [141] to correct the extra-articular and intra-articular impingements and reshaping of the femoral head to have a more spherical one. All of the corrections make it possible to have better containment and a more congruent joint to indicate PAO. In this situation a proximal femur correctional surgery may be advised at first and then PAO can be done in the same session to improve stability (for the technical details, please refer to Ref. [83]).

If congruous reduction could not be achieved but the articular cartilage is normal (no OA or early-stage OA), then salvage osteotomy like Chiari and a modified Codivilla–Hey Groves–Colonna procedure may be indicated in young patients to obtain joint stability.

Please refer to the algorithms (Figs. 21.32–21.34) in order to decide on the type of treatment.

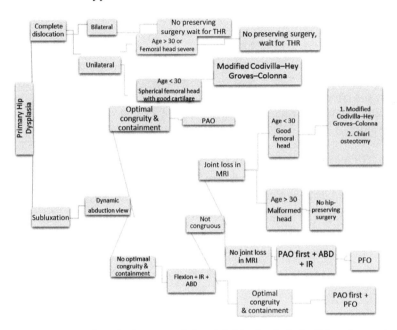

Figure 21.32 Algorithmic approach in adult hip dysplasia in dislocation and subluxation subtypes. In all cases the osteoarthritis grade should be <1. IR = internal rotation, ABD = abduction, PFO = proximal femoral osteotomy and PAO = periacetabular osteotomy.

Generally in many dysplastic patients (whether residual or adolescent onset) the first step is periacetabular osteotomy (PAO) [142], which is known as PAO-first principle. The reason is that a decrease in the femoral anteversion affects hip motions much more than the acetabular anteversion and most of the time the decision about femoral surgery is made during or after PAO surgery [57, 143].

Advances in Adult Dysplasia

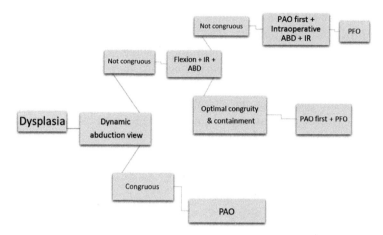

Figure 21.33 Algorithmic approach in adult hip dysplasia. In all cases the osteoarthritis grade should be < 1. IR = internal rotation, ABD = abduction, PFO = proximal femoral osteotomy and PAO = periacetabular osteotomy.

In patients where HD is secondary to Perthes disease and there is a clear pathology on the femoral side that prevents congruous reduction, the proximal femur should be addressed first and then PAO could be done in the same session (Figs. 21.35 and 21.36).

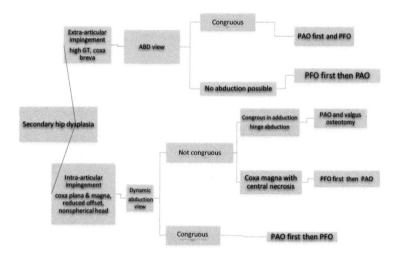

Figure 21.34 Algorithm for the management of secondary hip dysplasia due to Perthes and similar pathologies where the initial pathology starts on the femoral side. GT = greater trochanter, ABD = abduction and PFO = proximal femoral osteotomy.

Indications of Reorientation Osteotomy | 355

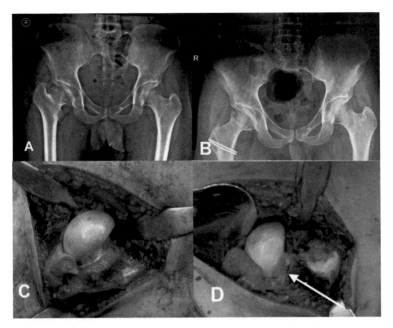

Figure 21.35 (A) Sequel of Perthes disease, including coxa magna, breva and plana, with greater trochanter (GT) overgrowth and sagging rope sign on the femoral side and secondary acetabular dysplasia with retroversion. (B) One year after relative femoral head lengthening, GT advancement and head trimming show correction of intra-articular and extra-articular impingement but joint instability still remains. An intraoperative photo (C) defines the short neck and high GT with an oval-shaped head. (D) After relative neck lengthening and head trimming. Arrow pointing to the superoretinacular flap containing vascular supply of the head.

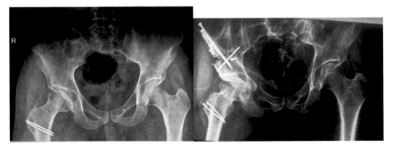

Figure 21.36 The same case as in Fig. 21.33 after PAO. Retroversion corrected (negative cross-over sign) and perfect coverage and joint medicalisation reflected by a normal acetabular index and LCEA.

21.15   Femoral Osteotomy

Intertrochanteric osteotomy used to be very common in the treatment of dysplasia up to the 1970s [144], but with a better understanding of the pathophysiology of dysplasia and the invention of new successful acetabular osteotomy techniques like PAO, their indications decreased significantly [145]. Currently the prevalence of proximal femoral osteotomy (PFO) has decreased significantly and dropped to less than 10% or even 5% [83, 142].

In addition to PFO (PFO: varus or valgus type or derotational), currently new techniques based on surgical hip dislocation and extended retinacular flaps are available that have increased the possibility of congruous reduction and provided a chance to use PAO instead of salvage surgeries [140, 141].

Indications of concomitant proximal femoral operation [83, 133].

- **Intra-articular pathology:** According to previous recommendations in dynamic views in addition to abduction, internal rotation and flexion must also be added. In some Perthes cases with hinge abduction, an adduction view may indicate valgus osteotomy.
- **Extra-articular impingement:** It is like a high-riding GT, which is seen in Perthes. It prevents abduction of the hip and prevents preoperative assessment by dynamic views; in these cases MR arthrography can be helpful.
- **Severe coxa valga with fovea alta**
- **Insufficient congruity:** If during PAO surgery, intraoperative radiography shows insufficient congruity after a well-oriented PAO (Fig. 21.37).
- Correction of the secondary coxa vara due to a previous varus osteotomy that causes anterolateral impingement after PAO.

In severe deformities, like Perthes, the head is nonspherical and the centre of rotation of the head can be at various points during joint motion (Fig. 21.38).

In these cases an adduction view may approve the need for valgus osteotomy or reshaping of the head by various techniques mentioned above to achieve better congruency. An important and useful intraoperative criterion of good correction is the possibility of internal rotation of 30° in hip flexion. If in the intraoperative radiography, correction of the acetabulum is appropriate and there is no retroversion (negative cross-over sign) but internal rotation is still limited, the femoral-side pathology (decreased head-neck offset) must be corrected by arthrotomy and trimming [133].

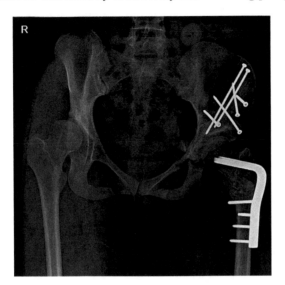

Figure 21.37 The same patient as in Fig. 21.30. Postoperative radiography of simultaneous PAO and intertrochanteric derotation–extension osteotomy. Although coverage is improved and the Shenton line is restored the cross-over sign is positive and acetabular reorientation is not optimal. One year after surgery pain is diminished but an anterior impingement test is positive and Merle d'Aubigne score = 15.

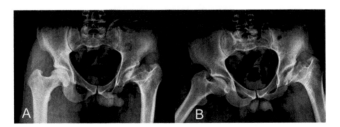

Figure 21.38 Secondary dysplasia due to Perthes disease. (A) Shallow and retroverted acetabulum with severe head flattening and deformity. (B) An abduction view confirms hinge abduction. It indicates valgus osteotomy for better containment.

21.16 Criteria of Correct Reorientation after PAO

The most important and most difficult part of PAO surgery is proper positioning of the acetabulum. In fact there is only one correct position, the best position, for the acetabulum. It has to be checked by standard AP radiography of the pelvis during surgery and not only with fluoroscopy. All of the following criteria must be checked [13, 133] (Fig. 21.39).

- LCEA = 20–35.
- Medial joint space of less than 10 mm must be created for medialisation of the head.
- The Shenton line must be intact.
- The sourcil angle should be 0°–10° and must be centred over the head.
- Anteversion of the acetabulum must be proper and defined by a negative cross-over sign that means the outlines of the anterior and posterior wall reach each other at the outer edge of the acetabulum.
- Proper congruency of the joint is a must.

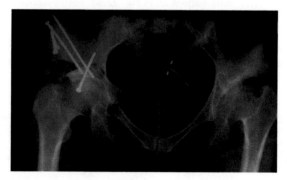

Figure 21.39 Optimal PAO correction. See the text for details.

During PAO there is a tendency to undercorrect the lateral coverage (LCEA undercorrection) and on the contrary there is a higher tendency for overcorrection of the anterior coverage (ACEA overcorrection), which results in acetabulum retroversion and a positive cross-over sign [146].

After obtaining satisfactory correction (all above-mentioned criteria), the intraoperative ROM must be checked. Sufficient internal rotation in flexion (compared to the other side and at least 15°–30°) reflects the absence of secondary FAI, either intra-articular or extra-articular. In a new study, the need for osteochondroplasty has been reported in up to 57% of PAO surgeries [147].

21.17 Outcome of Reorientation Osteotomy

Reorientational osteotomies can reduce the speed of joint destruction [148]. In all reorientation osteotomies that are close to the joint, the outcome after 10 and 20 years of follow-up shows a success rate of nearly 90% and 60%–70%, respectively [149–151].

Pain reduction and better functional and radiological scores have been reported in all studies.

On the other hand none of them showed improvements in the ROM and in fact they relatively reduce the ROM after surgery. This indicates a good preoperative range of motion as a prerequisite of an osteotomy [69].

The effects of PAO on the cartilage have been studied specifically. In a study using dGEMRIC before surgery and repeating it in the first and second years after PAO, Hingsammer et al. [152]. Demonstrated that although cartilage proteoglycan lowered in the first year, it became normal in the second year. It shows that normalisation of the biomechanics of the hip joint improves quality of the cartilage. This change was more prominent in the superior and anterosuperior region (WB area) of the acetabulum, where the loading changed by reorientation. Additionally, dGEMRIC before surgery is an imaging biomarker and can be considered as a prognostic factor before PAO surgery [153–155].

In recent studies, if patients with OA (Tonnis grade 2 and higher), older patients and patients with a history of previous surgery were ignored, a 100% success rate has been reported for PAO in a 10-year follow-up [76, 156].

The longest study on the outcome of PAO was done by Lerch et al. [157], and 30-year results of the first series of Ganz surgeries, including 75 hips, have been reported from Bern. In general one-third of the hips remained free of OA. Of course this group of patients were nonhomogeneous and in the learning curve period of the surgery in which 24% of them were suffering from advanced OA (grade ≥ 2) and 31% had a history of previous femoral or acetabular surgery. In patients with no OA (Tonnis 0), only 40% needed THR.

Factors that were associated with poor clinical or radiological prognosis include (i) ages of > 40 years at the time of surgery with a hazard ratio (HR) of 4.3, (ii) preoperative Merle d'Aubigne´–Postel score < 15 (HR = 4.1), (iii) preoperative Harris Hip Score (HHS) < 70 (HR = 5.8), (iv) limp before surgery (HR = 1.7), (v) preoperative positive anterior impingement test (HR = 3.6), (vi) preoperative positive posterior impingement test (HR = 2.5), (vii) preoperative internal rotation < 20° (HR = 4.3), (viii) preoperative Tonnis grade > 1 (HR = 5.7), (ix) postoperative retroversion (HR = 4.8) and (x) postoperative anterior coverage > 27% (HR = 3.2).

In a 20-year follow-up of the same group a 60% survival rate has been reported [148].

Although better long-term results were reported with Chiari and rotational osteotomy, that could be due to lower ages of the patients at the time of the surgery [158, 159]. Considering the survivorship of THR in patients under 50 years of age in the Swedish hip registry, the authors concluded that PAO can reduce the need for revision THA by at least one surgical intervention.

On the other hand, studies that have especially determined the effect of age on the outcome of PAO showed good short- to mid-term results in patients above 40 years if OA was negative or in the initial stages (Tonnis < 1) [160, 161].

Other criteria that are associated with poor prognosis are the severity of dysplasia and a high alpha angle, to indicate FAI (please refer to Chapter 15 of *The Hip Joint*). Low preoperative LCEA leads to postoperative lower LCEA and there is a higher risk of undercorrection [146].

In general the best candidate for PAO is a patient under 40 years of age who has little pain and no limp before surgery with negative impingement tests and no radiographic evidence of OA.

21.18 Salvage Osteotomy

Shelf operation: Because of low success rates of shelf surgery (46% after 10 years and 37% in 20 years) it has no indication in adult HD currently.

21.19 Chiari Osteotomy

Karl Chiari [119] introduced Chiari osteotomy in 1952, which is actually a capsular interposition arthroplasty in which the joint reaction force is reduced and stability increased by performing supra-acetabular osteotomy and joint medicalisation (for technical details, please look at the original paper). Eventually a while capsule transforms to fibrocartilage under load bearing. Because of the invention of new surgical techniques, on the basis of extended retinacular flap dissection that allows reshaping of the femoral head and expands the indications of the PAO, and high success rates of THR, the indications of Chiari osteotomy have significantly decreased. All in all regarding its good long-term results, Chiari has its own limited but special indications. One of the best indications is patients suffering from neuromuscular dysplasia that does not have the possibility of congruous reduction. Another indication is in patients suffering from severe dysplasia with a very small acetabular roof that does not allow congruous reduction and containment. However, in such cases if the patient is less than 25 years old and the surgeon is experienced enough, modified Colona surgery with surgical hip dislocation can be associated with very good results.

The prerequisites of Chiari surgery are [162, 163]:

- No OA or early stages of OA (Tonnis <= 1)
- Good ROM especially at least 90° flexion because patients will experience reduced flexion after surgery [164]
- Age less than 44 years

21.20 Results

After 20–34 years of follow-up in 236 patients, 51% of the patients had good to excellent results, 29% had fair results and 18% had poor outcomes. The higher the age the worse the outcome [133].

21.21 Neuromuscular Patients

HD in neuromuscular patients is acquired secondary to muscular imbalance. In cerebral palsy (CP) contractures of hip adductors and flexors and relative weakness of abductors and extensors and delayed walking of patients have a great role [165–167]. These factors result in impaired development of the hip joint leading to coxa valga and excessive femoral anteversion, which gradually results in an unstable joint due to high pressure on the outer edge of the acetabulum [168, 169]. Its prevalence is relatively high (about 35% in general) and it is the second deformity after equinus [170, 171]. Its severity increases with the severity of spasticity. If neuromuscular dysplastic hips are not addressed it usually becomes progressive and leads to subluxation or dislocation [172].

21.22 Symptoms

It can be asymptomatic in younger ages. Instability and subluxation can lead to functional shortening, limp and the limitation of abduction. In older ages pain and restricted ROM are associated with soft-tissue contractures. Subluxation or dislocation is painful and associated with abductor limp and disability, especially in walking and very painful sitting.

In the beginning of dysplasia in younger ages, perhaps the only necessary procedure is to release the soft-tissue contractures (adductor and flexor). With the presence of structural dysplasia in radiography, pelvic or femoral osteotomy or a combination of both, along with proper soft-tissue release, is necessary [173–175].

In patients who cannot walk, dislocation can be associated with severe pain and disability. However. these patients are not good candidates for complex hip-preserving surgery and usually undergo soft-tissue releases, valgus osteotomy or proximal femoral resection arthroplasty [176, 177].

In patients who are able to walk if congruous reduction is possible in dynamic views, PAO is performed along with PFO [178, 179] (Fig. 21.40).

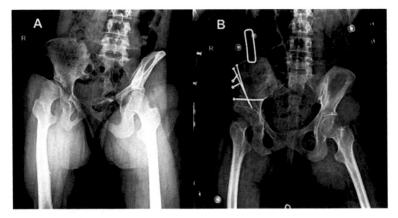

Figure 21.40 The same patient as in Fig. 21.7. A 22-year-old girl suffering from spastic CP. Three years after PAO and adductor release she had a stable joint with good coverage. In neurologic dysplasia proximal femoral osteotomy is usually required.

If congruous reduction is not possible, the Chiari procedure is better [180, 181]. In most cases PFO must also be performed in order to correct the associated coxa valga and excessive anteversion.

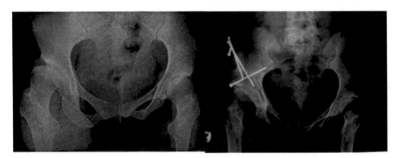

Figure 21.41 Dysplasia secondary to poliomyelitis (the same patient as in Fig. 21.8). Six years after optimal reorientation with PAO. Stability and congruity of the joint restored. Even in this flail limb the patient is satisfied because of pain relief and stability of the hip.

Secondary dysplasia is also seen in other neuromuscular patients, like poliomyelitis, although the number of patients has significantly decreased with the eradication of poliomyelitis. Sierra et al. [182] have reported the results of PAO in nine patients suffering from polio with successful results. The author performed PAO on a

35-year-old woman who was suffering from poliomyelitis. After five years satisfactory results were achieved in terms of pain reduction, functional score and joint stability (Fig. 21.41).

21.23 Arthroscopy

Several articles have emphasised the poor outcomes of arthroscopic capsulotomy and labral resection in DDH [183–185].

In a dysplastic patient compensatory labral hypertrophy has a great role in stability of the joint, so any debridement of labrum leads to instability and accelerates the degenerative process of the joint. In a study done on 48 patients with borderline dysplasia (LCEA: 20–25), 77.8% had labral and chondral lesions in the anterosuperior portion of the acetabulum (rim area). The results were good in the first year but declined in the second year [186].

The role of arthroscopy in the treatment of DDH is still unclear. In a recent study on 26 patients that underwent labral repair and capsular plication along with a capsular shift to inferior, good results have been reported after two years [187]. Because of the prevalence of labral damages in HD, arthroscopy may play a contributing role besides osteotomy. It could be used as a staging tool to quantify the damage to cartilage and to determinate the prognosis. It also can be used as a conjunctive therapeutic tool beside the PAO to address intra-articular FAI and labral repair [188, 189].

21.24 PAO Complications

PAO is a high-demand surgery with a long-term learning curve and can be associated with various complications [190–196]. The prevalence of complications has been reported to be around 6%–39% [2].

In a multicentred prospective study [197] on 205 PAO surgeries, complications in 34.6% of them have been reported in which around 28.7% were minor complications and 5.9% were major ones. The most common minor complications included superficial infection or stitch abscess, posterior column fracture, ramus osteotomy nonunion and heterotopic ossification (Fig. 21.42).

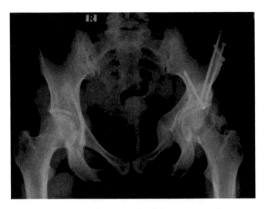

Figure 21.42 Heterotopic ossification (HO) four years after PAO in a 30-year-old lady. The patient is pain-free and asymptomatic.

Major complications were 12 in total, of which 9 cases required reoperation. They included migration or displacement of the osteotomy (Fig. 21.43), deep infection and peroneal palsy. Overall, with gaining more experience, complications have reduced and PAO can be considered as a safe procedure.

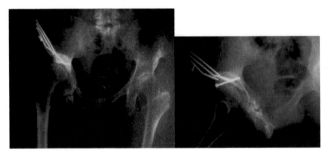

Figure 21.43 Loss of correction and ischial (posterior column) breakage. Eight weeks after surgery due to weight bearing and posterior column fracture, screw breakage and loss of correction happened.

21.25 Asymptomatic Dysplastic Patients

If radiologic criteria of dysplasia are clear and LCEA is less than 15–18 then PAO is indicated [198]. In borderline cases with LCEA from 18 to 25, the best method is to observe the asymptomatic patients and advise them to modify their daily life activities.

However, in symptomatic dysplastic patients, PAO is always recommended to correct dysplasia.

21.26 Summary

Adult HD is considered a wide-spectrum disorder with different aetiologies. It can present as undiagnosed mild dysplasia or be obvious with frank high dislocation. Treatment of HD in young patients still remains a challenge, and management should be based on a true understanding of the pathomorphology and available treatment options. Comprehensive evaluation of the patient by radiology, including dynamic radiography, and utilising new methods like MR arthrography and dGEMRIC can definitely be helpful in the decision-making process for hip-preserving surgeries.

Treatment should be tailored to the patient's age, symptoms and expectations. Although with an increased trend in successful results for arthroplasty its indication is expanding, a new era of hip-preserving surgery is opened for orthopaedic surgeons during last couple of years. In addition to PAO, advanced surgical techniques based on extended retinacular flap dissection of the femoral head have been presented that expand the indications of hip-preserving surgeries in young patients. It is not necessary to overemphasise the complexity of these high-demand operations that have a substantial learning curve and a potential for major complications.

References

1. Pun S. Hip dysplasia in the young adult caused by residual childhood and adolescent-onset dysplasia. *Curr Rev Musculoskelet Med*, 2016; **9**(4):427–434.
2. Clohisy JC, Schutz AL, St John LC, et al. Periacetabular osteotomy. A systemic literature review. *Clin Orthop Relat Res*, 2009; **467**:2041–2052.
3. Hodgson AR. Congenital dislocation of the hip. *Br Med J*, 1961; **2**:647.
4. Neel JV. A study on major congenital defects in Japanese infants. *Am J Hum Genet*, 1958; **10**:398.
5. Manaster BJ. From the RSNA Refresher Courses. Radiological Society of North America. Adult chronic hip pain: radiographic evaluation. *Radiographics*, 2000; **20**(Spec No):S3–S25.

6. Bracken J, Tran T, Ditchfield M. Developmental dysplasia of the hip: controversies and current concepts. *J Paediatr Child Health*, 2012; **48**(11):963–972; quiz 972–973.

7. Toennis D. *Congenital Dysplasia and Dislocation of the Hip* (New York, Berlin, Heidelberg: Springer), 1986; pp. 13–22.

8. Dunn PM. Perinatal observations on the etiology of congenital dislocation of the hip. *Clin Orthop Relat Res*, 1976; **119**:11–22.

9. Wynne-Davis R. Acetabular dysplasia and familial joint laxity: two etiological factors in congenital dislocation of the hip. *J Bone Joint Surg Br*, 1970; **52-B**:704–716.

10. Chan A, McCaul KA, Cundy PJ, Haan EA, Byron-Scott R. Perinatal risk factors for developmental dysplasia of the hip. *Arch Dis Child Fetal Neonatal Ed*, 1997; **76**(2):F94–F100.

11. Cady RB. Developmental dysplasia of the hip: definition, recognition, and prevention of late sequelae. *Pediatr Ann*, 2006; **35**(2):92–101.

12. Bache CE, Clegg J, Herron M. Risk factors for developmental dysplasia of the hip: ultrasonographic findings in the neonatal period. *J Pediatr Orthop B*, 2002; **11**(3):212–218.

13. Graf R. Fundamentals of sonographic diagnosis of infant hip dysplasia. *J Paediatr Orthop*, 1984; **4**(6):735–740.

14. Hoaglund FT, Shiba R, Newberg AH, Leung KY. Diseases of the hip: a comparative study of Japanese Oriental and American white patients. *J Bone Joint Surg Am*, 1985; **67-A**:1376–1383.

15. Aronson J. Osteoarthritis of the young adult hip: etiology and treatment. *Instr Course Lect*, 1986; **35**:119–128.

16. Harris WH. Etiology of osteoarthritis of the hip. *Clin Orthop Relat Res*, 1986; **213**:20–33.

17. Solomon L. Patterns of osteoarthritis of the hip. *J Bone Joint Surg Br*, 1976; **58**(2):176–183.

18. Gosvig KK, Jacobsen S, Sonne-Holm S, Palm H, Troelsen A. Prevalence of malformations of the hip joint and their relationship to sex, groin pain, and risk of osteoarthritis: a population-based survey. *J Bone Joint Surg Am*, 2010; **92**(5):1162–1169.

19. Jacobsen S, Sonne-Holm S. Hip dysplasia: a significant risk factor for the development of hip osteoarthritis. A cross-sectional survey. *Rheumatology (Oxford)*, 2005; **44**(2):211–218.

20. Watanabe RS. Embryology of the human hip. *Clin Orthop Relat Res*, 1974; (98):8–26.

21. Strayer LM Jr. Embryology of the human hip joint. *Clin Orthop Relat Res*, 1971; **74**:221–240.
22. Strayer LM Jr. The embryology of the human hip joint. *Yale J Biol Med*, 1943; **16**:13–26.6.
23. Cashin M, Uhthoff H, O'Neill M, et al. Embryology of the acetabular labral-chondral complex. *J Bone Joint Surg Br*, 2008; **90**:1019–1024.
24. Ponseti IV. Growth and development of the acetabulum in the normal child. Anatomical, histological, and roentgenographic studies. *J Bone Joint Surg Am*, 1978; **60**:575–585.
25. Liporace FA, Ong B, Mohaideen A, et al. Development and injury of the triradiate cartilage with its effects on acetabular development: review of the literature. *J Trauma*, 2003; **54**:1245–1249.
26. Tönnis D, Remus W. Development of hip dysplasia in puberty due to delayed ossification of femoral nucleus, growth plate and triradiate cartilage. *J Pediatr Orthop B*, 2004; **13**(5):287–292.
27. Lee CB, Mata-Fink A, Millis MB, Kim Y-J. Demographic differences in adolescent-diagnosed and adult-diagnosed acetabular dysplasia compared with infantile developmental dysplasia of the hip. *J Pediatr Orthop*, 2013; **33**(2):107–111. doi:10.1097/BPO.0b013e3182745456.
28. Ortiz-Neira CL, Paolucci EO, Donnon T. A meta-analysis of common risk factors associated with the diagnosis of developmental dysplasia of the hip in newborns. *Eur J Radiol*, 2012; **81**(3):e344–e351. doi:10.1016/j.ejrad.2011.11.003.
29. Okano K, Takaki M, Okazaki N, Shindo H. Bilateral incidence and severity of acetabular dysplasia of the hip. *J Orthop Sci*, 2008; **13**(5):401–404. doi:10.1007/s00776-008-1252-4.
30. Modaressi K, Erschbamer M, Exner GU. Dysplasia of the hip in adolescent patients successfully treated for developmental dysplasia of the hip. *J Child Orthop*, 2011; **5**(4):261–266. doi:10.1007/s11832-011-0356-0.
31. Okano K, Yamaguchi K, Ninomiya Y, et al. Relationship between developmental dislocation of the hip in infant and acetabular dysplasia at skeletal maturity. *Medicine (Baltimore)*, 2015; **94**(1):e268. doi:10.1097/MD.0000000000000268.
32. Wiberg G. Studies on dysplastic acetabula and congenital subluxation of the hip joint: with special reference to the complication of osteoarthritis. *Acta Chir Scand*, 1939; **83**(Suppl 58):5–135.
33. Murphy SB, Ganz R, Müller ME. The prognosis in untreated dysplasia of the hip: a study of radiographic factors that predict the outcome. *J Bone Joint Surg Am*, 1995; **77-A**:985–989.

34. Garabekyan T, Ashwell Z, Chadayammuri V, et al. Lateral acetabular coverage predicts the size of the hip labrum. *Am J Sports Med*, 2016; **44**(6):1582–1589. doi:10.1177/0363546516634058.

35. Gupta A, Chandrasekaran S, Redmond JM, et al. Does labral size correlate with degree of acetabular dysplasia? *Orthop J Sports Med*, 2015; **3**(2):2325967115572573. doi:10.1177/2325967115572573.

36. Sankar WN, Beaulé PE, Clohisy JC, et al. Labral morphologic characteristics in patients with symptomatic acetabular dysplasia. *Am J Sports Med*, 2015; **43**(9):2152–2156. doi:10.1177/0363546515591262.

37. Henak CR, Abraham CL, Anderson AE, Maas SA, Ellis BJ, Peters CL, Weiss JA. Patient-specific analysis of cartilage and labrum mechanics in human hips with acetabular dysplasia. *Osteoarthritis Cartilage*, 2014; **22**(2):210–217.

38. Ferguson SJ, Bryant JT, Ganz R, et al. The acetabular labrum seal: a poroelastic finite element model. *Clin Biomech*, 2000; **15**:463–468.

39. Parvizi J, Bican O, Bender B, et al. Arthroscopy for labral tears in patients with developmental dysplasia of the hip: a cautionary note. *J Arthroplasty*, 2009; **24**:110–113.

40. Klaue K, Durnin CW, Ganz R. The acetabular rim syndrome: a clinical presentation of dysplasia of the hip. *J Bone Joint Surg Br*, 1991; **73**:423–429.

41. Weinstein SL. Natural history of congenital hip dislocation (CDH) and hip dysplasia. *Clin Orthop Relat Res*, 1987; **225**:62–76.

42. Salter RB. Innominate osteotomy in the treatment of congenital dislocation and subluxation of the hip. *J Bone Joint Surg Br*, 1961; **43-B**:518–539.

43. Van Bosse H, Wedge JH, Babyn P. How are dysplastic hips different? A three-dimensional CT study. *Clin Orthop Relat Res*, 2015; **473**(5):1712–1723. doi:10.1007/s11999-014-4103-y.

44. Steppacher SD, Lerch TD, Gharanizadeh K, et al. Size and shape of the lunate surface in different types of pincer impingement: theoretical implications for surgical therapy. *Osteoarthr Cartil*, 2014; **22**(7):951–958. doi:10.1016/j.joca.2014.05.010.

45. Hipp JA, Sugano N, Millis MB, Murphy SB. Planning acetabular redirection osteotomies based on joint contact pressures. *Clin Orthop Relat Res*, 1999; **364**:134–143.

46. Mavcic B, Iglic A, Kralj-Iglic V, Brand RA, Vengust R. Cumulative hip contact stress predicts osteoarthritis in DDH. *Clin Orthop Relat Res*, 2008; **466**(4):884–891. doi:10.1007/s11999-008-0145-3.

47. Li PL, Ganz R. Morphologic features of congenital acetabular dysplasia. One in six is retroverted. *Clin Orthop Relat Res*, 2003; **416**:245-253.

48. Mast JW, Brunner RL, Zebrack J. Recognizing acetabular version in the radiographic presentation of hip dysplasia. *Clin Orthop Relat Res*, 2004; **418**:48-53.

49. Nepple JJ, Wells J, Ross JR, Bedi A, Schoenecker PL, Clohisy JC. Three patterns of acetabular deficiency are common in young adult patients with acetabular dysplasia. *Clin Orthop Relat Res*, 2016; **475**(4):1037-1044. doi:10.1007/s11999-016-5150-3.

50. Sankar WN, Flynn JM. The development of acetabular retroversion in children with Legg-Calvé-Perthes disease. *J Pediatr Orthop*, 2008; **28**:440-443.

51. Ezoe M, Naito M, Inoue T. The prevalence of acetabular retroversion among various disorders of the hip. *J Bone Joint Surg Am*, 2006; **88-A**:372-379.

52. Dora C, Durbach J, Hersche O, et al. Pathomorphological characteristics of posttraumatic acetabular dysplasia. *J Orthop Trauma*, 2000; **14**:483-489.

53. Dora C, Bühler C, Stover MD, et al. Morphological characteristics of acetabular dysplasia in proximal femoral focal deficiency. *J Pediatr Orthop B*, 2004; **13**:81-87.

54. Duncan ST, Bogunovic L, Baca G, Schoenecker PL, Clohisy JC. Are there sex-dependent differences in acetabular dysplasia characteristics? *Clin Orthop Relat Res*, 2015; **473**(4):1432-1439. doi:10.1007/s11999-015-4155-7.

55. Noble PC, Kamaric E, Sugano N, et al. Three-dimensional shape of the dysplastic femur: implications for THR. *Clin Orthop Relat Res*, 2003; (417):27-40.

56. Sugano N, Noble PC, Kamaric E, et al. The morphology of the femur in developmental dysplasia of the hip. *J Bone Joint Surg Br*, 1998; **80**:711-719.

57. Tönnis D, Heinecke A. Acetabular and femoral anteversion: relationship with osteoarthritis of the hip. *J Bone Joint Surg Am*, 1999; **81**:1747-1770.

58. McKibbin B. Anatomical factors in the stability of the hip joint in the newborn. *J Bone Joint Surg Br*, 1970; **52**:148-159.

59. Liu RY, Wang KZ, Wang CS, Dang XQ, Tong ZQ. Evaluation of medial acetabular wall bone stock in patients with developmental dysplasia

of the hip using a helical computed tomography multiplanar reconstruction technique. *Acta Radiol*, 2009; **50**:791–797.
60. Nötzli HP, Müller SM, Ganz R. The relationship between fovea capitis femoris and weight bearing area in the normal and dysplastic hip in adults: a radiologic study. *Z Orthop Ihre Grenzgeb*, 2001; **139**:502–506.
61. Steppacher SD, Tannast M, Werlen S, et al. Femoral morphology differs between deficient and excessive acetabular coverage. *Clin Orthop Relat Res*, 2008; (466):782–790.
62. Clohisy JC, Nunley RM, Carlisle JC, Schoenecker PL. Incidence and characteristics of femoral deformities in the dysplastic hip. *Clin Orthop Relat Res*, 2009; **467**(1):128–134.
63. Wells J, Nepple JJ, Crook K, Ross JR, Bedi A, Schoenecker P, Clohisy JC. Femoral morphology in the dysplastic hip: three-dimensional characterizations with CT. *Clin Orthop Relat Res*, 2017; **475**(4):1045–1054. doi: 10.1007/s11999-016-5119-2.
64. Argenson JN, Flecher X, Parratte S, Aubaniac JM. Anatomy of the dysplastic hip and consequences for total hip arthroplasty. *Clin Orthop Relat Res*, 2007; **465**:40–45.
65. Metcalfe JE, Banaszkiewicz P, Kapoor B, et al. Unexpected long femur in adults with acetabular dysplasia. *Acta Orthop Belg*, 2005; **71**:424–428.
66. Clohisy JC, Dobson MA, Robison JF, et al. Radiographic structural abnormalities associated with premature, natural hip-joint failure. *J Bone Joint Surg Am*, 2011; **93-A**(Suppl):3–9.
67. Sanchez-Sotelo J, Berry DJ, Trousdale RT, Cabanela ME. Surgical treatment of developmental dysplasia of the hip in adults: II. Arthroplasty options. *J Am Acad Orthop Surg*, 2002; **10**:334–344.
68. Kärrholm J. The swedish hip arthroplasty register. *Acta Orthop*, 2010; **81**:3–4.
69. Perry KI, Trousdale RT, Sierra RJ. Hip dysplasia in the young adult: an osteotomy solution. *Bone Joint J*, 2013; **95-B**(11 Suppl A):21–25.
70. Hartofilakidis G, Karachalios T, Stamos KG. Epidemiology, demographics, and natural history of congenital hip disease in adults. *Orthopedics*, 2000; **23**:823–827.
71. Weinstein SL. Natural history and treatment outcomes of childhood hip disorders. *Clin Orthop Relat Res*, 1997; **344**:227–242.
72. Crawford AH, Mehlman CT, Slovek RW. The fate of untreated developmental dislocation of the hip: long-term follow-up of eleven patients. *J Pediatr Orthop*, 1999; **19**:641–644.

73. Terjesen T. Residual hip dysplasia as a risk factor for osteoarthritis in 45 years follow-up of late-detected hip dislocation. *J Child Orthop*, 2011; **5**(6):425–431. doi:10.1007/s11832-011-0370-2.
74. Albinana J, Dolan LA, Spratt KF, Morcuende J, Meyer MD, Weinstein SL. Acetabular dysplasia after treatment for developmental dysplasia of the hip. Implications for secondary procedures. *J Bone Joint Surg Br*, 2004; **86**(6):876–886.
75. Cooperman DR, Wallensten R, Stulberg SD. Acetabular dysplasia in the adult. *Clin Orthop Relat Res*, 1983; **175**:79–85.
76. Wyles CC, Heidenreich MJ, Jeng J, Larson DR, Trousdale RT, Sierra RJ. The John Charnley award: redefining the natural history of osteoarthritis in patients with hip dysplasia and impingement. *Clin Orthop Relat Res*, 2017; **475**:336–350. doi:10.1007/s11999-016-4815-2.
77. Matheney T, Zaltz I, Kim YJ, Schoenecker P, Millis M, Podeszwa D, Zurakowski D, Beaulé P, Clohisy J. Activity level and severity of dysplasia predict age at bernese periacetabular osteotomy for symptomatic hip dysplasia. *J Bone Joint Surg Am*, 2016; **98**:665–671. http://dx.doi.org/10.2106/JBJS.15.00735.
78. Tönnis D. The prearthrotic deformity as origin of coxarthrosis: radiographic measurements and their value in the prognosis. *Z Orthop Ihre Grenzgeb*, 1978; **116**:444–446.
79. Anda S, Terjesen T, Kvistad KA, Svenningsen S. Acetabular angles and femoral anteversion in dysplastic hips in adults: CT investigation. *J Comput Assist Tomogr*, 1991; **15**:115–120.
80. Nakahara I, Takao M, Sakai T, Miki H, Nishii T, Sugano N. Three-dimensional morphology and bony range of movement in hip joints in patients with hip dysplasia. *Bone Joint J*, 2014; **96-B**(5):580–589.
81. Siebenrock KA, Steppacher SD, Haefeli PC, et al. Valgus hip with high antetorsion causes pain through posterior extraarticular FAI. *Clin Orthop Relat Res*, 2013; (471):3774–3780.
82. Lequesne MG, Laredo JD. The faux profil (oblique view) of the hip in the standing position. Contribution to the evaluation of osteoarthritis of the adult hip. *Ann Rheum Dis*, 1998; **57**:676–681.
83. Ganz R, Horowitz K, Leunig M. Algorithm for femoral and periacetabular osteotomies in complex hip deformities. *Clin Orthop Relat Res*, 2010; **468**:3168–3180. doi:10.1007/s11999-010-1489-z.
84. Tannas M, Siebenrock KA, Anderson SA. Femoroacetabular impingement: radiographic diagnosis—what the radiologist should know. *AJR Am J Roentgenol*, 2007; **188**:1540–1552.

85. Tonnis D. Surgical treatment of congenital dislocation of the hip. *Clin Orthop Relat Res*, 1990; **258**:33–40.
86. Sharp IK. Acetabular dysplasia: the acetabular angle. *J Bone Joint Surg Br*, 1961; **43**(2):268–272.
87. Tönnis D, Legal H, Graf R. *Congenital Dysplasia and Dislocation of the Hip in Children and Adults* (New York: Springer), 1987; pp. 116–121.
88. Nakamura S, Ninomiya S, Nakamura T. Primary osteoarthritis of the hip joint in Japan. *Clin Orthop Relat Res*, 1989; **241**:190–196.
89. Stulberg SD, Harris WH. Acetabular dysplasia and development of osteoarthritis of the hip. In *The Hip, Proceedings of the Second Open Scientific Meeting of The Hip Society*, Harris WH, ed. (St Louis: CV Mosby), 1974; pp. 82–93.
90. Toogood PA, Skalak A, Cooperman DR. Proximal femoral anatomy in the normal human population. *Clin Orthop Relat Res*, 2009; (467):876–885.
91. Hoaglund FT, Low WD. Anatomy of the femoral neck and head, with comparative data from Caucasians and Hong Kong Chinese. *Clin Orthop Relat Res*, 1980; (152):10–16.
92. Eijer H, Leunig M, Mohamed MN, et al. Cross-table lateral radiograph for screening of anterior femoral head-neck offset in patients with femoro-acetabular impingement. *Hip Int*, 2001; **11**:37–41.
93. Higgins SW, Spratley EM, Boe RA, et al. A novel approach for determining three-dimensional acetabular orientation: results from two hundred subjects. *J Bone Joint Surg Am*, 2014; **96**:1776–1784. doi:10.2106/JBJS.L.01141.
94. Laborie LB, Engesæter IØ, Lehmann TG, et al. Radiographic measurements of hip dysplasia at skeletal maturity—new reference intervals based on 2,038 19-year-old Norwegians. *Skeletal Radiol*, 2013; **42**(7):925–935. doi:10.1007/s00256-013-1574-y.
95. Yasunaga Y, Ochi M, Terayama H, et al. Rotational acetabular osteotomy for advanced osteoarthritis secondary to dysplasia of the hip. *J Bone Joint Surg Am*, 2006; **88**(9):1915–1919.
96. Clohisy JC, Barrett SE, Gordon JE, et al. Medial translation of the hip joint center associated with the Bernese periacetabular osteotomy. *Iowa Orthop J*, 2004; **24**:43–48.
97. Haddad FS, Garbuz DS, Duncan CP, et al. CT evaluation of periacetabular osteotomies. *J Bone Joint Surg Br*, 2000; **82**(4):526–531.

98. Nakamura S, Yorikawa J, Otsuka K, et al. Evaluation of acetabular dysplasia using a top view of the hip on three-dimensional CT. *J Orthop Sci*, 2000; **5**(6):533–539.

99. Ito H, Matsuno T, Hirayama T, et al. Three-dimensional computed tomography analysis of non-osteoarthritic adult acetabular dysplasia. *Skeletal Radiol*, 2009; **38**(2):131–139.

100. Mechlenburg I, Nyengaard JR, Rømer L, Søballe K. Changes in load-bearing area after Ganz periacetabular osteotomy evaluated by multi-slice CT scanning and stereology. *Acta Orthop Scand*, 2004; **75**(2):147–153.

101. Tallroth K, Lepistö J. Computed tomography measurement of acetabular dimensions: normal values for correction of dysplasia. *Acta Orthop*, 2006; **77**(4):598–602.

102. Nishii T, Tanaka H, Nakanishi K, et al. Fat suppressed 3D spoiled gradient-echo MRI and MDCT arthrography of articular cartilage in patients with hip dysplasia. *AJR Am J Roentgenol*, 2005; **185**(2):379–385.

103. James S, Miocevic M, Malara F, et al. MR imaging findings of acetabular dysplasia in adults. *Skeletal Radiol*, 2006; **35**(6):378–384.

104. Beltran LS, Mayo JD, Rosenberg ZS, et al. Fovea alta on MR images: is it a marker of hip dysplasia in young adults? *AJR Am J Roentgenol*, 2012; **199**(4):879–883.

105. Kim YJ, Jaramillo D, Millis MB, Gray ML, Burstein D. Assessment of early osteoarthritis in hip dysplasia with delayed gadolinium-enhanced magnetic resonance imaging of cartilage. *J Bone Joint Surg Am*, 2003; **85**(10):1987–1992.

106. Cunningham T, Jessel R, Zurakowski D, Millis MB, Kim YJ. Delayed gadolinium enhanced magnetic resonance imaging of cartilage to predict early failure of Bernese periacetabular osteotomy for hip dysplasia. *J Bone Joint Surg Am*, 2006; **88**(7):1540–1548.

107. Ganz R, Slongo T, Siebenrock KA, Turchetto L, Leunig M. Surgical technique: the capsular arthroplasty, a useful but abandoned procedure for young patients with developmental dysplasia of the hip. *Clin Orthop Relat Res*, 2012; **470**:2957–2967. doi:10.1007/s11999-012-2444-y.

108. Swedish Hip Arthroplasty Register. Annual Report 2007. Available at: http://www.shpr.se/en/default.aspx. Accessed May 23, 2012.

109. Codivilla A. [Operative treatment of congenital dislocation of the hip]. *Z Orthop Chir*, 1901; **9**:124–137 [in German].

110. Codivilla A. The operative treatment of congenital dislocations. *New York Med J*, 1902; **73**:741.

111. Colonna PC. Congenital dislocation of the hip in older subjects: based on a study of sixty-six open operations. *J Bone Joint Surg*, 1932; **14**:277–298.

112. Laurent LE. Capsular arthroplasty (Colonna's operation) for congenital dislocation of the hip: results of 102 operations. *Acta Orthop Scand*, 1964; **34**:66–86.

113. Pozo JL, Cannon SR, Catterall A. The Colonna-Hey Groves arthroplasty in the late treatment of congenital dislocation: a long-term review. *J Bone Joint Surg Br*, 1987; **69**:220–228.

114. Steel HH. Triple osteotomy of the innominate bone. *J Bone Joint Surg Am*, 1973; **55**(2):343–350.

115. Hasegawa Y, Iwase T, Kitamura S, Kawasaki M, Yamaguchi J. Eccentric rotational acetabular osteotomy for acetabular dysplasia and osteoarthritis: follow-up at a mean duration of twenty years. *J Bone Joint Surg Am*, 2014; **96**(23):1975–1982.

116. Ninomiya S, Tagawa H. Rotational acetabular osteotomy for the dysplastic hip. *J Bone Joint Surg Am*, 1984; **66**(3):430–436.

117. Ganz R, Klaue K, Vinh TS, Mast JW. A new periacetabular osteotomy for the treatment of hip dysplasias. Technique and preliminary results. *Clin Orthop Relat Res*, 1988; **232**:26–36.

118. Spitzy H. Künstliche Pfannendachbildung, Benutzung von Knochenbolzen zur temporären Fixation. *Z Orthop Chir*, 1923; **43**:284–294.

119. Chiari K. Medial displacement osteotomy of the pelvis. *Clin Orthop Relat Res*, 1974; **98**:55–71.

120. Leunig M, Siebenrock KA, Ganz R. Rationale of periacetabular osteotomy and background work. *J Bone Joint Surg Am*, 2001; **83-A**:438–448.

121. Tönnis D, Behrens K, Tscharani F. A modified technique of the triple pelvic osteotomy: early results. *J Pediatr Orthop*, 1981; **1**:241–249.

122. Carlioz H, Khouri N, Hulin P. Triple juxta-cotyloid osteotomy. *Rev Chir Orthop Reparstrice Appar Mot*, 1982; **68**:497–501.

123. Flückiger G, Eggli S, Kosina J, et al. Geburt nach periazetabulärer Osteotomie. *Orthopäde*, 2000; **29**:63–67.

124. Eppright RH. Abstract. Dial osteotomy of the acetabulum in the treatment of dysplasia of the hip. *J Bone Joint Surg Am*, 1975; **57-A**:1172.

125. Wagner H. Osteotomies for congenital hip dislocation. In *The Hip: Proceedings of the Fourth Open Scientific Meeting of the Hip Society* (St Louis, MO: CV Mosby), 1976; pp. 45–66.

126. Beck M, Leunig M, Ellis T, et al. The acetabular blood supply: implications for periacetabular osteotomies. *Surg Radiol Anat*, 2003; **25**:361–367.

127. Kalhor M, Beck M, Huff TW, et al. Capsular and pericapsular contributions to acetabular and femoral head perfusion. *J Bone Joint Surg Am*, 2009; **91-A**:409–418.

128. Siebenrock KA, Schöll E, Lottenbach M, Ganz R. Bernese periacetabular osteotomy. *Clin Orthop Relat Res*, 1999; (363):9–20.

129. Schramm M, Hohmann D, Radespiel-Troger M, Pitto RP. Treatment of the dysplastic acetabulum with Wagner spherical osteotomy. A study of patients followed for a minimum of twenty years. *J Bone Joint Surg Am*, 2003; **85**(5):808–814.

130. Hasegawa Y, Iwase T, Kitamura S, et al. Eccentric rotational acetabular osteotomy for acetabular dysplasia: follow-up of one hundred and thirty-two hips for five to ten years. *J Bone Joint Surg Am*, 2002; **84-A**(3):404–410.

131. Nakamura S, Ninomiya S, Takatori Y, et al. Long-term outcome of rotational acetabular osteotomy: 145 hips followed for 10–23 years. *Acta Orthop Scand*, 1998; **69**(3):259–265.

132. Maeyama A, Naito M, Moriyama S, Yoshimura I. Periacetabular osteotomy reduces the dynamic instability of dysplastic hips. *J Bone Joint Surg Br*, 2009; **91-B**:1438–1442.

133. Siebenrock KA, Steppacher SD, Albers ChE, et al. Diagnosis and management of developmental dysplasia of the hip from triradiate closure through young adulthood. An instructional course lecture, American Academy of Orthopaedic Surgeons. *Journal Bone Joint Surg Am*, 2013; **95-A**:749–755.

134. Siebenrock KA, Leunig M, Ganz R. Periacetabular osteotomy: the Bernese experience. *Instr Course Lect*, 2001; **50**:239–245.

135. Davey JP, Santore RF. Complications of periacetabular osteotomy. *Clin Orthop Relat Res*, 1999; **363**:33–37.

136. Peters CL, Beaulé PE, Beck M, et al. Report of breakout session: strategies to improve hip preservation training. *Clin Orthop Relat Res*, 2012; **470**(12):3467–3469.

137. Peters CL, Erickson JA, Hines JL. Early results of the Bernese periacetabular osteotomy: the learning curve at an academic medical center. *J Bone Joint Surg Am*, 2006; **88**(9):1920–1926.

138. Davey JP, Santore RF. Complications of periacetabular osteotomy. *Clin Orthop Relat Res*, 1999; **363**:33–37.

139. Tönnis D. Normal values of the hip joint for the evaluation of X-rays in children and adults. *Clin Orthop Relat Res*, 1976; **119**:39–47.

140. Ganz R, Gill TJ, Gautier E, et al. Surgical dislocation of the adult hip a technique with full access to the femoral head and acetabulum without the risk of avascular necrosis. *J Bone Joint Surg Br*, 2001; **83**:1119–1124.

141. Ganz R, Huff TW, Leunig M. Extended retinacular soft tissue flap for intra-articular hip surgery: surgical technique, indications, and results of application. *Instr Course Lect*, 2009; **58**:241–255.

142. Gala L, Clohisy JC, Beaulé PE. Hip dysplasia in the young adult. *J Bone Joint Surg Am*, 2016; **98**:63–73.

143. Clohisy JC, St John LC, Nunley RM, Schutz AL, Schoenecker PL. Combined periacetabular and femoral osteotomies for severe hip deformities. *Clin Orthop Relat Res*, 2009; **467**(9):2221–2227.

144. Santore RF, Turgeon TR, Phillips WF, III, Kantor SR. Pelvic and femoral osteotomy in the treatment of hip disease in the young adult. *Instr Course Lect*, 2006; **55**:131–144.

145. Turgeon TR, Phillips W, Kantor SR, Santore RF. The role of acetabular and femoral osteotomies in reconstructive surgery of the hip: 2005 and beyond. *Clin Orthop Relat Res*, 2005; **441**:188–199.

146. Novais EN, Duncan S, Nepple J, et al. Do radiographic parameters of dysplasia improve to normal ranges after Bernese periacetabular osteotomy? *Clin Orthop Relat Res*, 2016; **475**(4):1120–1127.

147. Albers CE, Steppacher SD, Ganz R, Tannast M, Siebenrock KA. Impingement adversely impacts 10-year survivorship after periacetabular osteotomy for DDH. *Clin Orthop Relat Res*, 2013; **471**(5):1602–1614. doi: 10.1007/s11999-013-2799-8.

148. Steppacher SD, Tannast M, Ganz R, Siebenrock KA. Mean 20-year follow-up of Bernese periacetabular osteotomy. *Clin Orthop Relat Res*, 2008; **466**(7):1633–1644.

149. Nakamura S, Ninomiya S, Takatori Y, Morimoto S, Umeyama T. Long-term outcome of rotational acetabular osteotomy: 145 hips followed for 10-23 years. *Acta Orthop Scand*, 1998; **69**(3):259–265.

150. Hellemondt GG, Sonneveld H, Schreuder MH, Kooijman MA, de Kleuver M. Triple osteotomy of the pelvis for acetabular dysplasia: results at a mean follow-up of 15 years. *J Bone Joint Surg Br*, 2005; **87**(7):911–915.

151. Janssen D, Kalchschmidt K, Katthagen BD. Triple pelvic osteotomy as treatment for osteoarthritis secondary to developmental dysplasia of the hip. *Int Orthop.* 2009; **33**(6):1555–1559.

152. Hingsammer AM, Kalish LA, Stelzeneder D, et al. Does periacetabular osteotomy for hip dysplasia modulate cartilage biochemistry? *J Bone Joint Surg Am*, 2015; **97**:544–550.

153. Bashir A, Gray ML, Boutin RD, Burstein D. Glycosaminoglycan in articular cartilage: in vivo assessment with delayed Gd(DTPA) (2-)-enhanced MR imaging. *Radiology*, 1997; **205**(2):551–558.

154. Bashir A, Gray ML, Burstein D. Gd-DTPA2- as a measure of cartilage degradation. *Magn Reson Med*, 1996; **36**(5):665–673.

155. Jessel RH, Zurakowski D, Zilkens C, et al. Radiographic and patient factors associated with pre-radiographic osteoarthritis in hip dysplasia. *J Bone Joint Surg Am*, 2009; **91**(5):1120–1129.

156. Reijman M, Hazes JM, Pols HA, Koes BW, Bierma-Zeinstra SM. Acetabular dysplasia predicts incident osteoarthritis of the hip: the Rotterdam study. *Arthritis Rheum*, 2005; **52**(3):787–793.

157. Lerch TD, Steppacher SD, Liechti EF, Tannast M, Siebenrock KA. One-third of hips after periacetabular osteotomy survive 30 years with good clinical results, no progression of arthritis, or conversion to THA. *Clin Orthop Relat Res*, 2017, **475**(4):1154–1168.

158. Guille JT, Forlin E, Kumar SJ, MacEwen GD. Triple osteotomy of the innominate bone in treatment of developmental dysplasia of the hip. *J Pediatr Orthop*, 1992; **12**:718–721.

159. Ohashi H, Hirohashi K, Yamano Y. Factors influencing the outcome of Chiari pelvic osteotomy: a long-term follow-up. *J Bone Joint Surg Br*, 2000; **82**:517–525.

160. Millis MB, Kain M, Sierra R, et al. Periacetabular osteotomy for acetabular dysplasia in patients older than 40 years: a preliminary study. *Clin Orthop Relat Res*, 2009; **467**:2228–2234.

161. Henak CR, Abraham CL, Anderson AE, et al. Patient-specific analysis of cartilage and labrum mechanics in human hips with acetabular dysplasia. *Osteoarthritis Cartilage*, 2014; **22**(2):210–217.

162. Lack W, Windhager R, Kutschera HP, et al. Chiari pelvic osteotomy for osteoarthritis secondary to hip dysplasia. Indications and long-term results. *J Bone Joint Surg Br*, 1991; **73**(2):229–234.

163. Ohashi H, Hirohashi K Yamano Y. Factors influencing the outcome of Chiari pelvic osteotomy: a long-term follow-up. *J Bone Joint Surg Br*, 2000; **82**(4):517–525.

164. Windhager R, Pongracz N, Schonecker W, et al. Chiari osteotomy for congenital dislocation and subluxation of the hip. Results after 20 to 34 years follow-up. *J Bone Joint Surg Br*, 1991; **73**(6):890–895.

165. Flynn JM, Miller F. Management of hip disorders in patients with cerebral palsy. *J Am Acad Orthop Surg*, 2002; **10**:198–209.

166. Root L. Surgical treatment for hip pain in the adult cerebral palsy patient. *Dev Med Child Neur*, 2009; **52**(Suppl 4):84–91.

167. Renshaw TS, Green NE, Griffin PP, et al. Cerebral palsy: orthopaedic management. *Instr Course Lect*, 1996; **45**:457–490.

168. Abel MF, Wenger DR, Mubarak SJ, et al. Quantitative analysis of hip dysplasia in cerebral palsy: a study of radiographs and 3-D reformatted images. *J Pedatr Orthop*, 1994; **14**:283–289.

169. Laplaza FJ, Root L. Femoral anteversion and neck-shaft angles in hip instability in cerebral palsy. *J Pediatr Orthop*, 1994; **14**:719–723.

170. Cook PH, Cole WG, Carney RPL. Dislocation of the hip in cerebral palsy. Natural history and predictability. *J Bone Joint Surg Br*, 1989; **71**:441–446.

171. Root L, Laplaza FJ, Brourman SN, et al. The severely unstable hip in cerebral palsy. Treatment with open reduction, pelvic osteotomy, and femoral osteotomy with shortening. *J Bone Joint Surg Am*, 1995; **77**:703–712.

172. McHale KA, Bagg M, Nason SS. Treatment of the chronically dislocated hip in adolescents with cerebral palsy with femoral head resection and subtrochanteric valgus osteotomy. *J Pediatr Orthop*, 1990; **10**:504–509.

173. Gordon JE, Capelli AM, Strecker WB, et al. Pemberton pelvic osteotomy and varus rotational osteotomy in the treatment of acetabular dysplasia in patients who have static encephalopathy. *J Bone Joint Surg Am*, 1996; **78**:1863–1871.

174. McNerney NP, Mubarak SJ, Wenger DR. One-stage correction for the dysplastic hip in cerebral palsy with the San Diego acetabuloplasty: results and complications in 104 hips. *J Pediatr Orthop*, 2000; **20**:93–103.

175. Al-Ghadir M, Masquijo JJ, Guerra LA, et al. Combined femoral and pelvic osteotomies vs. femoral osteotomy alone in the treatment of hip dysplasia in children with cerebral palsy. *J Pediatr Orthop*, 2009; **29**:779–783.

176. Castle ME, Schneider C. Proximal femoral resection-interposition arthroplasty. *J Bone Joint Surg Am*, 1978; **60**:1051–1054.

177. McCarthy RE, Simon S, Douglas B, et al. Proximal femoral resection to allow adults who have severe cerebral palsy to sit. *J Bone Joint Surg Am*, 1988; **70**:1011–1016.

178. Clohisy JC, Barrett SE, Gordon JE, et al. Periacetabular osteotomy in the treatment of severe acetabular dysplasia. Surgical technique. *J Bone Joint Surg Am*, 2006; **88**(Suppl 1 Pt 1):65–83.

179. Millis MB, Murphy SB. Periacetabular osteotomy. In *The Adult Hip*, Vol I, 2nd ed., Callaghan JJ, Rosenburg AG, Rubash HE, et al. eds. (Philadelphia, PA: Lippincott, Williams & Wilkins), 2007; pp. 795–825.

180. Dietz FR, Knutson LM. Chiari pelvic osteotomy in cerebral palsy. *J Pediatr Orthop*, 1995; **15**:372–380.

181. Fong HC, Lu W, Li YH. Chiari osteotomy and shelf augmentation in the treatment of hip dysplasia. *J Pediatr Orthop*, 2000; **20**:740–744.

182. Sierra RJ, Schoeniger SR, Millis M, Ganz R. Periacetabular osteotomy for containment of the nonarthritic dysplastic hip secondary to poliomyelitis. *J Bone Joint Surg Am*, 2010; **92**:2917–2923.

183. Kain MS, Novais EN, Vallim C, Millis MB, Kim YJ. Periacetabular osteotomy after failed hip arthroscopy for labral tears in patients with acetabular dysplasia. *J Bone Joint Surg Am*, 2011; **93**(Suppl 2):57–61.

184. Yamamoto Y, Ide T, Nakamura M, Hamada Y, Usui I. Arthroscopic partial limbectomy in hip joints with acetabular hypoplasia. *Arthroscopy*, 2005; **21**(5):586–591.

185. Parvizi J, Bican O, Bender B, et al. Arthroscopy for labral tears in patients with developmental dysplasia of the hip: a cautionary note. *J Arthroplasty*, 2009; **24**(6 Suppl):110–113.

186. Byrd JW, Jones KS. Hip arthroscopy in the presence of dysplasia. *Arthroscopy*, 2003; **19**(10):1055–1060.

187. Domb BG, Stake CE, Lindner D, El-Bitar Y, Jackson TJ. Arthroscopic capsular plication and labral preservation in borderline hip dysplasia: two-year clinical outcomes of a surgical approach to a challenging problem. *Am J Sports Med*, 2013; **41**(11):2591–2598.

188. Fujii M, Nakashima Y, Noguchi Y, et al. Effect of intra-articular lesions on the outcome of periacetabular osteotomy in patients with symptomatic hip dysplasia. *J Bone Joint Surg Br*, 2011; **93-B**:1449–1456.

189. Ross JR, Zaltz I, Nepple JJ, Schoenecker PL, Clohisy JC. Arthroscopic disease classification and interventions as an adjunct in the treatment of acetabular dysplasia. *Am J Sports Med*, 2011; **39**(Suppl):72–78.

190. Hussell JG, Mast JW, Mayo KA, Howie DW, Ganz R. A comparison of different surgical approaches for the periacetabular osteotomy. *Clin Orthop Relat Res*, 1999; (363):64–72.

191. Crockarell J Jr, Trousdale RT, Cabanela ME, Berry DJ. Early experience and results with the periacetabular osteotomy. The Mayo Clinic experience. *Clin Orthop Relat Res*, 1999; (363):45–53.

192. Davey JP, Santore RF. Complications of periacetabular osteotomy. *Clin Orthop Relat Res*, 1999; (363):33–37.

193. Trumble SJ, Mayo KA, Mast JW. The periacetabular osteotomy. Minimum 2 year follow-up in more than 100 hips. *Clin Orthop Relat Res*, 1999; (363):54–63.

194. McKinley TO. The Bernese periacetabular osteotomy: review of reported outcomes and the early experience at the University of Iowa. *Iowa Orthop J*, 2003; **23**:23–28.

195. Peters CL, Erickson JA, Hines JL. Early results of the Bernese periacetabular osteotomy: the learning curve at an academic medical center. *J Bone Joint Surg Am*, 2006; **88**(9):1920–1926.

196. Biedermann R, Donnan L, Gabriel A, et al. Complications and patient satisfaction after periacetabular pelvic osteotomy. *Int Orthop*, 2008; **32**(5):611–617.

197. Zaltz I, Baca G, Kim YJ, et al. Complications associated with the periacetabular osteotomy: a prospective multicenter study. *J Bone Joint Surg Am*, 2014; **96**:1967–1974.

198. Wenger DR, Is there a role for acetabular dysplasia correction in an asymptomatic patient? *J Pediatr Orthop*, 2013; **33**(1 Suppl):S8–S12.

Chapter 22

Advances in Avascular Necrosis of the Hip joint

Kaveh Gharanizadeh

Shafa Yahyaeian Hospital, Baharestan Square, Tehran, Iran
Orthopedic Bone and Joint Research Center,
Iran University of Medical Sciences (IUMS), Iran
kaveh.gharani@gmail.com

22.1 Femoral Head Osteonecrosis

22.1.1 Introduction

Femoral head osteonecrosis (FHON) is a progressive disease that usually happens in the third to the fifth decades of life, and in case it is not managed properly, up to 80% of the cases would be symptomatic and two-thirds of the patients would experience head collapse [1]. Haenisch [2] called it hip necrosis in 1925, and after that it came under the names of atraumatic necrosis, idiopathic necrosis and avascular necrosis (AVN); however, nowadays osteonecrosis is the preferred name. Currently it includes 10%–12% of all primary total hip replacements (THRs) in the United States, which would

---

The Hip Joint in Adults: Advances and Developments
Edited by K. Mohan Iyer
Copyright © 2018 Pan Stanford Publishing Pte. Ltd.
ISBN 978-981-4774-72-7 (Hardcover), 978-1-351-26244-6 (eBook)
www.panstanford.com

be around 20,000 cases annually [1, 3, 4]. In spite of all the various studies being carried out, the basic reasons of this disorder still remain unclear, and various aetiologies have been suggested [5, 6]. Since this disorder affects a very active age group of population, considering its disabling effects can impose a huge socioeconomic burden. In the meantime its prevalence is also increasing [7, 8]. During the last decade various hip-preserving techniques have been provided, which also brought about positive results [9].

22.1.2 Pathology and Aetiology

In this chapter we will talk about nontraumatic FHON. Different and variable mechanisms would result in necrosis that are not completely clear. However, ultimately after necrosis starts it would always follow through a common pathway [5, 10]. Necrosis caused by apoptosis without ischaemia, resulting from corticosteroids and alcohol usage, has also been strongly suggested recently [11, 12].

Arterial obstruction can be caused by internal and external reasons. Internal obstructions can be caused by local thrombosis, fat emboli, nitrogen bubbles or deformed and abnormal red blood cells [13–15].

External arterial obstruction in the bones can be caused by increased bone marrow pressure (marrow packing) due to increased fat or accumulation of cellular elements in the tissue [13, 16, 17]; or it may be due to increased pressure on the retinacular arteries of the femoral neck caused by intra-articular effusion leading to increased intra-articular pressure.

Vasculitis, irradiation and chemical toxicity can directly damage arteries [16, 18, 19]. In case the severity of ischaemia is higher than the osteocyte tolerance, irreversible ischaemic changes occur in cells, which is common disregarding the initial causing factor. Histologically this injury can be seen as dead osteocyte cells and increased empty lacunas after 24–72 hours, which can be identified by an inflammatory reaction in the healthy surrounding tissue and bone oedema in MRI [20, 21]. The healing process begins with an inflammatory cascade, including fibrovascular tissue. This tissue penetrates into dead tissue and in addition to resorption of dead trabecula, it starts to produce woven bone using primitive mesenchymal cells. The dead tissue can then be absorbed through a creeping substitution process [20, 22]; however, unfortunately the new bone lacks the necessary strength and rigidity to resist physiologic loads and results in collapses in the weight-bearing (WB) area, which would ultimately lead to a flat and deformed femoral head and secondary osteoarthritis.

Different reasons have been suggested for the basic cause of the necrosis. Various studies have been carried out about the relationship between the genetic background and hypercoagulation. Hereditary thrombophilia, impaired fibrinolysis, antiphospholipid antibodies, mutation of Factor V Leiden and lack of proteins S and C have been suggested in different studies [23–26].

Acquired coagulation disorders such as pregnancy, hyperlipidaemia, malignant tumours and inflammatory bowel disease (IBD) have also been proposed. Obviously patients who have an increased coagulation status can be potential candidates for AVN. One of the other theories that refer to genetic influence is that all the people who use alcohol or corticosteroids don't experience FHON. According to studies, this can be related to variation in metabolism and enzymes [27, 28]. About corticosteroids, they found the P. glycoprotein (Pqp) gene responsible for resistance against the creation of FHON secondary to corticosteroids.

High doses of corticosteroids [29–35] and alcohol abuse [36] are among known reasons of FHON, in which various reasons are suggested for their causes. Forty per cent of corticosteroid receivers may experience some degree of FHON and this is considered as the most common cause of nontraumatic FHON [37, 38].

Corticosteroids can cause impaired lipid metabolism, hyperlipidaemia, fatty liver, diffused fat embolism and increased bone marrow fat. In addition they inhibit osteogenesis and affect mesenchymal stem cells to differentiate into adipocyte and increase bone marrow fat. They have also been known as a reason for apoptosis of osteocyte cells [39].

Alcohol can also cause apoptosis [12, 39], lipid metabolism disorder and fat accumulation in the bone marrow. Besides, alcohol and corticosteroids can activate immune mechanisms that can be related to inflammatory factors such as interleukin-1 (IL-1), interleukin-10 (IL-10), tumour necrosis factor–alpha (TNF-α) and transforming growth factor–beta (TGF-β) [9].

Alcohol usage of 400 cc a week increases the risk of FHON by 11 times; smoking 20 packs of cigarette annually can also be a risk factor [40].

There are other risk factors related to FHON, which can be seen in Table 22.1.

Table 22.1 Risk factors associated with osteonecrosis

- Corticosteroid usage
- Alcohol usage
- Smoking cigarette
- HIV
- SLE

Coagulopathy
- Deficiency of antithrombin III
- Proteins C and S deficiency
- Thrombocytosis
- Decreased fibrinolysis: Increase in plasminogen activator inhibitor (PAI)

Haemoglobinopathy
- Thalassemia
- Polycythemia

Metabolic disorders
- Gaucher disease
- Necrosis

Rare disorders
- Dysbaric osteonecrosis (Giessen disease)
- Hyperlipidaemia

Other factors
- Radiotherapy
- Pregnancy
- Chemotherapy

Trauma
- Hip dislocation
- Femoral neck fracture

Systemic lupus erythematosus (SLE) patients using corticosteroids experienced a higher rate of FHON compared to non-SLE corticosteroid users (37% versus 20%, respectively) (P = 0.001) [41].

22.1.3 Other Diseases

Sickle cell disease: FHON can happen due to the closure of bone marrow arteries, the resulting pressure increase and the collapse of arteries. **Gaucher disease** can also lead to FHON by a similar mechanism (marrow packing) [42].

HIV can also cause FHON due to chemotherapy and the usage of corticosteroids and antiretrovirus medications.

Finally the multiple hit theory has been proposed in the incidence of nontraumatic FHON, where creating interruption in bone homeostasis, direct cell damage and indirect impaired blood supply can cause FHON, which is similar to what has been explained about corticosteroids.

Despite all studies idiopathic FHON constitutes about 30% of cases and no risk factors can be found [43]. Ollivier et al. [44] in a comparison of idiopathic FHON patients with healthy people using a computed tomography (CT) scan have shown that a neck–shaft angle less than 129° with a likelihood ratio (LR) of 3.6, anteversion of the femoral neck > 17 with LR = 3.8, lateral centre edge angle (LCEA) < 32 with LR = 5.7 and acetabular version < 19 with LR = 1.38 are associated with FHON where the combination of three out of four of these measurements is seen in 73% of patients suffering from FHON whereas this ratio is only 11% in the healthy group (LR = 6) [44]. In a different study, the decrease in the head-neck offset in the FHON group was significantly increased compared to the control group (63° versus 47°) ($p < 0.0001$). The above findings demonstrate the likelihood of the influence of hip morphology and their biomechanical effects, which would require further evaluation.

Despite all extended studies the exact mechanism and the common justifying reason of osteonecrosis of the femoral head are still unclear. However, the bone healing process after cellular necrosis and the subsequent mechanisms are common in all types. Small lesions that involve less than 15% of the femoral head are usually asymptomatic, and there is a chance of recovery [45]. Larger lesions enter the healing cycle but due to weaknesses in the initial immature bone, microfractures occur, which would eventually lead to a collapse in the subchondral bone or on the border between healthy and necrotic bone [11].

22.1.4 Clinical Presentation and Diagnosis

Typically, patients present with acute onset of pain in the groin, which would get aggravated during walking and sitting. However, 10%–15% of the patients refer with nonspecific pain. The pain is mostly described as deep and sharp as it can spread to the anterior thigh or around the knee. Patients usually suffer from an antalgic limp. They also usually complain about pain during physical examination, especially in flexion and during internal rotation (anterior impingement sign). They may also complain about episodes of locking or click (usually after the femoral head collapse). A log roll test can be positive in such patients. Due to secondary synovitis and joint effusion, a change in position from sitting to standing and vice versa can be associated with worsening pain.

A careful history of past illnesses and the possibility of having used various drugs, such as corticosteroids, alcohol, chemotherapy and drug addiction, is necessary.

Duration of the disease and whether or not the patient is symptomatic can also affect the outcome of treatment. Beaulé et al. [46] report much better results in patients who referred within 12 months of their symptomatic disorder and underwent total resurfacing arthroplasty [47]. Similarly Mont et al., in a research of 45 cases of THR after FHON, reported a much lower revision rate and a higher Harris Hip Score (HHS) in patients who referred within less than six months of being symptomatic [48].

The author's recommendation is to consider each patient according to his or her age, amount of activity, general health and underlying diseases and set diagnostic and therapeutic methods in accordance with that patient's requirements. For example, in a patient suffering from leukaemia, who is not expected to live very long, no surgical procedures can probably be appropriate.

22.1.5 Imaging

22.1.5.1 Plain X-ray

An anteroposterior view of the pelvis and lateral hip radiographs are the main steps for the primary diagnosis. Major and minor diagnostic criteria have been described previously [49]. Major criteria include defined sclerosis and subchondral collapse (crescent sign) in radiography. Minor criteria include narrowing of the joint space with femoral head collapse, mottled sclerosis and involvement of the acetabular side. The most common radiographic view to see the subchondral fracture or crescent sign is in the lateral frog leg position, because the common place of the necrosis is in the superolateral part of the anterior half of the head [40]. The main weakness of plain radiography is in its false negativity during the first stages of FHON (Fig. 22.1A).

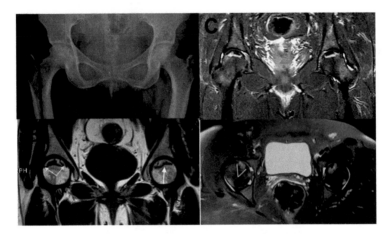

Figure 22.1 Bilateral femoral head osteonecrosis stage I. (A) Plain radiography is almost normal. (B) Bilateral femoral head osteonecrosis stage I – a T1 coronal MRI shows the typical well-defined 'band-like lesion'. Minor criteria are lesions with hypointensity without well-demarcated banding in T1 [49]. (C) Bilateral femoral head osteonecrosis stage I; a coronal T2 MRI shows the double line sign. (D) Axial fat suppression; a T2 cut of MRI shows a small anterior location. (B, D) Karboul angle measurement depicted with small involvement < 200.

In the early stages of the disease the first change is mild sclerosis around the osteopenic area can be completely hidden (Fig. 22.2).

As the disease progresses, sclerosis and osteopenic areas will appear in a geographic pattern and after that the head collapses, widens and becomes flat (Figs. 22.3–22.5).

Finally, degenerative changes start to involve on the acetabular side, too.

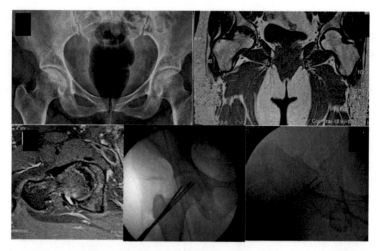

Figure 22.2 Stage II osteonecrosis of the right hip. (A) Minor radiolucency, especially in the weight-bearing area of the head without any collapse. (B) T1 coronal image (no collapse). (C) Axial MRI showing involvement of the anterior half of the head. (D, E) Multiple drilling with exact locations of the small drills towards the necrotic area.

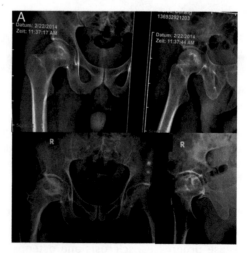

Figure 22.3 Right hip osteonecrosis following corticosteroid; evolution in time. (A) AP and lateral views show a large radiolucent lesion with sclerosis (major criteria) with no obvious collapse. On the basis of ARCO classification it is stage II, location C, >30%. Five months later a subchondral fracture and depression happened (arrow), and the head is flat. Stage III late.

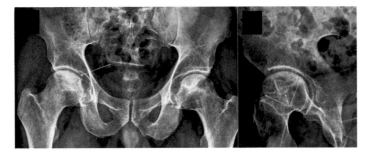

Figure 22.4 Bilateral hip osteonecrosis in a 28-year-old man following corticosteroid therapy. Typical radiolucent and sclerotic lesions scattered in a geographic pattern in both hips. Careful focus on the pelvic anteroposterior (A) view shows a fracture line (arrow) with minimal depression. Also in the lateral view (B) you can see the subtle subchondral fracture starting from the central area (right arrow). Three arrows depict the outline of the necrosis. The overall stage is early III, location C, depression <2 mm and extension C.

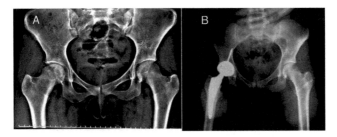

Figure 22.5 Right hip osteonecrosis in a 20-year-old medical student. She is a known case of sickle cell anaemia. Stage III late with head flattening. THR selected with a short stem and ceramic bearing because any hip-preserving surgery has a poor prognosis at this stage.

22.1.5.2 MRI

MRI is the gold standard and the method of choice in the diagnosis of FHON in the precollapse stages of the femoral head [9]. There are numerous advantages, including the fact that it's noninvasive and provides the ability to diagnose the asymptomatic lesions of the opposite side and determine the extent of the involvement and quantitatively measure the necrotic area [50–53]. A major diagnosis criterion in MRI is the existence of a band-like lesion with a low intensity in T1 around the area of necrosis (Fig. 22.1B).

In T2, a double line sign includes an external rim with a low intensity and an internal rim with a high intensity [52, 53] (Fig. 22.1C).

22.1.5.3 CT scan

A CT scan is the most sensitive test for the diagnosis of a small subchondral fracture [54] that cannot be very clear in radiography or MRI. The main disadvantage of this method is the high radiation exposures.

22.1.5.4 Nuclear bone scan

In the early stages of the disease the affected area is seen as a cold signal due to lack of absorption of TC 99. However, the rim around it is seen as hot. As the disease progresses and in the incidence of collapse, a hot signal would be dominant in most parts of the image. Due to its low specificity and inability to quantify the extent of involvement it is now rarely used.

Other methods, like positron emission tomography (PET) and dual-energy X-ray absorptiometry (DEXA), are not useful in the diagnosis or classification of FHON [22].

22.1.6 Differential Diagnosis

Transient osteoporosis is a rare disorder that presents with severe hip pain with no history of trauma or extreme activity. Transient osteoporosis is self-limited, and its spontaneous recovery happens within some months (6–12 months) [40]. Usually middle-aged men are involved although it can be seen in women, especially in the late stages of pregnancy. In the radiography only some osteopenia can be seen [54–57]. MRI shows a homogeneous oedema in the bone marrow of the femoral head and neck that is extended into intertrochanteric area and proximal femur without having any focal lesions, with a low intensity. In a bone scan an extended increased uptake is seen in the proximal part of the femur.

Another differential diagnosis is a subchondral insufficiency fracture of the femoral head that can be seen in elderly people, usually followed by a small trauma such as a hip sprain or leaning forwards or after a long walk. The reason for the fracture is osteoporosis or osteopenia, which can happen in elderly women or organ transplantation patients [58, 59]. It would be difficult to diagnose from FHON. The most important findings are in MRI. In the area of the fracture there is a band-like lesion with a low-intensity signal. However, unlike FHON, this lesion is irregular, disconnected, parallel and convex to the surface of cartilage, while in FHON the band is continuous and completely clear and is shaped concavely to the cartilage surface and is getting away from it. Diagnosis can sometimes be difficult, although the treatment of choice in both cases would be arthroplasty.

Neoplasm: Tumoural lesions that can be in differential diagnosis are clear cell chondrosarcoma. Among epiphyseal benign tumours, chondroblastoma and giant cell tumour of the bone can be classified as differential diagnosis.

22.1.7 Classification

Various classifications have been noted for FHON. The most common classification is Ficat & Arlet [60]. It was introduced in 1960 using plain radiography, measurement of bone marrow pressure and biopsy (Table 22.2). With the application of MRI in the diagnosis, in all new classifications a major part was given to MRI. Other common classifications are Steinberg [61] and Association Research Circulation Osseous (ARCO) [62, 63] (Table 22.2).

All classifications are determined on the basis of some common prognostic factors.

22.1.7.1 Collapse

The most important prognostic factor is collapse, which is identified by a crescent sign in radiography. Chondral collapse happens at the weakened subchondral bone that usually occurs at the lateral border between healthy and necrotic lesions. Collapses can appear in larger lesions that are more medially drawn than fovea of the femoral head. It can happen in the depth of necrosis or even at the junction of the necrosis with metaphysis (Fig. 22.3).

However, in smaller lesions that are located lateral to fovea, there is always a subchondral collapse [64].

Although a large collapse is visible in plain radiology and MRI, some smaller hidden collapses can simply be missed and are best shown in CT scans. Therefore the author advises CT scans in all cases that the surgeon decides for hip preservation surgery in order to make sure there is no collapse before the surgery. Hips that are in precollapse stages may be treated with preserving methods but in postcollapse stages, arthroplasty is generally known as the best treatment [1, 65].

Table 22.2 Classifications of osteonecrosis

Stage		Ficat & Arlet [60]	Steinberg et al. [61]	ARCO[a] [62, 63]
0		Normal or nondiagnostic radiographic, bone scan and MRI findings	Normal or nondiagnostic radiographic, bone scan and MRI findings	Normal or nondiagnostic radiographic, bone scan and MRI findings
I		Normal radiographs Bone scan: Cold spot on the femoral head Pathology: Infarction of weight-bearing area, abundant dead marrow cells and osteoblasts	Normal radiographic findings, abnormal bone scan and/or MRI findings A: Mild (<15% of head affected) B: Moderate (15%–30% affected) C: Severe (>30% affected)	Normal radiography Positive findings on bone scan, MRI or both
II	A	Sclerotic or cystic lesions No collapse	Lucent and sclerotic changes in the femoral head A: Mild (<15%) B: Moderate (15%–30%) C: Severe (>30%)	Radiography: Mottled appearance of the femoral head, osteosclerosis, cyst formation and osteopenia Bone scan, MRI and CT scan: Positive No collapse
	B	Subchondral collapse (crescent sign) without femoral head flattening		

III	Femoral head flattening	Subchondral fracture (crescent sign) without flattening A: Mild (<15% of articular surface) B: Moderate (15%–30%) C: Severe (>30%)	Early	Subchondral fracture (crescent sign) without flattening
			Late	Flattening of the femoral head
IV	Joint space narrowing and osteoarthritis	Flattening of the femoral head A: Mild (<15% of surface or <2 mm depression) B: Moderate (15%–30% of surface or 2–4 mm depression) C: Severe (>30% of surface or >4 mm depression)	Joint space narrowing; change in the acetabulum and osteoarthritis	
			Modifiers in ARCO classification 1. Location A. Medial third B. Central third C. Lateral third 2. Area of involvement or length of crescent A: Mild (<15%) B: Moderate (15%–30%) C: Severe (>30%) 3. % surface collapse and dome depression A: <2 mm, <15% B: 2–4 mm, 15%–30% C: >4 mm, >30%	
V		Joint narrowing and/or acetabular changes A: Mild (average of femoral head involvement as in stage IV and estimated acetabular involvement) B: Moderate involvement C: Severe involvement		
VI		Advanced degenerative changes		

[a]ARCO: Association Research Circulation Osseous

22.1.7.2 The lesion extent

Studies have approved the effects of the lesion size on the prognosis [1, 66]. Kerboul et al. [67] have used anterior and lateral radiography for its measurement (Fig. 22.1).

The sums of the arc angles in both views are added together and angles less than 150° are equivalent to a small lesion (15%), 150°–250° are medium (15%–30%) and more than 250° are considered as big lesions (more than 30%) [68]. it seems this method cannot be very accurate in the 3D volume measurement of the lesion and mostly shifts lesions to larger ones. Different studies have used MRI in midcoronal and midsagittal views and calculated the largest angle and proved that there is a significant relationship between the size of the lesion and collapse [69, 70].

22.1.7.3 Location of the lesion

For the first time, a Japanese investigation committee found the importance of the location of the lesion and noticed the probability of the incidence of collapse is a lot more in the lesions that are located in the lateral part of the WB area. Three regions, A, B and C, were identified in the WB area as medial, central and lateral, respectively. Association of research circulation osseous (ARCO) added this modifier to their classification in 1992 [62].

22.1.7.4 Depression rate of the head and head flattening

The femoral head can be not flat despite the existence of a subchondral fracture (crescent sign). The depression rate is classified as mild (<2 mm), moderate (2–4 mm) and severe (>4 mm), which has the worst prognosis [71].

22.1.7.5 Author's preference

The author prefers to use ARCO classification. However, in the case of a decision for hip preservation surgeries, the precollapse stage should be confirmed by a CT scan. Perhaps one of the reasons for inconsistent and contradicted results of preserving surgeries in different literatures is failure to detect small lesions of collapse preoperatively due to lack of CT scanning.

Overall, the evidence shows asymptomatic small lesions, especially in the medial, can be followed but larger symptomatic lesions require surgical procedures, which can differ on the basis of the severity of the collapse.

22.1.8 Treatment

Various surgical treatments have been proposed and reviewed for FHON. Conservative treatments include pharmacologic and nonpharmacologic methods and surgical treatments are basically classified in the two groups of joint preservation and joint replacement.

22.1.8.1 Nonsurgical treatment

Due to a lack of understanding of the pathogenesis of this disorder, there is a lot of controversy in the existing treatment methods. On the other hand since this disorder mostly involves young and active people (25% of the patients are less than 25 years old) [72], there is a great tendency to do nonarthroplasty treatments and possibly find a way to control the disorder.

Maybe the most important way to choose the right treatment method is knowing about the natural history. However, interestingly, results of various studies in this field are also different. Nishii et al. [73] followed precollapse patients for five years and 50% of them experienced collapse. However, the collapse did not progress and stopped in 54% of these patients. Cease of the progression of collapse depends on the extent of the lesion. The collapse stopped in 81% of the hips where less than two-thirds of the WB area was involved. Of these patients, 64% become asymptomatic eventually.

On the other hand Hernigou et al. [74] did a study on 40 patients with small asymptomatic lesions in stage I that could be found in MRI during a 15-year period and found that 88% of the hips started to become symptomatic and 73% of those ended up having a collapse and required a surgery.

Finally the lesions that affect more than two-thirds of the WB area [75] and usually present with severe pain will lead to collapse.

22.1.8.1.1 *Pharmacological methods*

Bisphosphonates inhibit osteoclasts, which would result in the inhibition of bone resorption and decrease remodelling and may eventually reduce the incidence of collapse. The longest study, of 10 years, showed that 87% of the cases treated with alendronate did not need surgical interventions and that the best time for starting medications is before the collapse stage. On the contrary, in a new multicentre randomised controlled trial (RCT) study [76] they found that there is no difference in the treatment of patients with alendronate or a placebo. Studies about the effects of other medications, such as statin [77–80], anticlotting drugs (including enoxaparin) [77], lipoic acid [78], antiapoptotic factors [79], growth factors [80] and herbal remedies [81], have been done, but most studies have low evidence (level IV) and the results are limited; therefore their results cannot be recommended for treatment.

22.1.8.1.2   Nonpharmacologic methods

These types of treatments are biophysical techniques, including pulsed electromagnetic field (PEMF) and extracorporeal shockwave therapy (ESWT) [82–84]. They have been studied and resulted in lowering pain and improvements in function. In an RCT study with hyperbaric oxygen therapy on 10 patients in a period of seven years, none of them underwent joint replacement [85].

Overall, in all level I and II studies, there have been variable and inconsistent results of nonsurgical treatments [86–88]. Furthermore their detailed indications are not clear, and they require further studies to investigate the effects.

22.1.8.1.3   Author's preference

All in all, nonsurgical treatments can be recommended for patients with small and limited lesions, especially those who are asymptomatic (stage I), but if the lesion size is increased and is symptomatic, the effects of these methods are uncertain and the indications are still unclear.

22.1.8.2   Core decompression

It includes a set of operations that is performed based on decompression of avascular bone marrow area. This operation can be combined with other techniques, such as adding a bone graft (BG), bone marrow aspiration (BMA) and cell-based therapy, bone morphogenic protein (BMP), calcium bone substitution and platelet-rich plasma (PRP) [89].

The aim of this operation, whether it would be using a single core or multiple drilling [89–92], is to reduce the internal bone pressure, which would result in lowering of the patient's pain immediately after the surgery (Fig. 22.2).

In a survey done on multiple drilling, 78% (120 hips out of 193) were preserved after five years of follow-up and after 7.2 years, 88% of the hips that had small to medium lesions didn't require surgical treatments [90].

In a systematic review [93] it was shown that core decompression is the most effective method in the first stages of FHON and smaller lesions. Patients with smaller lesions experience about 14%–25% failure but in larger lesions 42%–84% of patients may experience treatment failure [93, 94].

The location of the lesion can also be effective in the outcome of the surgery. Small lesions that involve the medial part of the WB area (stage IIA) rarely fail in treatment [93], but Yoon et al. [95] showed that 17 out of 23 hips with lateral lesions experienced collapse.

The combination of core decompression with a tantalum rod [96, 97] and calcium phosphate cement [98, 99] was reviewed in different studies, and variable results were obtained.

For example, in a prospective study on 27 patients with a 38-month follow-up of patients who underwent core decompression with a tantalum rod, the lesion progressed in 50% of the cases and eventually 70% survived [100].

A different study on 15 cases of THR who had undergone decompression with a tantalum rod previously showed that bone growth happened in 13 cases but it was very limited and only existed on 1.9% of the surface of the rod and wear debris could be seen in 13 cases [101].

In the case of BMA injection during core decompression, although there are promising results [102], there are no specific criteria about the sufficient number of colony forming units in BMA. The other problem is that in patients with alcohol or corticosteroids usage, the number of these units is very low and their differentiation into bone tissue is very little.

22.1.8.2.1 Core decompression with bone graft

Phemister introduced this method for the first time in 1949 [103]. Currently this operation can be done as both nonvascularised bone graft (NVBG) and vascularised bone graft (VBG). BG includes osteoinductive and osteoconductive properties along with structural strength. In a study done on 28 hips in the stages of flat II–IV all surgical procedures on patients with a collapse or those with corticosteroid usage failed but there was 73% success rate in stage II.

VBG was introduced by Urbaniak [104] using fibular vascularised graft (FVG), and after that methods like muscle pedicle graft and iliac crest were also reported later [105–107].

Recent studies on VBG show that this surgical procedure can be more successful in the precollapse stages. Several studies show that it had a much better result compared to core decompression alone or with NVBG [107, 108].

However, none of the studies were randomised and the complications of vascular grafts were relatively higher, including motor weakness, discomfort of the graft donor place and sensory impairments [108, 109]. Specific complications of FVG are the creation of clawing in the big toe and anterior tibial muscle weakness [110, 111].

22.1.8.2.2 Core decompression complications

- Femoral fracture may happen during or after the surgery. The incidence rate used to be 2.5% in the past, which is now reduced to less than 1% [112]. The reason of fracture is technical and in case the entrance of the core is above the lesser trochanter it can reduce significantly.
- Previously use of tantalum rod or fibula during THR surgery could be technically difficult, with more blood loss and longer operation time [112, 113]. However recent studies show that they do not affect the survival of the prostheses.

22.1.8.2.3 Author's preference

Multiple core decompression (Fig. 22.2) is currently known as the preferred treatment method for FHON in the precollapse stages and in smaller to medium symptomatic lesions (stages I or II). It should not be done in larger lesions or in collapse stages. Adding BMP, stem cell or other medications may improve the results but requires RCT and well-designed level I studies.

22.1.8.3 Bone graft

22.1.8.3.1 *Open bone graft*

It can be VBG or NVBG. The goal is to debride and remove the dead bone, preserve the head, create a protective scaffold resistant to collapse, accelerate bone repair and create an environment with osteoconductive properties while decompressing the head. It can be done in three ways.

The first method is the Phemister technique, which is similar to single core decompression and after the evacuation of the necrotic area it will be filled with a variety of BG.

The second method is trapdoor, which requires anterolateral dislocation of the head and opening a window from inside the cartilage of the head. Due to the possibility of iatrogenic damage to the cartilage and intensifying AVN, it is not favoured by the author [114] (Fig. 22.6).

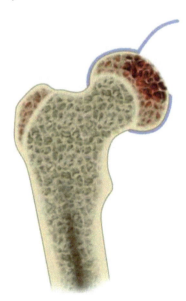

Figure 22.6 Trapdoor technique. A window created in the femoral head cartilage after anterolateral dislocation of the head.

In the lightbulb technique with the anterior or anterolateral approach, the capsule is opened and a window is created from the head-neck junction. After debridement, the defective part should be filled with a BG (Fig. 22.7).

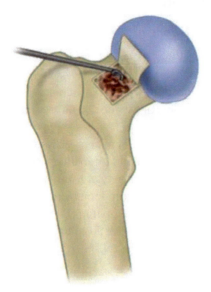

Figure 22.7 Lightbulb technique. A window created at the head-neck junction in the metaphysis after anterior capsulotomy.

The advantage of this method is that there is no more iatrogenic injury to cartilage or impaired blood supply to the head but there is a possibility of femoral neck fracture [115].

Various studies suggest that the best results can be obtained with smaller lesions and before collapse. The total success rate was reported to be 55%–87% [116–120].

In a recent study [121] about the lightbulb technique and the use of a combination of allograft and BMP 7, 67% of the 28 hips survived after 36 months of follow-up. The most successful results were in the small- to medium-sized lesions.

22.1.8.3.2 *Vascularised bone graft*

In addition to the advantages of the NVBG method, in this method the blood flow in the necrotic area can contribute to faster healing and union of the live graft [122, 123].

FVG is still the most commonly used method. It is relatively easy to harvest. The desired length of the bone is removed, and there is reliable blood supply and vascular pedicle. There is also the possibility of doing the operation with two operating teams in order to reduce the time of surgery. Other methods, including muscle pedicle graft from tensor fascia lata (TFL), quadratus femoris and enhanced iliac BG, have been reported [122–125].

Results from different studies, including the one done by Urbaniak et al. [104], show that the most effective time is the precollapse stage (11% failure in precollapse versus 65% failure in postcollapse). In a study of 224 hips in the postcollapse stage that were treated with vascularised fibular grafting (VFG), there was 65% success rate after 4.3 years of follow-up.

For the surgical technique, please refer to Ref. [40].

Complications: In a five-year survey done on 247 surgical cases, complications were reported [126] in 24% of them, of which the most common ones included ankle pain, motor weakness, sensory impairment and femoral fracture.

22.1.8.3.3 Author's preference

NVBG in the lightbulb method is recommended in patients with smaller lesions and in precollapse stages. VBG is not very often used due to technical problems and extensive necessary resources. However, it can still be used in facilitated centres with experienced surgical teams for larger lesions in the precollapse stage. There have been promising results in up to 65% of the cases of postcollapse patients, but considering the extent of the surgery and its complications, the author would only recommend it to the young, motivated patients in stage II and early III lesion.

22.1.8.4 Proximal femoral osteotomy

It can be done in two ways: angular and rotational [127–135]. The aim is changing biomechanics, removing the WB load from the necrotic area and finally transferring the healthy cartilage under the WB area [136]. The best cases are young patients (<40 years old) with BMI < 25 and initial stages of FHON where after the operation at least 33% of the WB area will be covered with healthy cartilage.

22.1.8.4.1 *Indication*

The indication is stage II or early III with smaller lesions where there is a possibility to transfer the healthy cartilage to the WB area. Decisions about the osteotomy method can be made according to the necrotic area.

22.1.8.4.2 *Contraindication*

Contraindications are underlying medical conditions that prevent natural bone healing, like kidney disorders, long-term use of corticosteroids and smoking.

Sugioka et al. reported successful results of rotational osteotomy [137]. However, their successful results were not reported by others [138, 139]. Most of the successful results are from Japan, and they didn't gain much popularity in Western countries.

Success rates of 82%–100% have been reported for rotational osteotomy [140–143] and 82%–98% have been reported for angular osteotomy [144–148]. In the case of failure due to malunion or nonunion the conversion to THR would have technical problems, with much more blood loss and a longer operation time [149, 150].

22.1.8.4.3 *Author's preference*

In case the surgeon is familiar with the technique it can have favourable outcomes in young patients who have been selected carefully and the lesion is in stage II or early III with involvement of less than 30% and depression less than 2 mm. For the surgical technique please refer to the original article in Ref. [151].

22.1.8.5 Hip replacement surgery

In the late postcollapse stage (late stage III) or stage IV and the involvement of the acetabulum, arthroplasty is the treatment of choice.

22.1.8.5.1 *Hemiresurfacing*

Although this method used to be practiced in order to maintain bone stock and reduce complications of joint instability in younger patients, it is rarely used nowadays [152, 153]. The main reason is degradation of acetabular cartilage due to exposure to the metal surface in young and active patients. Amstutz [154] did a study follow-up of 27 years and showed that it has a 63% survival rate after 10 years and 36% after 15 years.

22.1.8.5.2   Total resurfacing

This method is used in order to preserve proximal femoral bone stock and lower the rates of dislocation after surgery in younger patients. Its popularity rate has decreased significantly because of major complications of metal-on-metal [155] debris and lymphocytic reactions. Furthermore, other reasons of failure are progression of the necrotic area and loosening of the prosthesis [156, 157], thermal damages [158, 159] and increased risk of periprosthetic fractures [160]. Nevertheless, Amstutz [161] published the results of 92 hips that had been followed-up for 10.8 years in 2016 and showed a 90.3% survival rate, which is comparable to THR results. Good results are disregarding the extension of the necrosis or femoral head defects, and the author notes that the most important factor is the correct technique and the use of cement in metaphysis and normal distribution of pressure in the proximal part of the femur.

In case of consideration of total resurfacing for FHON, it is very important to pay attention to the technique and choose the patient carefully. The best cases are men under the age of 50 with a femoral head diameter of >50 mm and smaller lesions (2 cm cube) [162].

22.1.8.5.3   Total hip arthroplasty

Historically, THR results were disappointing up to 1990 [163]. But results after 1990 were much better, and they all reported over 90% survival rates, which is comparable with THR after osteoarthritis [164–172].

The underlying reasons for improvements in bearing surfaces and tribology are the use of ultra-high-molecular-weight polyethylene (UHMWPE) and ceramic, which have significantly reduced wearing [168–171]. In a research study with a midterm follow up to 8.5 years, 100% survival rates (ignoring the infection) were reported in prostheses with ceramic on polyethylene bearing [169]. In an analysis of 3277 THR cases following FHON, the highest failure rates were reported in patients with Gaucher disease, sickle cell anaemia and kidney transplant. On the contrary idiopathic patients, SLE cases and heart transplantation were reported to have better results [172]. In a study done by Kim et al. [171] stem and cup survival after 7.3 years was reported 98% and 85%, respectively, and the main reasons for cup failure were wearing and loosening. Similar results can also be seen in non-FHON patients. Porous metal technology like tantalum or highly porous titanium cups have had excellent short-term results for non-FHON patients, which would require the passage of time for long-term review of results.

Another issue is the effect of previous surgical procedures like proximal femur osteotomy (PFO) and resurfacing in FHON patients on the results of hip arthroplasty. Recent studies with midterm follow-ups indicated that PFO [173] or resurfacing [174–176] has no effects on THR survival rates. However, it is technically possible to experience bleeding or extended operation time during surgery [177, 178].

22.1.8.5.4 *Author's preference*

Cementless THRs are recommended for patients in stage IV or late stage III and also in patients in early stage III with large lesions. New bearings like ceramics are preferred in young patients (Fig. 22.5).

22.1.9 Overview

FHON still remains a real challenge for orthopaedic surgeons to manage. FHON usually happens in active and high-demand patients. The accurate aetiology is still unclear. The anatomy of the hip joint and femoral head blood supply as well as the forces that act on the femoral head and the minimal available bone stock of the femoral head, all have made surgical treatment in this area much more difficult. The decision to treat each patient needs to be individualised in a way that considers the patient's general health and accompanying disorders in addition to the exact classification of the stages of the disorder. The flow chart in Fig. 22.8 shows our preferred methods of treatment.

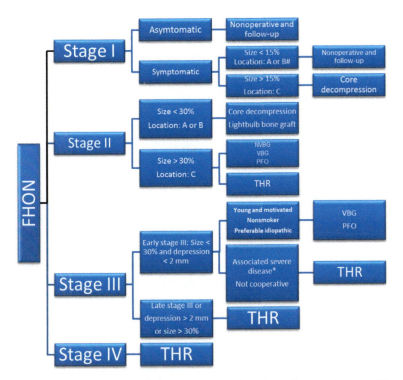

Figure 22.8 Recommended treatment algorithm for patients with osteonecrosis. FHON = femoral head osteonecrosis, NVBG = nonvascularised bone grafting, VBG = vascularised bone grafting, THA = total hip arthroplasty and PFO = proximal femur osteotomy. # location of osteonecrosis: medial = A, central = B, lateral = C. * like sickle cell disease, chronic corticosteroid use and renal failure: Confirm stage II with CT scan.

References

1. Mont MA, Hungerford DS. Non-traumatic avascular necrosis of the femoral head. *J Bone Joint Surg Am*, 1995; **77**(3):459–474.
2. Haenisch H. Arthritis dissecans der hufte. *Zentralblatt*, 1925; **52**:1.
3. Lavernia CJ, Sierra RJ, Grieco FR. Osteonecrosis of the femoral head. *J Am Acad Orthop Surg*, 1999; **7**:250–261.
4. Mont MA, Jones LC, Sotereanos DG, Amstutz HC, Hungerford DS. Understanding and treating osteonecrosis of the femoral head. *Instr Course Lect*, 2000; **49**:169–185.

5. Mont MA, Hungerford DS. Non-traumatic avascular necrosis of the femoral head. *J Bone Joint Surg Am*, 1995; **77-A**:459–474.
6. Mont MA, Jones LC, Hungerford DS. Nontraumatic osteonecrosis of the femoral head: ten years later. *J Bone Joint Surg Am*, 2006; **88-A**:1117–1132.
7. Cooper C, Steinbuch M, Stevenson R, Miday R, Watts NB. The epidemiology of osteonecrosis: findings from the GPRD and THIN databases in the UK. *Osteoporos Int*, 2010; **21**(4):569–577.
8. Kang JS, Park S, Song JH, Jung YY, Cho MR, Rhyu KH. Prevalence of osteonecrosis of the femoral head: a nationwide epidemiologic analysis in Korea. *J Arthroplasty*, 2009; **24**(8):1178–1183.
9. Mont MA, Cherian JJ, DO, Sierra RJ, Jones, LC, Lieberman JR. Nontraumatic osteonecrosis of the femoral head: where do we stand today? A ten-year update. *J Bone Joint Surg Am*, 2015; **97**:1604–1627.
10. Orth P, Anagnostakos K. Coagulation abnormalities in osteonecrosis and bone marrow edema syndrome. *Orthopedics*, 2013; **36**:290–300.
11. Calder JD, Buttery L, Revell PA, Pearse M, Polak JM. Apoptosis – a significant cause of bone cell death in osteonecrosis of the femoral. *J Bone Joint Surg Br*, 2004; **86-B**:1209–1213.
12. Wang J, Kalhor A, Lu S, Crawford R, Ni JD, Xiao Y. iNOS expression and osteocyte apoptosis in idiopathic, nontraumatic osteonecrosis. *Acta Orthop*, 2015; **86**(1):134–141.
13. Jones JP Jr. Fat embolism, intravascular coagulation, and osteonecrosis. *Clin Orthop Relat Res*, 1993; (292):294–308.
14. Jones JP Jr. Intravascular coagulation and osteonecrosis. *Clin Orthop Relat Res*, 1992; (277):41–53.
15. Jones JP Jr. Alcoholism, hypercortisonism, fat embolism and osseous avascular necrosis. 1971. *Clin Orthop Relat Res*, 2001; (393):4–12.
16. Hungerford DS, Lennox DW. The importance of increased intraosseous pressure in the development of osteonecrosis of the femoral head: implications for treatment. *Orthop Clin North Am*, 1985; **16**(4):635–654.
17. Wang Y, Li Y, Mao K, Li J, Cui Q, Wang GJ. Alcohol-induced adipogenesis in bone and marrow: a possible mechanismfor osteonecrosis. *Clin Orthop Relat Res*, 2003; (410):213–224.
18. Arlet J. Nontraumatic avascular necrosis of the femoral head: past, present, and future. *Clin Orthop Relat Res*, 1992; (277):12–21.
19. Jacobs B. Epidemiology of traumatic and nontraumatic osteonecrosis. *Clin Orthop Relat Res*, 1978; **130**:51–67.

20. Day S, Ostrum RF, Chao EYS, Rubin CT, Aro HT, Einhorn TA. Bone injury, regeneration and repair. In *Orthopaedic Basic Science: Biology and Biomechanics of the Musculoskeletal System*, Buckwalter J, Einhorn TA, Simon SR, eds. (Rosemont, USA: American Academy of Orthopaedic Surgeons), 2000; pp. 372–375.

21. Glimcher MJ, Kenzora JE. The biology of osteonecrosis of the human femoral head and its clinical implications: II. The pathological changes in the femoral head as an organ and in the hip joint. *Clin Orthop Relat Res*, 1979; **139**:283–312.

22. Glimcher MJ, Kenzora JE. The biology of osteonecrosis of the human femoral head and its clinical implications: III. Discussion of the etiology and genesis of the pathological sequelae; comments on treatment. *Clin Orthop Relat Res*, 1979; **140**:273–312.

23. Rueda JC, Duque MAQ, Mantilla RD, Iglesias-Gamarra A. Osteonecrosis and antiphospholipid syndrome. *J Clin Rheumatol*, 2009; **15**(3):130–132.

24. Lieberman JR, Berry DJ, Mont MA, et al. Osteonecrosis of the hip: management in the 21st century. *Instr Course Lect*, 2003; **52**:337–355.

25. Jones LC, Hungerford DS. Osteonecrosis: etiology, diagnosis, and treatment. *Curr Opin Rheumatol*, 2004; **16**(4):443–449.

26. Chao YC, Wang SJ, Chu HC, Chang WK, Hsieh TY. Investigation of alcohol metabolizing enzyme genesin Chinese alcoholics with avascular necrosis of hip joint, pancreatitis and cirrhosis of the liver. *Alcohol Alcohol*, 2003; **38**(5):431–436.

27. Asano T, Takahashi KA, Fujioka M, et al. ABCB1 C3435T and G2677T/A polymorphism decreased the risk for steroid-induced osteonecrosis of the femoral head after kidney transplantation. *Pharmacogenetics*, 2003; **13**(11):675–682.

28. Asano T, Takahashi KA, Fujioka M, et al. Genetic analysis of steroid-induced osteonecrosis of the femoral head. *J Orthop Sci*, 2003; **8**(3):329–333.

29. Klingenstein G, Levy RN, Kornbluth A, Shah AK, Present DH. Inflammatory bowel disease related osteonecrosis: report of a large series with a review of the literature. *Aliment Pharmacol Ther*, 2005; **21**(3):243–249.

30. Zhao FC, Guo KJ, Li ZR. Osteonecrosis of the femoral head in SARS patients: seven years later. *Eur J Orthop Surg Traumatol*, 2013; **23**(6):671–677.

31. Griffith JF, Antonio GE, Kumta SM, Hui DS, Wong JK, Joynt GM, Wu AK, Cheung AY, Chiu KH, Chan KM, Leung PC, Ahuja AT. Osteonecrosis of hip and knee in patients with severe acute respiratory syndrome treated with steroids. *Radiology*, 2005; **235**(1):168–175.
32. Lieberman JR, Roth KM, Elsissy P, Dorey FJ, Kobashigawa JA. Symptomatic osteonecrosis of the hip and knee after cardiac transplantation. *J Arthroplasty*, 2008; **23**(1):90–96.
33. Sahraian MA, Yadegari S, Azarpajouh R, Forughipour M. Avascular necrosis of the femoral head in multiple sclerosis: report of five patients. *Neurol Sci*, 2012; **33**(6):1443–1446.
34. Talamo G, Angtuaco E, Walker RC, Dong L, Miceli MH, Zangari M, Tricot G, Barlogie B, Anaissie E. Avascular necrosis of femoral and/or humeral heads in multiple myeloma: results of a prospective study of patients treated with dexamethasone-based regimens and high-dose chemotherapy. *J Clin Oncol*, 2005; **23**(22):5217–5223.
35. Renedo RJ, Sousa MM, Pérez SF, Zabalbeascoa JR, Carro LP. Avascular necrosis of the femoral head in patients with Hodgkin's disease. *Hip Int*, 2010; **20**(4):473–481.
36. Kamal D, Trăistaru R, Alexandru DO, Kamal CK, Pirici D, Pop OT, Mălăescu DG. Morphometric findings in avascular necrosis of the femoral head. *Rom J Morphol Embryol*, 2012; **53**(3 Suppl):763–767.
37. Weinstein RS, Nicholas RW, Manolagas SC. Apoptosis of osteocytes in glucocorticoid-induced osteonecrosis of the hip. *J Clin Endocrinol Metab*, 2000; **85**(8):2907–2912.
38. Koo KH, Kim R, Kim YS, et al. Risk period for developingosteonecrosis of the femoral head in patients on steroid treatment. *Clin Rheumatol*, 2002; **21**(4):299–303.
39. Seamon J, Keller T, Saleh J, Cui Q. The pathogenesis of nontraumatic osteonecrosis. *Arthritis*, 2012; **2012**: Article ID 601763, 11 pages.
40. Clohisy JC, Beaulé PE, Della Valle CJ, Callaghan JJ, Rosenberg AG, Rubash HE. *The Adult Hip: Hip Preservation Surgery* (Philadelphia: Wolters Kluwer), 2015; Chapter 21, pp. 224–231.
41. Shigemura T, Nakamura J, Kishida S, Harada Y, Ohtori S, Kamikawa K, et al. Incidence of osteonecrosis associated with corticosteroid therapy among different underlying diseases: prospective MRI study. *Rheumatology*, 2011; **50**(11):2023–2028.
42. Rodrigue SW, Rosenthal DI, Barton NW, Zurakowski D, Mankin HJ. Risk factors for osteonecrosis in patients with type 1 Gaucher's disease. *Clin Orthop Relat Res*, 1999; **362**:201–207.

43. Assouline-Dayan Y, Chang C, Greenspan A, Shoenfeld Y, Gershwin ME. Pathogenesis and natural history of osteonecrosis. *Semin Arthritis Rheum*, 2002; **32**(2):94–124.

44. Ollivier M, Lunebourg M, Abdel MP, Parratt S, Argenson JN. Anatomical findings in patients undergoing total hip arthroplasty for idiopathic femoral head osteonecrosis. *J Bone Joint Surg Am*, 2016; **98**:672–676.

45. Nam KW, Kim YL, Yoo JJ, et al. Fate of untreated asymptomatic osteonecrosis of the femoral head. *J Bone Joint Surg Am*, 2008; **90**(3):477–484.

46. Fackson M, Wang H, Johnson AJ, et al. Abnormal vascular endothelial growth factor expression in mesenchymal stem cells from both osteonecrotic and osteoarthritic hips. *Bull NYU Hosp Jt Dis*, 2011; **69**(Suppl 1):S56–S61.

47. Beaulé PE, Schmalzried TP, Campbell P, et al. Duration of symptoms and outcome of hemiresurfacing for hip osteonecrosis. *Clin Orthop Relat Res*, 2001; (385):104–117.

48. Mont MA, Ragland PS, Etienne G. Core decompression of the femoral head for osteonecrosis using percutaneous multiple small-diameter drilling. *Clin Orthop Relat Res*, 2004; (429):131–138.

49. Sugano N, Kubo T, Takaoka K, et al. Diagnostic criteria for non-traumatic osteonecrosis of the femoral head. A multicentre study. *J Bone Joint Surg Br*, 1999; **81**(4):590–595.

50. Hauzeur JP, Pasteels JL, Schoutens A, Hinsenkamp M, Appelboom T, Chochrad I, et al. The diagnostic value of magnetic resonance imaging in non-traumatic osteonecrosis of the femoral head. *J Bone Joint Surg Am*, 1989; **71**(5):641–649.

51. Markisz JA, Knowles RJ, Altchek DW, Schneider R, Whalen JP, Cahill PT. Segmental patterns of avascular necrosis of the femoral heads: early detection with MR imaging. *Radiology*, 1987; **162**(3):717–720.

52. Mitchell DG, Rao VM, Dalinka MK, Spritzer CE, Alavi A, Steinberg ME, et al. Femoral head avascular necrosis: correlation of MR imaging, radiographic staging, radionuclide imaging, and clinical findings. *Radiology*, 1987; **162**(3):709–715.

53. Steinberg DR, Steinberg ME. The University of Pennsylvania classification of osteonecrosis. In *Osteonecrosis*, Koo KH, Mont MA, Jones LC, eds. (Heidelberg: Springer), 2014; pp. 201–206.

54. Stevens K, Tao C, Lee SU, Salem N, Vandevenne J, Cheng C, et al. Subchondral fractures in osteonecrosis of the femoral head:

comparison of radiography, CT, and MR imaging. *Am J Roentgenol*, 2003; **180**(2):363-368.
55. Korompilias AV, Karantanas AH, Lykissas MG, Beris AE. Transient osteoporosis. *J Am Acad Orthop Surg*, 2008; **16**(8):480-489.
56. Patel S. Primary bone marrow oedema syndromes. *Rheumatology*, 2014; **53**(5):785-792.
57. Szwedowski D, Nitek Z, Walecki J. Evaluation of transient osteoporosis of the hip in magnetic resonance imaging. *Pol J Radiol*, 2014; **79**:36-38.
58. Ikemura S, Yamamoto T, Motomura G, Nakashima Y, Mawatari T, Iwamoto Y. The utility of clinical features for distinguishing subchondral insufficiency fracture from osteonecrosis of the femoral head. *Arch Orthop Trauma Surg*, 2013; **133**(12):1623-1627.
59. Yamamoto T. Subchondral insufficiency fractures of the femoral head. *Clin Orthop Surg*, 2012; **4**(3):173-180.
60. Ficat RP. Idiopathic bone necrosis of the femoral head. Early diagnosis and treatment. *J Bone Joint Surg Br*, 1985; **67**(1):3-9.
61. Steinberg ME, Brighton CT, Steinberg DR, et al. Treatment of avascular necrosis of the femoral head by a combination of bone grafting, decompression, and electrical stimulation. *Clin Orthop Relat Res*, 1984; (186):137-153.
62. Gardeniers JWM. ARCO committee on terminology and staging (report from the Nijmegen meeting). *ARCO News Lett*, 1992; **4**:6.
63. Gardeniers JWM. Report of the committee of staging and nomenclature. *ARCO News Lett*, 1993; **5**:4.
64. Motomura G, Yamamoto T, Yamaguchi R, Ikemura S, Nakashima Y, Mawatari T, Iwamoto Y. Morphological analysis of collapsed regionsin osteonecrosis of the femoral. *Bone Joint Surg Br*, 2011; **93-B**:184-187.
65. Schmalzried TP, Tiberi JV. Metal-metal reactivity: Houston, we have a problem! *Orthopedics*. 2010; **33**(9):647.
66. Saito S, Ohzono K, Ono K. Joint-preserving operations for idiopathic avascular necrosis of the femoral head. Results of core decompression, grafting and osteotomy. *J Bone Joint Surg Br*, 1988; **70**(1):78-84.
67. Kerboul M, Thomine J, Postel M, et al. The conservative surgical treatment of idiopathic aseptic necrosis of the femoral head. *J Bone Joint Surg Br*, 1974; **56**(2):291-296.
68. Chen CH, Chang JK, Huang KY, et al. Core decompression for osteonecrosis of the femoral head at pre-collapse stage. *Kaohsiung J Med Sci*, 2000; **16**(2):76-82.

69. Koo KH, Kim R. Quantifying the extent of osteonecrosis of the femoral head. A new method using MRI. *J Bone Joint Surg Br*, 1995; **77**(6):875–880.
70. Cherian SF, Laorr A, Saleh KJ, Kuskowski MA, Bailey RF, Cheng EY. Quantifying the extent of femoral head involvement in osteonecrosis. *J Bone Joint Surg Am*, 2003; **85-A**(2):309–315.
71. Berend KR, Gunneson E, Urbaniak JR, et al. Hip arthroplasty after failed free vascularized fibular grafting for osteonecrosis in young patients. *J Arthroplasty*, 2003; **18**(4):411–419.
72. Hungerford DS. Osteonecrosis: avoiding total hip arthroplasty. *J Arthroplasty*, 2002; **17**:121–124.
73. Nishii T, Sugano N, Ohzono K, et al. Progression and cessation of collapse in osteonecrosis of the femoral head. *Clin Orthop Relat Res*, 2002; (400):149–157.
74. Hernigou P, Poignard A, Nogier A, et al. Fate of very small asymptomatic stage-I osteonecrotic lesions of the hip. *J Bone Joint Surg Am*, 2004; **86-A**:2589–2593.
75. Sugano N, Atsumi T, Ohzono K, et al. The 2001 revised criteria for diagnosis, classification, and staging of idiopathic osteonecrosis of the femoral head. *J Orthop Sci*, 2002; **7**:601–605.
76. Chen CH, Chang JK, Lai KA, et al. Alendronate in the prevention of collapse of the femoral head in non-traumatic osteonecrosis: a two-year multicenter, prospective, randomized, double-blind, placebo-controlled study. *Arthritis Rheum*, 2012; **64**(5):1572–1578.
77. Glueck CJ, Freiberg RA, Sieve L, Wang P. Enoxaparin prevents progression of stages I and II osteonecrosis of the hip. *Clin Orthop Relat Res*, 2005; (435):164–170.
78. Lu BB, Li KH. Lipoic acid prevents steroid-induced osteonecrosis in rabbits. *Rheumatol Int.* 2012; **32**(6):1679–1683.
79. Park BH, Jang KY, Kim KH, Song KH, Lee SY, Yoon SJ, Kwon KS, Yoo WH, Koh YJ, Yoon KH, Son HH, Koh GY, Kim JR. COMP-Angiopoietin-1 ameliorates surgeryinduced ischemic necrosis of the femoral head in rats. *Bone*, 2009; **44**(5):886–892.
80. Kuroda Y, Akiyama H, Kawanabe K, Tabata Y, Nakamura T. Treatment of experimental osteonecrosis of the hip in adult rabbits with a single local injection of recombinant human FGF-2 microspheres. *J Bone Miner Metab*, 2010; **28**(6):608–616.

81. Tian L, Dang XQ, Wang CS, Yang P, Zhang C, Wang KZ. Effects of sodium ferulate on preventing steroid-induced femoral head osteonecrosis in rabbits. *J Zhejiang Univ Sci B*, 2013; **14**(5):426–437.

82. Kong XY, Wang RT, Tian N, Li L, Lin N, Chen WH. Effect of Huogu II Formula (II) with medicinal guide Radix Achyranthis Bidentatae on bone marrow stem cells directional homing to necrosis area after osteonecrosis of the femoral head in rabbit. *Chin J Integr Med*, 2012; **18**(10):761–768.

83. Chen JM, Hsu SL, Wong T, Chou WY, Wang CJ, Wang FS. Functional outcomes of bilateral hip necrosis: total hip arthroplasty versus extracorporeal shockwave. *Arch Orthop Trauma Surg*, 2009; **129**(6):837–841.

84. Vulpiani MC, Vetrano M, Trischitta D, Scarcello L, Chizzi F, Argento G, Saraceni VM, Maffulli N, Ferretti A. Extracorporeal shock wave therapy in early osteonecrosis of the femoral head: prospective clinical study with long-term follow-up. *Arch Orthop Trauma Surg*, 2012; **132**(4):499–508.

85. Wang CJ, Wang FS, Yang KD, Huang CC, Lee MS, Chan YS, Wang JW, Ko JY. Treatment of osteonecrosis of the hip: comparison of extracorporeal shockwave with shockwave and alendronate. *Arch Orthop Trauma Surg*, 2008; **128**(9):901–908.

86. Camporesi EM, Vezzani G, Bosco G, Mangar D, Bernasek TL. Hyperbaric oxygen therapy in femoral head necrosis. *J Arthroplasty*, 2010; **25**(6 Suppl):118–123.

87. Mont MA, Zywiel MG, Marker DR, McGrath MS, Delanois RE. The natural history of untreated asymptomatic osteonecrosis of the femoral head: a systematic literature review. *J Bone Joint Surg Am*, 2010; **92**(12):2165–2170.

88. Nakamura J, Harada Y, Oinuma K, Iida S, Kishida S, Takahashi K. Spontaneousrepair of asymptomatic osteonecrosis associated with corticosteroid therapy insystemic lupus erythematosus: 10-year minimum follow-up with MRI. *Lupus*, 2010; **19**(11):1307–1314.

89. Petrigliano FA, Lieberman JR. Osteonecrosis of the hip: novel approaches to evaluation and treatment. *Clin Orthop Relat Res*, 2007; **465**:53.

90. Al Omran A. Multiple drilling compared with standard core decompression for avascular necrosis of the femoral head in sickle cell disease patients. *Arch Orthop Trauma Surg*, 2013; **133**(5):609–613.

91. Song WS, Yoo JJ, Kim YM, Kim HJ. Results of multiple drilling compared with those of conventional methods of core decompression. *Clin Orthop Relat Res*, 2007; **454**:139–146.
92. Israelite C, Nelson CL, Ziarani CF, Abboud JA, Landa J, Steinberg ME. Bilateral core decompression for osteonecrosis of the femoral head. *Clin Orthop Relat Res*, 2005; **441**:285–290.
93. Lieberman JR, Engstrom SM, Meneghini RM, et al. Which factors influence preservation of the osteonecrotic femoral head. *Clin Orthop Relat Res*, 2011; **470**:525–534.
94. Soohoo NF, Vyas S, Manunga J, et al. Cost-effectiveness analysis of core decompression. *J Arthroplasty*, 2006; **21**(5):670.
95. Yoon TR, Song EK, Rowe SM, et al. Failure after core decompression in osteonecrosis of the femoral head. *Int Orthop.* 2001;24(6):316.
96. Floerkemeier T, Thorey F, Daentzer D, Lerch M, Klages P, Windhagen H, von Lewinski G. Clinical and radiological outcome of the treatment of osteonecrosis of the femoral head using the osteonecrosis intervention implant. *Int Orthop*, 2011; **35**(4):489–495.
97. Liu G, Wang J, Yang S, Xu W, Ye S, Xia T. Effect of a porous tantalum rod on early and intermediate stages of necrosis of the femoral head. *Biomed Mater*, 2010; **5**(6):065003.
98. Jiang HJ, Huang XJ, Tan YC, Liu DZ, Wang L. Core decompression and implantation of calcium phosphate cement/Danshen drug delivery system for treating ischemic necrosis of femoral head at Stages I, II and III of antigen reactive cell opsonization. *Chin J Traumatol*, 2009; **12**(5):285–290.
99. Civinini R, De Biase P, Carulli C, Matassi F, Nistri L, Capanna R, Innocenti M. The use of an injectable calcium sulphate/calcium phosphate bioceramic in the treatment of osteonecrosis of the femoral head. *Int Orthop*, 2012; **36**(8):1583–1588.
100. Varitimidis SE, Dimitroulias AP, Karachalios TS, et al. Outcome after tantalum rod implantation for treatment of femoral head osteonecrosis: 26 hips followed for an average of 3 years. *Acta Orthop*, 2009; **80**(1):20.
101. Tanzer M, Bobyn JD, Krygier JJ, et al. Histopathologic retrieval analysis of clinically failed porous tantalum osteonecrosis implants. *J Bone Joint Surg Am*, 2008; **90**(6):1282.
102. Hernigou P, Poignard A, Zilber S, et al. Cell therapy of hip osteonecrosis with autologous bone marrow grafting. *Indian J Orthop*, 2009; **43**(1):40.

103. Phemister DB. Treatment of the necrotic head of the femur in adults. *J Bone Joint Surg Am*, 1949; **31-A**(1):55.

104. Urbaniak JR, Coogan PG, Gunneson EB, et al. Treatment of osteonecrosis of the femoral head with free vascularized fibular grafting. A long-term follow-up study of one hundred and three hips. *J Bone Joint Surg Am*, 1995; **77**(5):681.

105. Babhulkar S. Osteonecrosis of femoral head: treatment by core decompression and vascular pedicle grafting. *Indian J Orthop*, 2009; **43**(1):27.

106. Iwata H, Torii S, Hasegawa Y, et al. Indications and results of vascularized pedicle iliac bone graft in avascular necrosis of the femoral head. *Clin Orthop Relat Res*, 1993; (295):281.

107. Chen CC, Lin CL, Chen WC, et al. Vascularized iliac bone-grafting for osteonecrosis with segmental collapse of the femoral head. *J Bone Joint Surg Am*, 2009; **91**(10):2390.

108. Berend KR, Gunneson EE, Urbaniak JR. Free vascularized fibular grafting for the treatment of postcollapse osteonecrosis of the femoral head. *J Bone Joint Surg Am*, 2003; **85-A**(6):987.

109. Nagoya S, Nagao M, Takada J, et al. Predictive factors for vascularized iliac bone graft for nontraumatic osteonecrosis of the femoral head. *J Orthop Sci*, 2004; **9**(6):566.

110. Yoo MC, Chung DW, Hahn CS. Free vascularized fibula grafting for the treatment of osteonecrosis of the femoral head. *Clin Orthop Relat Res*, 1992; (277):128.

111. Yoo MC, Kim KI, Hahn CS, et al. Long-term follow-up of vascularized fibular grafting for femoral head necrosis. *Clin Orthop Relat Res*, 2008; **466**(5):1133.

112. Fairbank AC, Bhatia D, Jinnah RH, et al. Long-term results of core decompression for ischaemic necrosis of the femoral head. *J Bone Joint Surg Br*, 1995; **77**(1):42.

113. Camp JF, Colwell CW Jr. Core decompression of the femoral head for osteonecrosis. *J Bone Joint Surg Am*, 1986; **68**(9):1313.

114. Merle d'Aubigne R, Postel M, Mazabraud A, et al. Idiopathic necrosis of the femoral head in adults. *J Bone Joint Surg Br*, 1965; **47**(4):612–633.

115. Clohisy JC, Beaulé PE, Della Valle CJ, Callaghan JJ, Rosenberg AG, Rubash HE. *The Adult Hip: Hip Preservation Surgery* (Philadelphia: Wolters Kluwer), 2015; Chapter 57, pp. 681–691.

116. Saito S, Ohzono K, Ono K. Joint-preserving operations for idiopathic avascular necrosis of the femoral head. Results of core decompression, grafting and osteotomy. *J Bone Joint Surg Br.* 1988; **70**(1):78–84.

117. Rosenwasser MP, Garino JP, Kiernan HA, et al. Long term followup of thorough debridement and cancellous bone grafting of the femoral head for avascular necrosis. *Clin Orthop Relat Res.* 1994; (306):17–27.

118. Hsu JE, Wihbey T, Shah RP, Garino JP, Lee GC. Prophylactic decompression and bone grafting for small asymptomatic osteonecrotic lesions of the femoral head. *Hip Int*, 2011; **21**(6):672–677.

119. Zhang HJ, Liu YW, Du ZQ, Guo H, Fan KJ, Liang GH, Liu XC. Therapeutic effect of minimally invasive decompression combined with impaction bone grafting on osteonecrosis of the femoral head. *Eur J Orthop Surg Traumatol*, 2013; **23**(8):913–919.

120. Helbig L, Simank HG, Kroeber M, Schmidmaier G, Grützner PA, Guehring T. Core decompression combined with implantation of a demineralised bone matrix for non-traumatic osteonecrosis of the femoral head. *Arch Orthop Trauma Surg*, 2012; **132**(8):1095–1103.

121. Seyler TM, Marker DR, Ulrich SD, et al. Nonvascularized bone grafting defers joint arthroplasty in hip osteonecrosis. *Clin Orthop Relat Res,* 2008; **466**(5):1125–1132.

122. Ali SA, Christy JM, Griesser MJ, Awan H, Pan X, Ellis TJ. Treatment of avascular necrosis of the femoral head utilising free vascularised fibular graft: a systematic review. *Hip Int*, 2014; **24**(1):5–13.

123. Yin S, Zhang C, Jin D, Chen S, Sun Y, Sheng J. Treatment of osteonecrosis of the femoral head in lymphoma patients by free vascularised fibular grafting. *Int Orthop*, 2011; **35**(8):1125–1130.

124. Wang YS, Zhang Y, Li JW, Yang GH, Li JF, Yang J, Yang GH. A modified technique of bone grafting pedicled with femoral quadratus for alcohol-induced osteonecrosis of the femoral head. *Chin Med J (Engl)*, 2010; **123**(20):2847–2852.

125. Baksi DP, Pal AK, Baksi DD. Long-term results of decompression and musclepedicle bone grafting for osteonecrosis of the femoral head. *Int Orthop*, 2009; **33**(1):41–47.

126. Vail TP, Urbaniak JR. Donor-site morbidity with use of vascularized autogenous fibular grafts. *J Bone Joint Surg Am.* 1996; **78**(2):204–211.

127. Hamanishi M, Yasunaga Y, Yamasaki T, Mori R, Shoji T, Ochi M. The clinical and radiographic results of intertrochanteric curved varus osteotomy for idiopathic osteonecrosis of the femoral head. *Arch Orthop Trauma Surg*, 2014; **134**(3):305–310.

128. Ito H, Tanino H, Yamanaka Y, Nakamura T, Takahashi D, Minami A, Matsuno T. Long-term results of conventional varus half-wedge proximal femoral osteotomy for the treatment of osteonecrosis of the femoral head. *J Bone Joint Surg Br*, 2012; **94**(3):308–314.

129. Zhao G, Yamamoto T, Ikemura S, Motomura G, Mawatari T, Nakashima Y, Iwamoto Y. Radiological outcome analysis of transtrochanteric curved varus osteotomy for osteonecrosis of the femoral head at a mean follow-up of 12.4 years. *J Bone Joint Surg Br*, 2010; **92**(6):781–786.

130. Ikemura S, Yamamoto T, Jingushi S, Nakashima Y, Mawatari T, Iwamoto Y. Leg-length discrepancy after transtrochanteric curved varus osteotomy for osteonecrosis of the femoral head. *J Bone Joint Surg Br*, 2007; **89**(6):725–729.

131. Park KS, Tumin M, Peni I, Yoon TR. Conversion total hip arthroplasty after previous transtrochanteric rotational osteotomy for osteonecrosis of the femoral head. *J Arthroplasty*, 2014; **29**(4):813–816.

132. Zhao G, Yamamoto T, Motomura G, Iwasaki K, Yamaguchi R, Ikemura S, Iwamoto Y. Radiological outcome analyses of transtrochanteric posterior rotational osteotomy for osteonecrosis of the femoral head at a mean follow-up of 11 years. *J Orthop Sci*, 2013; **18**(2):277–283.

133. Zhao G, Yamamoto T, Ikemura S, Motomura G, Iwasaki K, Yamaguchi R, Nakashima Y, Mawatari T, Iwamoto Y. Clinico-radiological factors affecting the joint space narrowing after transtrochanteric anterior rotational osteotomy for osteonecrosis of the femoral head. *J Orthop Sci*, 2012; **17**(4):390–396.

134. Ha YC, Kim HJ, Kim SY, Kim KC, Lee YK, Koo KH. Effects of age and body mass index on the results of transtrochanteric rotational osteotomy for femoral head osteonecrosis: surgical technique. *J Bone Joint Surg Am*, 2011; **93**(Suppl 1):75–84.

135. Yamamoto T, Ikemura S, Iwamoto Y, Sugioka Y. The repair process of osteonecrosis after a transtrochanteric rotational osteotomy. *Clin Orthop Relat Res*, 2010; **468**(12):3186–3191.

136. Motomura G, Yamamoto T, Suenaga K, Nakashima Y, Mawatari T, Ikemura S, Iwamoto Y. Long-term outcome of transtrochanteric anterior rotational osteotomy forosteonecrosis of the femoral head in patients with systemic lupus erythematosus. *Lupus*, 2010; **19**(7):860–865.

137. Sugioka Y, Hotokebuchi T, Tsutsui H. Transtrochanteric anterior rotational osteotomy for idiopathic and steroid-induced necrosis of

the femoral head. Indications and long-term results. *Clin Orthop Relat Res*, 1992; (277):111–120.

138. Tanaka S, Fukuda K, Tomihara M. Simulation by stereographic processing of computed tomography for transtrochanteric rotation osteotomy in necrosis of the femoral head. *Int Orthop*, 1998; **22**(2):116–121.

139. Dean MT, Cabanela ME. Transtrochanteric anterior rotational osteotomy for avascular necrosis of the femoral head. Long-term results. *J Bone Joint Surg Br*, 1993; **75**(4):597–601.

140. Tooke SM, Amstutz HC, Delaunay C. Hemiresurfacing for femoral head osteonecrosis. *J Arthroplasty*, 1987; **2**(2):125–133.

141. Zhao G, Yamamoto T, Motomura G, Iwasaki K, Yamaguchi R, Ikemura S, Iwamoto Y. Radiological outcome analyses of transtrochanteric posterior rotational osteotomy for osteonecrosis of the femoral head at a mean follow-up of 11 years. *J Orthop Sci*, 2013; **18**(2):277–283.

142. Ha YC, Kim HJ, Kim SY, Kim KC, Lee YK, Koo KH. Effects of age and body mass index on the results of transtrochanteric rotational osteotomy for femoral head osteonecrosis: surgical technique. *J Bone Joint Surg Am*, 2011; **93**(Suppl 1):75–84.

143. Hiranuma Y, Atsumi T, Kajiwara T, Tamaoki S, Asakura Y. Evaluation of instability after transtrochanteric anterior rotational osteotomy for nontraumatic osteonecrosis of the femoral head. *J Orthop Sci*, 2009; **14**(5):535–542.

144. Gallinaro P, Masse A. Flexion osteotomy in the treatment of avascular necrosis of the hip. *Clin Orthop Relat Res,* 2001; (386):79–84.

145. Jacobs MA, Hungerford DS, Krackow KA. Intertrochanteric osteotomy for avascular necrosis of the femoral head. *J Bone Joint Surg Br,* 1989; **71**(2):200–204.

146. Hamanishi M, Yasunaga Y, Yamasaki T, Mori R, Shoji T, Ochi M. The clinical and radiographic results of intertrochanteric curved varus osteotomy for idiopathic osteonecrosis of the femoral head. *Arch Orthop Trauma Surg*, 2014; **134**(3):305–310.

147. Ito H, Tanino H, Yamanaka Y, Nakamura T, Takahashi D, Minami A, Matsuno T. Long-term results of conventional varus half-wedge proximal femoral osteotomy for the treatment of osteonecrosis of the femoral head. *J Bone Joint Surg Br*, 2012; **94**(3):308–314.

148. Zhao G, Yamamoto T, Ikemura S, Motomura G, Mawatari T, Nakashima Y, Iwamoto Y. Radiological outcome analysis of transtrochanteric curved varus osteotomy for osteonecrosis of the femoral head at a

mean follow-up of 12.4 years. *J Bone Joint Surg Br*, 2010; **92**(6):781–786.

149. Ikemura S, Yamamoto T, Jingushi S, Nakashima Y, Mawatari T, Iwamoto Y. Leg-length discrepancy after transtrochanteric curved varus osteotomy for osteonecrosis of the femoral head. *J Bone Joint Surg Br*, 2007; **89**(6):725–729.

150. Park KS, Tumin M, Peni I, Yoon TR. Conversion total hip arthroplasty after previous transtrochanteric rotational osteotomy for osteonecrosis of the femoral head. *J Arthroplasty*, 2014; **29**(4):813–816.

151. Sugioka Y. Transtrochanteric anterior rotational osteotomy of the femoral head in the treatment of osteonecrosis affecting the hip: a new osteotomy operation. *Clin Orthop Relat Res*, 1978; (130):191–201.

152. Ullmark G, Sundgren K, Milbrink J, Nilsson O, Sörensen J. Osteonecrosis following resurfacing arthroplasty. *Acta Orthop*, 2009; **80**(6):670–674.

153. Thorey F, Reck F, Windhagen H, von Lewinski G. Influence of bone density on total hip resurfacing arthroplasty in patients with osteonecrosis of the femoral head - a radiological analysis. *Technol Health Care*, 2008; **16**(3):151–158.

154. Amstutz HC, Le Duff MJ. Current status of hemi-resurfacing arthroplasty for osteonecrosis of the hip: a 27-year experience. *Orthop Clin North Am*, 2009; **40**(2):275–282.

155. De Smet KA, Van Der Straeten C, Van Orsouw M, Doubi R, Backers K, Grammatopoulos G. Revisions of metal-on-metal hip resurfacing: lessons learned and improved outcome. *Orthop Clin North Am*, 2011; **42**(2):259–269, ix.

156. Dy CJ, Thompson MT, Usrey MM, Noble PC. The distribution of vascular foramina at the femoral head/neck junction: implications for resurfacing arthroplasty. *J Arthroplasty*, 2012; **27**(9):1669–1675.

157. Kabata T, Maeda T, Tanaka K, Yoshida H, Kajino Y, Horii T, Yagishita S, Tsuchiya H. Hemi-resurfacing versus total resurfacing for osteonecrosis of the femoral head. *J Orthop Surg (Hong Kong)*, 2011; **19**(2):177–180.

158. Hsieh PH, Tai CL, Liaw JW, Chang YH. Thermal damage potential during hip resurfacing in osteonecrosis of the femoral head: an experimental study. *J Orthop Res*, 2008; **26**(9):1206–1209.

159. Baker R, Whitehouse M, Kilshaw M, Pabbruwe M, Spencer R, Blom A, Bannister G. Maximum temperatures of 89°C recorded during the mechanical preparation of 35 femoral heads for resurfacing. *Acta Orthop*, 2011; **82**(6):669–673.

160. Beaulé PE, Campbell PA, Hoke R, Dorey F. Notching of the femoral neck during resurfacing arthroplasty of the hip: a vascular study. *J Bone Joint Surg Br*, 2006; **88**(1):35–39.

161. Amstutz HC, Le Duff MJ. Hip resurfacing for osteonecrosis: two- to 18-year results of the Conserve Plus design and technique. *Bone Joint J*, 2016; **98-B**:901–999.

162. Johnson AJ, Zywiel MG, Hooper H, Mont MA. Narrowed indications improve outcomes for hip resurfacing arthroplasty. *Bull NYU Hosp Jt Dis*, 2011; **69**(Suppl 1):S27–S29.

163. Johannson HR, Zywiel MG, Marker DR, Jones LC, McGrath MS, Mont MA. Osteonecrosis is not a predictor of poor outcomes in primary total hip arthroplasty: a systematic literature review. *Int Orthop*, 2011; **35**(4):465–473.

164. Zustin J, Sauter G, Morlock MM, Rüther W, Amling M. Association of osteonecrosis and failure of hip resurfacing arthroplasty. *Clin Orthop Relat Res*, 2010; **468**(3):756–761. Epub 2009 Jul 14.

165. Bedard NA, Callaghan JJ, Liu SS, Greiner JJ, Klaassen AL, Johnston RC. Cementless THA for treatment of osteonecrosis at 10-year follow-up: have we improved compared to cemented THA? *J Arthroplasty*, 2013; **28**(7):1192–1199. Epub 2013 Feb 12.

166. Bergh C, Fenstad AM, Furnes O, Garellick G, Havelin LI, Overgaard S, Pedersen AB, Mäkelä KT, Pulkkinen P, Mohaddes M, Kärrholm J. Increased risk of revision in patients with non-traumatic femoral head necrosis. Acta Orthop, 2014; **85**(1):11–17.

167. Goffin E, Baertz G, Rombouts JJ. Long-term survivorship analysis of cemented total hip replacement (THR) after avascular necrosis of the femoral head in renal transplant recipients. *Nephrol Dial Transplant*, 2006; **21**(3):784–788.

168. Nich C, Courpied JP, Kerboull M, Postel M, Hamadouche M. Charnley-Kerboull total hip arthroplasty for osteonecrosis of the femoral head a minimal 10-year follow-up study. *J Arthroplasty*, 2006; **21**(4):533–540.

169. Kim YH, Choi Y, Kim JS. Cementless total hip arthroplasty with alumina-on-highly cross-linked polyethylene bearing in young patients with femoral head osteonecrosis. *J Arthroplasty*, 2011; **26**:218–223.

170. Min BW, Lee KJ, Song KS, Bae KC, Cho CH. Highly cross-linked polyethylene in total hip arthroplasty for osteonecrosis of the femoral

head: a minimum 5-year follow-up study. *J Arthroplasty*, 2013; **28**:526–530.

171. Kim YH, Kim JS, Park JW, Joo JH. Contemporary total hip arthroplasty with and without cement in patients with osteonecrosis of the femoral head: a concise follow-up, at an average of seventeen years, of a previous report. *J Bone Joint Surg Am*, 2011; **93-A**:1806–1810.

172. Johannson HR, Zywiel MG, Marker DR, Jones LC, McGrath MS, Mont MA. Osteonecrosis is not a predictor of poor outcomes in primary total hip arthroplasty: a systematic literature review. *Int Orthop*, 2011; **35**(4):465–473.

173. Kawasaki M, Hasegawa Y, Sakano S, Masui T, Ishiguro N. Total hip arthroplasty after failed transtrochanteric rotational osteotomy for avascular necrosis of the femoral head. *J Arthroplasty*, 2005; **20**:574–579.

174. McGrath MS, Marker DR, Seyler TM, Ulrich SD, Mont MA. Surface replacement is comparable to primary total hip arthroplasty. *Clin Orthop Relat Res*, 2009; **467**:94–100.

175. Beaulé PE, Schmalzried TP, Campbell P, Dorey F, Amstutz HC. Duration of symptoms and outcome of hemiresurfacing for hip osteonecrosis. *Clin Orthop Relat Res*, 2001; **385**:104–117.

176. Cuckler JM, Moore KD, Estrada L. Outcome of hemiresurfacing in osteonecrosis of the femoral head. *Clin Orthop Relat Res*, 2004; **429**:146–150.

177. Gilbert RE, Cheung G, Carrothers AD, Meyer C, Richardson JB. Functional results of isolated femoral revision of hip resurfacing arthroplasty. *J Bone Joint Surg Am*, 2010; **92-A**:1600–1604.

178. Ball ST, Le Duff MJ, Amstutz HC. Early results of conversion of a failed femoral component in hip resurfacing arthroplasty. *J Bone Joint Surg Am*, 2007; **89-A**:735–741.

Index

abduction, 6, 38, 57, 91, 126–28, 132, 160–61, 163, 177, 180, 320–22, 335–36, 350–54, 356–57, 362
abnormal anatomies, 150, 231, 310
abnormalities, 147, 150–51, 157
ACEA, *see* anterior centre edge angle
acetabular anteversion, 353
acetabular cartilage delamination, 170
acetabular chondral lesions, 174, 187
acetabular component, 33–34, 37, 60–62, 88–93, 95, 230, 262, 310
acetabular defects, 231, 327
acetabular deficiency, 62, 330
acetabular depth-to-width index, 334
acetabular depth–width ratio (ADR), 337
acetabular dysplasia, 148, 174, 231, 321, 324, 327, 330, 355
acetabular erosion, 133, 135
acetabular fractures, 26–27, 29, 55, 61, 169, 315
acetabular inclination angle (AIA), 35, 37
acetabular labral tears, 112, 142, 169–70
acetabular labrum lesions, 152
acetabular reorientation, 177, 357
acetabular retroversion, 149, 156, 172–73, 322, 329
acetabular rim, 37, 54, 145–46, 166, 168, 171, 173, 186–87, 321, 347
acetabulum
  false, 333, 336, 343–44
  retroverted, 330, 357
  steep, 330
ACI, *see* autologous chondrocyte implantation
acute slip, 11, 14
AD-SVF, *see* adipose-derived stromal vascular fraction
adduction, 109, 127, 144, 158, 160, 162, 165, 320, 335–36, 356
adipose-derived stromal vascular fraction (AD-SVF), 197–98
ADR, *see* acetabular depth–width ratio
adult hip dysplasia, 343–44, 348, 353–54, 361, 366
adverse events, 80, 197–98, 274, 278–79
adverse local tissue reaction (ALTR), 38
adverse reactions to metal debris (ARMD), 39–41, 43
aetiology, 24, 29, 115, 178, 227, 290, 294, 366, 384, 406
aetiopathogenesis, 290
AIA, *see* acetabular inclination angle
algorithm, 353–54
alignments, 14, 93, 127, 219, 310
allograft, 59, 171, 173, 175, 198, 402
allograft–prosthesis composite (APC), 59

ALTR, *see* adverse local tissue reaction
ALVAL, *see* aseptic lymphocytic vasculitis associated lesion
ambulatory setting, 268, 280
AMIC, *see* autologous matrix-induced chondrogenesis
anaesthesia, 26, 77, 166, 197, 248, 268–70, 278
  local infiltration, 270–71, 313
  optimised, 313
  regional, 166
  spinal, 270
  total intravenous, 270
analgesia, 77, 197, 278, 296, 300
anterior approach, 14, 52, 75, 102, 265, 270, 276, 278
anterior centre edge angle (ACEA), 339–40, 358
anterior impingement, 162, 335, 357, 360, 388
anterior insufficiency, 330
anterior pelvic plane (APP), 93–94
anterior-superior iliac spine (ASIS), 103–4, 147, 159–61, 257–58
anterolateral approach, 128, 135, 402
anteroposterior (AP) 6–8, 11–12, 14, 17, 29, 126, 128, 147–50, 218, 232, 244, 336, 350, 388, 390–91
anteversion, 33, 38, 88, 90–91, 93–95, 149–50, 218, 323, 327, 330, 335, 343, 350, 358, 363
antitubercular chemotherapy (ATT), 120
AP, *see* anteroposterior
APC, *see* allograft–prosthesis composite
apoptosis, 49, 235, 384–85
APP, *see* anterior pelvic plane
approach
  anterior single-incision, 276
  anterior tissue-sparing, 278
  endoscopic, 113
  laparoscopic-assisted single-port, 103
  lateral, 75
  posterior, 37, 75–76, 81
  surgical, 74, 78, 94
  tissue regeneration, 196
  traditional transgluteal, 75, 82
  trochanteric slide, 37
  two-incision muscle-sparing, 75
ARCO classification, 390, 393–96
ARMD, *see* adverse reactions to metal debris
Arnold–Hilgartner classification, 245
arthrography, 156, 342
arthropathy, 143
  advanced, 241
  crystalline hip, 142
  haemophilic hip, 242
arthroplasty, 34, 49, 54, 73, 76, 92, 133, 137, 171, 174, 344, 346, 366, 393–94, 404
  abrasion, 188
  capsular, 345, 374–75
  shelf, 229
  total, 302
  vitallium-mould, 256
arthroscopy, 142, 164, 172, 174–78, 187, 189, 295, 343, 364
arthrotomy, 175, 348, 357
articular cartilage, 145, 147, 156, 164, 174, 176, 187, 189, 195, 199–200, 243, 287–88, 292–95, 351, 353
articular hyaline cartilage lesions, 187
articular surface, 14, 28, 37, 145, 166, 171, 176, 193, 329, 395
articular surface replacement (ASR), 39

aseptic lymphocytic vasculitis associated lesion (ALVAL), 39
ASIS, *see* anterior-superior iliac spine
ASR, *see* articular surface replacement
asymptomatic lesions, 391, 397
ATT, *see* antitubercular chemotherapy
autograft, 59, 173, 175
autologous chondrocyte implantation (ACI), 175, 188–89, 195
autologous matrix-induced chondrogenesis (AMIC), 189, 202
avascular necrosis (AVN), 10–11, 13–17, 23, 25–26, 28–29, 35, 37, 126, 131, 136, 227, 243, 297, 326, 383, 385, 401
avascular osteonecrosis, 157
AVN, *see* avascular necrosis

ball-and-socket type, 143
bearing surfaces, 33–34, 302, 305, 308–9, 316–17, 344, 405
Bernhardt–Roth syndrome, 101, 104
BG, *see* bone graft
BHR, *see* Birmingham hip resurfacing
bilateral hip osteonecrosis, 391
bilateral hip subluxation, 321, 351
bilateral severe osteoarthritis, 324
bilateral severe Perthes disease, 330
biomechanics, 213, 314, 326, 345, 359, 403
bipolar hemiarthroplasty, 133, 138
bipolar prosthesis, 133
Birmingham hip resurfacing (BHR), 36, 38
BMA, *see* bone marrow aspiration
BMI, *see* body mass index

BMP, *see* bone morphogenic protein
body mass index (BMI), 48, 60, 231, 291, 334, 350, 403
bone defects, 49, 51, 54, 58–59, 194
bone deformity, 170, 295, 297
bone formation, 227, 233, 235, 257
bone graft (BG), 15, 53, 131, 137, 348, 398, 400–402
bone grafting, 59, 62, 407, 412, 417
bone loss, 54–55, 59–60, 315
bone marrow aspiration (BMA), 398–99
bone marrow transplantation, 190
bone morphogenic protein (BMP), 235, 399, 401–2
bone remodelling, 49, 301
bone stock, 32–34, 36, 48, 52, 54, 57–59, 61, 88, 216, 248, 405–6
bony augmentation osteotomies, 346
bursitis, 107–10, 151, 158, 160

cable-locking plates, long, 52
callus formation, 18, 77, 124
cam-type lesions, 172
cancellous screws, 127, 129–30, 132, 137–38
CAOS, *see* computer-assisted orthopaedic surgery
capsular plication, 167, 174, 364, 380
capsulotomy, anterior, 166–67, 402
caput-collum-diaphyseal (CCD), 215
cartilage damage, 17, 187, 295, 342, 351
cartilage defects, 188, 194–95
cartilage disease, 186, 200
cartilage injury, 171–72
cartilage lesions, 156, 175, 231
cartilage loss, 157, 175, 245, 350

cartilage repair, 190, 192, 194–96, 199
Catterall classifications, 229
Catterall stages, 229
CCD, *see* caput-collum-diaphyseal
CEA, *see* centre edge angle
cemented total hip arthroplasty, 135
cementless implants, 231
cementless stems, 48, 57, 120
centre edge angle (CEA), 338, 344
ceramic-on-ceramic, 34, 41, 92, 309, 316
ceramic-on-conventional-polyethylene, 316
ceramic-on-HCLPE, 316
ceramic-on-polyethylene, 34
cerebral palsy (CP), 323, 362
Charnley curette, 263
Charnley monobloc stem, 314
Chiari osteotomy, 229, 346–47, 350, 353, 360–61, 363
chondral lesions, 78, 156, 176, 185–88, 193, 196, 199, 364
chondrodysplasia, 290
chondrolysis, 5, 10–12, 14–17
chondropathy, 170–71, 343
chronic regional pain syndrome (CRPS), 169
classic extreme anteverted acetabulum, 330
clinical outcomes, 36, 78, 89, 91, 189, 278, 310, 345
closed reduction, 13–14, 16, 23, 27, 128, 317
Codivilla–Hey Groves–Colonna procedure, 345–46, 353
cold welding, 315
collapse, 37, 39, 176, 233, 387, 389–90, 392–94, 396–402
combined anteversion, 37
comminution, 126, 137
comorbidities, 23, 36, 48–49, 78, 278, 313

complication rates, 10, 14, 23, 33, 58–59, 179, 222, 248
component anteversion, 89, 310
component implantation, 102
component insertion, 61
component positioning, 88, 92–93, 310
compression, 23, 28, 101–3, 128–29, 146, 170, 179, 248, 306, 312
computed tomography (CT), 9, 27, 29, 40, 55, 90–91, 109, 126, 131, 147, 150–51, 230, 297, 310, 329, 342, 371, 374, 387, 394, 396, 407
computer-assisted orthopaedic surgery (CAOS), 310–11
concomitant reduction, 212, 220
congruency, 150, 162, 228, 340–43, 350–52, 356, 358
congruity, 27, 145, 322, 345, 350–51, 356, 363
congruous reduction, 346, 353–54, 356, 361–63
contact patch-to-rim distance (CPRD), 37–38
containment, 229, 321–22, 350–52, 357, 361
contractures, 248, 327, 362
contraindications, 29, 120, 142, 152, 177, 221, 257, 265, 345, 404
contralateral hip, 16, 162
contrecoup mechanism, 156
core decompression, 399–401
correction, 15, 93, 171–72, 313, 345–49, 351–52, 355–57, 359, 365
correctional osteotomy, 14
corrosion, 215, 312, 314–15
cortical allografts, 52–53
cortical perforations, 48–49, 51, 53, 129
cortical strut graft/plate, 53

corticosteroids, 35, 49, 246, 384–88, 399–400, 404
coxa magna, 322, 352, 355
coxa profunda, 149, 156
coxa valga, 49, 150, 323, 331–32, 335, 338, 351, 356, 362–63
coxa vara, 150, 356
CP, *see* cerebral palsy
CPRD, *see* contact patch-to-rim distance
Craig's test, 163
C-reactive protein (CRP), 76
CRP, *see* C-reactive protein
CRPS, *see* chronic regional pain syndrome
CT, *see* computed tomography
cumulative contact stresses, 328–29
cup deflection, 39
cup orientation, 38, 91, 310
cup placement, 89, 95

DAA, *see* direct anterior approach
DDH, *see* developmental dysplasia of the hip
debridement, 120, 146, 175, 188–89, 327, 364, 402
decompression, 79, 127, 176, 179, 398–99
deep vein thrombosis (DVT), 39, 249, 257
defects, 57, 62, 188, 194, 325, 327, 329
   acetabular chondral, 167
   anterior, 329
   fascial, 160
   full-thickness, 187, 203
   global, 329
   osseous, 194
   osteochondral, 188, 194, 202
   posterior, 329
   posterosuperior, 335
   rare, 290
   small segmental, 173

degenerative joint disease (DJD), 174, 186, 287
delamination, 170, 187–88
delayed gadolinium-enhanced MRI of cartilage (dGEMRIC), 342–43, 359, 366
depression, 243, 390–91, 395–96, 404
developmental dysplasia of the hip (DDH), 35, 37, 49, 152, 158, 162, 322–26, 331–34, 345–46, 350, 364
DEXA, *see* dual-energy X-ray absorptiometry
DEXRIT, *see* dynamic external rotator impingement test
dGEMRIC, *see* delayed gadolinium-enhanced MRI of cartilage
DHS, *see* dynamic hip screw
direct anterior approach (DAA), 37, 255, 257
DIRI, *see* dynamic internal rotatory impingement
disease-modifying antirheumatic drugs (DMARDs), 116
dislocation, 28–29, 31–32, 37, 88–89, 92, 133, 135–36, 310, 316–17, 319, 322, 332, 335–36, 343–44, 362
   anterior, 28, 310
   anteroinferior, 28
   anterolateral, 401
   anterosuperior, 28
   low, 322, 332, 343
   posterior, 26, 28, 310
   safe, 232
   true, 319, 321, 324, 335
   unilateral, 327, 344
displacement, 1, 5, 9–11, 22, 57, 126–27, 136, 349, 365
disruption
   iatrogenic, 16
   vascular, 13–14
distraction, 142, 164–66, 193

DJD, see degenerative joint disease
DMARDs, see disease-modifying antirheumatic drugs
DNA, 116
Dorr classification, 135
Drennan's sign, 6
drilling, 51, 188, 390, 399
dual-energy X-ray absorptiometry (DEXA), 221, 392
Durom acetabular component, 39
DVT, see deep vein thrombosis
dynamic external rotator impingement test (DEXRIT), 162
dynamic hip screw (DHS), 23, 127, 130, 132
dynamic internal rotatory impingement (DIRI), 162
dysplasia, 89–90, 175, 177, 319, 322, 326, 332–34, 336, 338, 341–42, 344–45, 356, 360, 362–63, 365–66
  adolescent onset, 322, 325
  bilateral, 325
  borderline, 174, 333, 364
  multiple epiphyseal, 227
  neuromuscular, 361
  residual, 325–26
  secondary, 329, 357, 363
  severe, 142, 334, 361
dysplastic hips, 168, 329, 333
  bilateral, 324
  borderline, 177
  neuromuscular, 362
dysplastic patients, 333, 335, 349, 353, 364, 366

EBRA-FCA, see Ein-Bild-Roentgen-Analyse-femoral component analysis
Ein-Bild-Roentgen-Analyse-femoral component analysis (EBRA-FCA), 221
electromyography (EMG), 102
electroneuromyography, 179
EMG, see electromyography
endoprosthesis, 276, 284
endoscopic bursectomy, 110
end-stage haemophiliac arthropathy, 243
enhanced recovery programmes (ERPs), 305, 313
E-Poly, see vitamin E–doped polyethylene
ERPs, see enhanced recovery programmes
ESWT, see extracorporeal shockwave therapy
extension, 111, 127, 152, 160–62, 177–78, 257, 263–64, 335–36, 357, 391, 405
external rotation, 4, 6, 26, 126, 128, 144, 160, 162–63, 167, 177, 264, 320, 335
extracorporeal shockwave therapy (ESWT), 398

FAAH, see fatty acid amide hydrolase
FADDIR, see flexion, adduction and internal rotation
FAI, see femoroacetabular impingement
failure, mechanical, 39, 78, 250, 291, 314, 316
failure rates, 39, 77, 250, 405
fatigue strength, 306, 312
fatty acid amide hydrolase (FAAH), 290
FDA, see Food and Drug Administration
femoral anteversion, 147, 320, 327, 335, 353, 362
femoral component, 38, 41, 48, 51–53, 88, 91, 93, 230, 263, 310
femoral deformities, 51, 330

femoral fractures, 24, 49, 52, 55, 77, 400
femoral head, 7-8, 11-18, 23-29, 34-39, 124, 127-28, 142-43, 148-50, 155-57, 172-74, 231-34, 324-27, 331-33, 338-46, 391-96
femoral head collapse, 34, 41, 228-29, 233, 388-89
femoral head deformity, 233-34, 239, 326, 329, 345, 352
femoral head extrusion index, 148, 150, 334, 338-39
femoral head osteonecrosis (FHON), 124, 127, 137, 234, 383-89, 391-93, 397, 399-400, 403, 405-7
femoral neck, 7, 22-23, 33-34, 41, 128-29, 137-39, 155-56, 213, 215, 217-18, 222, 224, 259, 261-62, 340-41
femoral neck anteversion, 163
femoral neck fracture (FNF), 33, 37, 39, 93, 124, 127-28, 133, 135, 169, 386
femoral osteotomy, 51, 229, 246, 353-54, 356, 362-63
femoral retroversion, 2
femoral stem, 51-53, 55, 57, 212, 220, 316
femoral varus osteotomies, 229-30
femoroacetabular impingement (FAI), 5, 15, 18, 78-79, 142-43, 146, 149, 152, 154, 156, 158, 162, 167, 170, 172, 174, 177, 186-87, 231, 297, 334, 351, 359
FHON, *see* femoral head osteonecrosis
fibular graft, 131
fibular vascularised graft (FVG), 400, 403
Ficat & Arlet, 393-94

fixation, 11-13, 23-24, 33, 47, 49-50, 56, 77-78, 88, 124, 127, 187, 311, 315, 349
 arthroscopic-assisted fracture, 176
 biological, 50, 311, 315
 cementless acetabular, 315
 cementless stem, 50
 diaphyseal, 212, 220
 intramedullary, 58
 press-fit, 315
flexion, adduction and internal rotation (FADDIR), 335
FNF, *see* femoral neck fracture
Food and Drug Administration (FDA), 38
fracture healing, 57, 59, 77
fracture line, 50, 58, 125-26, 129, 137, 391
fractures, 13, 21-23, 25-26, 34, 38, 41, 47-62, 124-29, 132, 136-37, 263, 265, 306, 308, 392-93
 avulsion, 104
 basicervical, 129
 clamshell, 53
 complete, 22
 controlled, 349
 displaced, 53, 61, 126, 136
 extracapsular, 21, 25
 femoral shaft, 27-28, 124
 impacted, 22
 intertrochanteric, 24
 isolated, 60
 nondisplaced transcervical, 129
 pertrochanteric, 77
 posterior column, 364-65
 primary, 50
 proximal, 51, 58
 reverse oblique, 24
 shaft, 51
 stress, 126, 142
 subtrochanteric, 24
 total neck, 123

transverse, 60
trochanteric avulsion, 24
undisplaced, 22–23, 61
unstable trochanteric, 78
vertebral compression, 124
free vascularised fibular grafting (FVFG), 137
FVFG, *see* free vascularised fibular grafting
FVG, *see* fibular vascularised graft

gadolinium-enhanced MRI (Gd-MRI), 233, 342
GAG, *see* glycosaminoglycan
gait, 75, 159, 180
　antalgic, 5
　circumduction, 159
　short-limb, 159
　waddling, 336
Galeazzi test, 161
Garden's alignment index, 128
Garden's classification, 125, 129, 132–33
Gaucher disease, 386–87, 406
Gd-MRI, *see* gadolinium-enhanced MRI
Giessen disease, 386
glycosaminoglycan (GAG), 293
grafts, 171
　cancellous, 131
　cortical strut, 53, 58
　free vascularised/nonvascularised fibula, 137
　live, 402
　muscle pedicle, 137, 400, 403
　rectus auto, 171
　vascular, 400
greater trochanter (GT), 50, 53–54, 107–10, 112, 126, 129, 158, 178, 218–19, 263, 265, 335, 340, 354–55
greater trochanteric pain syndrome (GTPS), 109–10
grit-blasting, 311

growth plates, 3–5, 7, 325, 348
GT, *see* greater trochanter
GTPS, *see* greater trochanteric pain syndrome

haemarthrosis, 13, 16, 241–44, 246
haematopoietic stem cells (HSCs), 190
haemophilia, 4, 241–43, 246–50
haemophilic arthropathy, 241–43, 245, 250
HAGOS, *see* Hip and Groin Outcome Score
Harris Hip Score (HHS), 75, 120, 171, 174, 181, 230–31, 346, 360, 388
HCLPE, *see* highly cross-linked polyethylene 306–7
HD, *see* hip dysplasia
HD acetabular cartilage
HDP, *see* high-density polyethylene
head–neck junction, 154, 172, 186, 314, 348, 402
head–neck offset, 3, 332, 335, 357, 387
head sphericity, 149–50, 234, 344
hemiarthroplasty, 23–24, 54–55, 133, 135, 137
HES, *see* Hospital Episode Statistics
heterotopic ossification (HO), 15, 235, 364–65
HHS, *see* Harris Hip Score
high-density polyethylene (HDP), 306
highly cross-linked polyethylene (HCLPE), 306–7
Hip and Groin Outcome Score (HAGOS), 181
hip arthroplasty, 31, 34, 49, 52, 74, 88, 91–92, 176, 179, 230–31, 247, 250, 256, 284, 305
hip arthroscopy, 74, 78–80, 117, 141–42, 176–77, 181, 186, 200

hip deformities, 142, 231–32, 296–97, 372
hip disease, 92, 243
hip dislocations, 26–28, 50, 231, 309–10, 327–28, 345–46, 352, 356, 361, 386
Hip Dysfunction and Osteoarthritis Outcome Score (HOOS), 197–98
hip dysplasia (HD), 158, 292, 310, 319, 322, 324–27, 329–30, 332, 335–38, 343, 345, 350, 353–54, 362, 364, 366
hip fractures, 5, 7, 13, 21, 23, 49, 80
hip osteoarthritis, 142–43, 197, 290–92, 296, 300, 324
Hip Outcome Score (HOS), 181
hip pain, 102, 107, 109–10, 145–46, 157–58, 181, 186, 392
hip-preserving surgeries, 346, 351, 362, 366, 391
hip replacement, 50, 55, 74–75, 119, 197, 248, 258, 268, 273–74, 276, 278, 280, 284, 305, 345
hip resurfacing, 31–32, 34, 43, 91–93, 102, 231
hip resurfacing arthroplasty (HRA), 33–34, 36, 88, 93
hiPSCs, see human induced pluripotent stem cells
HIV, 247, 386–87
HO, see heterotopic ossification
HOOS, see Hip Dysfunction and Osteoarthritis Outcome Score
HOS, see Hip Outcome Score
Hospital Episode Statistics (HES), 79
HRA, see hip resurfacing arthroplasty
HSCs, see haematopoietic stem cells

human induced pluripotent stem cells (hiPSCs), 199

IBD, see inflammatory bowel disease
IHOT, see International Hip Outcome Tool
iliotibial band, see ITB
impingement, 15, 18, 33–34, 79–80, 88, 110, 162, 170, 172, 174, 176, 316, 335, 338, 355–56
implantation, 48, 50, 211–13, 215, 218–21, 279
implants, 36–37, 41–42, 50–51, 53–55, 57–61, 77–78, 199, 211–13, 215, 218, 222, 230–31, 248–50, 275–76, 309–12
induced pluripotent stem cells (iPSCs), 196, 199
inflammatory bowel disease (IBD), 385, 409
INR, see international normalised ratio
in situ fixation, 16
internal fixation, 13, 23, 25, 57, 131–32
internal rotation, 6, 127–28, 144, 158, 160, 165, 172, 297, 335, 350, 353–54, 356–57, 359–60, 388
internal snapping, 111, 178–79
International Hip Outcome Tool (IHOT), 181
international normalised ratio (INR), 247
International Society for Cell Therapy (ISCT), 191–92
interventions, 166, 178, 230, 298, 300
  mechanical, 302
  operative, 27, 57
  surgical, 109, 178, 296, 360, 398

Index

therapeutic, 294
intra-articular fractures, 291
intra-articular injury, 297
intra-articular pathology, 151, 158, 356
intra-articular snapping hip, 112
intracapsular fractures, 22, 24–26
intracapsular pressure, 124, 127–28, 243
intraoperative fractures, 51, 58, 71
intravenous (IV), 270–71, 276, 280–81, 284
invasive surgery, 52
iPSCs, *see* induced pluripotent stem cells
ISCT, *see* International Society for Cell Therapy
ITB (iliotibial band), 110, 112, 151, 160, 171, 173, 178, 180
IV, *see* intravenous

jigs, traditional, 88
joint arthroplasty, total, 246
joint effusion, 9, 40, 244–45, 296, 388
joint replacement, 28, 54, 117, 120, 247, 249–50, 267, 276, 289, 302, 345, 397–98
joint space, 17, 27, 171, 174, 288–89, 297, 340–41, 343–44, 346, 350–51, 389
joint space narrowing, 297, 350, 395
joint stability, 145–46, 327, 345, 353, 364

Karboul angle measurement, 389
keyhole surgery, 110
Klein's line, 7
knee arthroplasty, 87
knee replacement, 55, 87
knee valgus, 333
Konan grades, 188

Konan's classification, 188
Kyle classification, 24

labral damage, 168, 173, 335, 364
labral pathology, 18, 151, 158, 297
labral repair, 167–68, 171, 173–74, 177, 364
labrum, 143, 145–46, 150, 152–53, 164, 166, 170–72, 174, 326–27, 335, 347, 364
  abnormal, 170
  anterior, 145
  degenerated, 327
  fibrocartilaginous, 143, 145–46
  healthy, 170
  hypertrophic, 326, 342
  hypertrophied, 342
  posterior, 145
  repaired, 177
  unstable, 170
laparoscopic-assisted single-port appendectomy, 103
lateral centre edge angle (LCEA), 174, 333–34, 338–39, 342, 355, 358, 360, 364–65, 387
lateral femoral circumflex artery (LFCA), 124
lateral femoral cutaneous nerve (LFCN), 37, 101–4, 142, 164, 168
lateral transgluteal approach, 75–76
LCEA, *see* lateral centre edge angle
LCPD, *see* Legg–Calvé–Perthes disease
Leadbetter technique, 127
Legg–Calvé–Perthes disease (LCPD), 227–36, 243, 322
leg length, 28, 32, 88, 148, 150, 159, 161, 215–16, 218, 230–31, 263, 296–97, 314
lesions, 120, 146, 156, 171, 187, 295, 387, 389, 391, 393, 396–99, 401–6

band-like, 389
bony, 18
cam, 5, 154–55, 162, 173, 332, 334, 338, 342
cartilaginous, 151
countercoup, 331
cystic, 394
diffuse, 175
femoral, 172
focal, 392
lateral, 399
necrotic, 393
osteochondral, 194, 202
osteolytic, 56
pincer, 173
radiolucent, 390
sclerotic, 391
symptomatic, 400
tong-type, 172
lesser trochanter (LT), 21, 53–54, 67, 129, 148, 179, 261, 263, 340, 400
LFCA, see lateral femoral circumflex artery
LFCN, see lateral femoral cutaneous nerve
LIA, see local infiltration anaesthesia
ligamentum teres, 4, 23, 79, 142–43, 146, 157, 167, 177, 262, 331
lightbulb method, 403
likelihood ratio (LR), 387
limb malalignment, 49
limp, 5, 119, 335, 360, 362
  abductor, 362
  antalgic, 388
  reduced, 230
local infiltration anaesthesia (LIA), 270–71, 277, 281, 313
locking, 158, 243, 296, 315, 335, 388
Loder classification, 10–12
LR, see likelihood ratio

LT, see lesser trochanter
lumbar hyperlordosis, 327, 333, 336

MACI, see matrix-induced autologous chondrocyte implantation
magnetic resonance imaging (MRI), 9, 22, 40, 79, 89–90, 102, 109–10, 119, 126, 150–52, 154–56, 181–83, 230, 245, 295, 326–27, 342, 391–94, 396–97, 411–14
malalignment, 288, 290–91, 297, 336
MAP, see modified anterior portal
MARS, see metal artefact reduction sequence
matrix-induced autologous chondrocyte implantation (MACI), 189, 202
MDCT, see multidetector computed tomographic
medial femoral circumflex artery (MFCA), 37, 124, 144
Medicines and Healthcare Regulatory Agency (MHRA), 41
meralgia paresthetica, 101–4
Merle d'Aubigne score, 181, 350, 352, 357
mesenchymal stem cells (MSCs), 186, 189, 191–93, 195–97, 235, 385
meta-analysis, 78, 91, 93, 135, 230
metal artefact reduction sequence (MARS), 40, 42
metal-on-metal (MOM), 32, 34–35, 39, 42, 88, 309–10, 316
metal-on-metal hip resurfacing arthroplasty (MOM HRA), 31, 36, 39–41, 43
metaphysis, 3–4, 6, 8, 14, 49, 216, 219, 232, 393, 402, 405

MFA, *see* musculoskeletal function assessment
MFCA, *see* medial femoral circumflex artery
MHRA, *see* Medicines and Healthcare Regulatory Agency
MI, *see* minimally invasive
MI hip fracture plating, 78, 80
microfracture, 167, 175, 188–89, 193, 295, 387
minimally invasive (MI), 73–77, 175, 200, 211–12, 219, 270, 278, 284, 300, 302
Minimally Invasive Screw System (MISS), 78
minimally invasive surgery (MIS), 52, 73–78, 80, 82, 84, 95, 223, 273, 278, 284
minimally invasive total hip arthroplasty (MITHA), 74
MIS, *see* minimally invasive surgery
MISS, *see* Minimally Invasive Screw System
MITHA, *see* minimally invasive total hip arthroplasty
mobilisation, 16, 23, 26, 74, 92, 103, 248, 257, 259, 265, 271, 300, 313–14
mobility, 23, 135, 144, 249, 262, 265, 316
modified anterior portal (MAP), 165–66
modular stems, 58, 230
MOM, *see* metal-on-metal
MOM HRA, see metal-on-metal hip resurfacing arthroplasty
morbidity, 73, 94, 230, 313
magnetic resonance arthrography, 152, 154, 156–57, 326, 342, 351, 366
MRI, *see* magnetic resonance imaging
MSCs, *see* mesenchymal stem cells

multidetector computed tomographic (MDCT), 156–57
musculoskeletal function assessment (MFA), 181

National Health Service (NHS), 79, 87, 94
National Joint Registry (NJR), 48, 87
neck–shaft angle (NSA), 124, 148, 150, 331, 338, 387
necrosis, 77, 221, 243, 352, 383–85, 387, 389, 391, 393, 405
nerve growth factor (NGF), 302
nerve injury, 27, 30, 74, 166, 168, 179
NGF, *see* nerve growth factor
NHS, *see* National Health Service
NJR, *see* National Joint Registry
nonoperative treatment, 23, 57, 109–10, 228
nonsteroidal anti-inflammatory drug (NSAID), 273, 299–300
nonsurgical treatments, 181, 230, 398
nonunion, 23–24, 26, 37, 49, 57, 59, 74, 126, 133, 136–37, 162, 364, 404
nonvascularised bone graft (NVBG), 400–403, 407
NSA, *see* neck–shaft angle
NSAID, *see* nonsteroidal anti-inflammatory drug
NVBG, *see* nonvascularised bone graft

OA, *see* osteoarthritis
Ober's test, 161
offset, 33, 150, 177, 215–18, 310, 335, 338, 342, 348
open reduction and internal fixation (ORIF), 52–53, 56–57, 60

ORIF, *see* open reduction and internal fixation
osteoarthritis (OA), 5, 18, 36, 38, 79, 90, 158, 170–71, 186, 190, 193, 195–98, 211, 221, 243, 287–95, 297–99, 301–02, 324–25, 327, 332–34, 338, 343–45, 347–49, 351, 353–54, 359–61, 395, 405
  advanced, 142, 296, 333, 350, 360
  early, 79, 186, 295–96, 325, 333, 343, 353
  end-stage, 36, 188
  late-stage OA, 295–96
  posttraumatic, 48–49
  primary, 230, 250
  primary hip, 290–91, 332
osteoarthrosis, 229
osteointegration, 36, 212, 220
osteolysis, 33–34, 39–40, 48, 50, 53, 219, 306, 316
osteonecrosis, 10, 12, 15–16, 28, 41, 90, 124, 127, 136, 142, 169, 175–76, 186, 235, 348, 383, 386–87, 394, 407
osteopenia, 392–94
osteoporosis, 21, 35, 48–49, 60, 123, 222, 244–45, 393
osteosclerosis, 157, 394
osteotomy, 13–18, 58, 74, 229–30, 261, 263, 343, 345, 347–49, 352–53, 356, 359, 361, 364–65, 404
outcomes
  long-term, 43, 178, 212, 221, 228
  poor, 48, 361, 364

PAI, *see* plasminogen activator inhibitor
pain, joint, 87, 142–43, 295–96, 300

pain management, 270, 277–78, 299, 314
pain relief, 105, 109, 188, 250, 262, 363
PAO, *see* periacetabular osteotomy
Paprosky classification, 61
pathomorphology, 171, 336, 366
pathophysiology, 228, 235–36, 290, 295, 324, 356
patient position, 149–50
patients
  active, 33–36, 211, 219, 268, 405
  asymptomatic, 40, 42, 153, 324, 365
  older, 23, 48, 123, 136–37, 359
pelvic osteotomy, 229, 347
PEMF, *see* pulsed electromagnetic field
perfusion MRI, 232–33, 238
periacetabular osteotomy, 142, 173, 177, 231, 334, 346–66
peripheral compartment, 164, 167–68, 172–73, 175
periprosthetic fractures, 18, 47–48, 52, 55, 60, 135, 249, 405
Perthes sequelae, 177
PET, *see* positron emission tomography
PFFD, *see* proximal femoral focal deficiency
PFF, *see* proximal femur fracture
PFO, *see* proximal femoral osteotomy
physical therapy, 179, 248, 273, 285, 299
pincer-type lesions, 172
piriformis syndrome, 179
plasminogen activator inhibitor (PAI), 386
platelet-rich plasma (PRP), 197–98, 299
polyethylene, 33–34, 53, 56, 88, 306–9, 316, 405

PONV, *see* postoperative nausea and vomiting
portal placement, 142, 164, 166, 168
portals, 110, 165–68, 172
positron emission tomography (PET), 392
postcollapse, 394, 403–4
postoperative nausea and vomiting (PONV), 270–71
posterior impingement, 162, 335, 360
posterior superior iliac spine (PSIS), 160
postoperative fractures, 38, 48, 50–51
preoperative planning, 40, 55, 90, 94, 150, 215, 217, 230, 276, 310
preoperative preparation, 268, 275
prognosis, 5, 175, 232, 364, 396
 long-term, 178
 poor, 171, 329, 360, 391
 radiological, 360
 worst, 396
prosthesis, 29, 47, 50–52, 54, 59, 61, 74, 88, 133, 263, 265, 305, 311–12, 400, 405
proximal femoral focal deficiency (PFFD), 329, 370
proximal femoral osteotomy (PFO), 230, 353–54, 356, 362–63, 406–7
proximal femur fracture (PFF), 47–49, 50–52, 55–56, 58
PRP, *see* platelet-rich plasma
PSIS, *see* posterior superior iliac spine
pulsed electromagnetic field (PEMF), 398

QoL, *see* quality of life
quadratus femoris, 131, 137, 144, 403

quality of life (QoL), 31, 177, 198, 200, 243, 246, 288–89, 296, 299–300

RA, *see* rheumatoid arthritis
radiofrequency (RF), 166, 168–69, 188
radiographs, 4, 6–9, 11, 14, 17, 27–28, 55, 59, 112, 126, 130, 147–48, 150, 177, 244
radiography, 40, 147, 234, 331, 336, 342, 351, 358, 362, 366, 389, 392–94, 396
radiostereometric analysis (RSA), 221, 224
radiotherapy, 2, 190, 386
randomised control trial (RCT), 316, 398, 401
RCT, *see* randomised control trial
rectus femoris, 28, 102, 144, 151, 255–59
reduction, 11, 13–14, 24, 26–29, 75–76, 124, 127–28, 197, 199, 273, 299, 306, 309, 312, 316
 open, 13–14, 57, 128
regenerative medicine, 185, 196, 200
rehabilitation, 16, 26–27, 47, 77, 179–80, 188, 246, 248, 270, 272, 278, 283–84, 314, 348
remodelling, 50, 228–29, 232, 326, 397
resection, 34, 41, 92, 170–71, 174, 213, 215, 217
resection arthroplasty, 59, 120, 255, 362
resurfacing, 32–35, 41, 231, 257, 405–6
resurfacing arthroplasty, 31–32, 34, 388
retroversion, 15, 149, 173, 329, 355, 357

revision, 26, 33, 37, 39, 41–42, 48, 51–53, 61, 92, 231, 250, 309, 315
revision arthroplasty, 47, 57
revision surgery, 36, 41, 48, 50, 57–59, 220, 305, 309–10, 314
RF, *see* radiofrequency
rheumatoid arthritis (RA), 48–49, 71, 108, 115–17, 175, 221, 245, 296
risk factors, 2, 4, 16, 21, 35–36, 48–49, 51, 60, 102, 278, 288, 324, 367, 386–87
robotic surgery, 95
rotational osteotomy, 347, 360, 404
RSA, *see* radiostereometric analysis

sagging rope sign, 355
SARI, *see* Surface Arthroplasty Risk Index
scaffold, 189, 194, 200, 401
scanning electron microscopy, 145
Scarpa's triangle, 111
SCFE, *see* slipped capital femoral epiphysis
sclerosis, 244, 297, 321, 389–90
scoliosis, 108, 159
screw fixation, 13, 78, 315
screws, 11–14, 17, 26, 52–53, 60, 62, 129, 131, 133, 315, 348
  antirotatory, 129–30
  bicortical, 60
  cannulated, 11, 23, 129
  conventional hip, 78
  dynamic compression, 26
  inferior, 129
  minifragment, 265
  posterior, 129
  sliding hip, 77, 129
  superior, 129
  threaded, 11–14
  unicortical, 60

severe subluxation, 321–22, 335, 345
Shenton line, 7, 321–22, 341, 351, 357–58
shockwave therapy, 110, 398
shortening, 6, 15, 49, 119, 126, 213, 307, 333, 335, 346, 362
short-stem designs, 212, 215–16
short stems, 212–13, 215–22, 391
  calcar-guided, 215–18
  modern, 213, 215, 220–21
  neck-preserving, 276–78
  neck-retaining, 214
  rounded, 218
SHS, *see* sliding hip screw
SLE, *see* systemic lupus erythematosus
sliding hip screw (SHS), 77–78, 129
slipped capital femoral epiphysis (SCFE), 1, 157–58, 175, 177, 186
slipped upper femoral epiphysis (SUFE), 1–13, 15–18, 292
  bilateral, 2, 5
  displaced, 5, 9
  stable, 11–12, 16
Smith-Petersen approach, 255
snapping hip, 111–12, 142–43
soft tissues, 13, 39–40, 74, 76, 93, 108, 151, 162, 168, 222, 259, 262, 265, 284–85, 314
sourcil, 149, 331, 336–40, 358
Southwick's angle, 8–9
Southwick's classification, 10–12
stem cells, 175, 185, 190, 196, 198, 200, 401
stem designs, 221, 314–15
straight stems, 215–16, 219, 221–22
strain theory, 77
stress shielding, 33, 50, 58, 219, 312, 316

stromal vascular fraction (SVF), 197–98
Stulberg classification, 228
subchondral fracture, 389–92, 395
subluxation, 33, 146, 153, 168, 319–23, 326, 330, 332–35, 341, 343–45, 350, 353, 362
subsidence, 41, 57–59, 222
SUFE, *see* slipped upper femoral epiphysis
Surface Arthroplasty Risk Index (SARI), 36
surface replacement, 39, 54
surgical dislocation, 14, 37, 141, 172, 232, 346
surgical management, 18, 56–57, 61, 171
surgical procedures, 94–95, 120, 211, 228–29, 234, 246, 273, 276, 388, 397, 400, 406
surgical techniques, 43, 48, 77, 174, 248, 273, 284, 349, 361, 403–4
   advanced, 366
   conventional, 76
   invasive, 284
   minimal invasive, 279
   standardised, 313
surgical treatment, 32, 144, 172, 187, 246, 288, 296, 397, 399, 406
survival rate, 38, 231, 250, 284, 360, 405
survivorship, 37, 58, 92, 231, 316, 360
SVF, *see* stromal vascular fraction
synovectomy, 117, 246
synovial chondromatosis, 79, 142–43, 175
synovitis, 233, 235, 245, 302
   hypertrophic, 157
   secondary, 388

transient, 158
systemic lupus erythematosus (SLE), 386, 406

TB, *see* trochanteric bursitis
tensor fascia lata (TFL), 102, 137, 144, 151, 255–59, 264, 277, 403
TFL, *see* tensor fascia lata
TGF-ß, *see* transforming growth factor–beta
THA, *see* total hip arthroplasty
therapeutic strategies, 186, 188, 191, 194, 196, 302
thermotherapy, 299
THR, *see* total hip replacement
titanium modular fluted tapered (TMFT), 56, 58–59
TIVA, 270
TMFT, *see* titanium modular fluted tapered
TNF-α, *see* tumour necrosis factor–alpha
Tonnis angle, 334, 337
Tonnis classification, 334, 343–44
Tonnis grade, 349, 351, 359
Tonnis method, 347
Tonnis operations, 347
total hip arthroplasty (THA), 31–32, 47–48, 74–75, 87–89, 91–95, 102, 120, 133, 135–37, 211, 267–68, 278, 280, 300–301, 305
   cemented, 50, 421
   cementless, 92, 120, 268
   conventional, 32–33, 95, 231
   conventional straight-stem, 218
   primary, 48, 51, 120
   revision, 48–49, 51, 62, 360
   robotic-assisted, 95
   same-day-discharge, 278
   short-stem, 212, 220–22

total hip replacement (THR), 23–24, 31–33, 74, 119–20, 248–50, 257–58, 296, 325, 332–33, 344, 360–61, 383, 388, 391, 400, 404–6
  bilateral, 54, 325, 333, 405
  cementless, 406
  revision, 87
traction, 13–14, 27–29, 57, 120, 127–28, 164–68, 170, 257, 263
transforming growth factor–beta (TGF-ß), 385
transplantation, 166, 188, 190, 195, 406
trauma, 2, 5, 28–29, 47–49, 51, 73–74, 77, 80, 243, 273, 278, 284, 288, 291, 392
traumatic hip dislocation, 176
Trendelenburg limp, 159, 178, 335
Trendelenburg test, 159
triple osteotomy, 345, 347, 352
trochanteric bursitis (TB), 107–9, 169, 178
  chronic, 178
trochanteric osteotomy, 15, 74
true acetabulum, 320, 343, 346
tumour necrosis factor–alpha (TNF-α), 235, 385

UCS, *see* unified classification system
UHMWPE, see ultra-high-molecular-weight polyethylene 306, 317, 405
ultra-high-molecular-weight polyethylene (UHMWPE), 306, 317, 405
ultrasonography (USG), 119, 151
ultrasound, 9, 40, 42, 93, 110, 116, 151, 249, 324
uncemented femoral stems, 309
uncemented shafts, 276
uncemented stems, 50

uncemented total hip arthroplasty, 136
unified classification system (UCS), 54–55, 61, 67
union, 16, 58, 143, 402
  delayed, 49
  femoral, 58
Universal Data Collection Project, 242
unstable slips, 12, 16
USG, *see* ultrasonography

valgus, 331–32, 338
valgus alignment, 217, 247
valgus intertrochanteric osteotomy, 136, 138
valgus osteotomy, 132, 356–57, 362
Vancouver classification, 52–53
Vancouver classification system (VCS), 52, 55, 67
varus, 15, 26, 38, 41, 93, 216–17, 229, 265, 331–32, 338, 345, 351, 356
VAS, *see* visual analogue scale
vascularised bone graft (VBG), 400–401, 403, 407
vascularised fibular grafting (VFG), 403
VBG, *see* vascularised bone graft
VCS, *see* Vancouver classification system
vertical mattress suture technique, 171
VFG, *see* vascularised fibular grafting
visual analogue scale (VAS), 197
visual analogue score, 181
vitamin E–doped polyethylene (E-Poly), 306–7

Waldenstrom stage, 232
Watson-Jones approach, 75

WB, *see* weight-bearing
WB area, 331, 336, 338–39, 347, 359, 396–97, 399, 403–4
wear, 32, 34, 37–40, 53, 231, 249, 287, 290, 306–7, 315–17, 399
wear and tear, 193
wear rates, 35, 37, 92, 306, 308–9
weight-bearing (WB), 16, 159, 188, 194, 235, 271, 291, 331, 335, 385, 390, 394
WFH, *see* World Federation of Hemophilia

Whitman's method, 128
World Federation of Hemophilia (WFH), 245

X-linked recessive condition, 241
X-rays, 13, 102, 109, 147–48, 197, 275, 277, 297, 339

Yasunaga grading, 343

zone of hypertrophy, 3
zone of Ranvier, 3

PGMO 04/17/2018